Advances in Natural and Technological Hazards Research

Volume 49

More information about this series at http://www.springer.com/series/6362

Nirupama Agrawal

Natural Disasters and Risk Management in Canada

An Introduction

Springer

Nirupama Agrawal
York University
Toronto, ON, Canada

ISSN 1878-9897 ISSN 2213-6959 (electronic)
Advances in Natural and Technological Hazards Research
ISBN 978-94-024-1281-9 ISBN 978-94-024-1283-3 (eBook)
https://doi.org/10.1007/978-94-024-1283-3

Library of Congress Control Number: 2018937409

Printed on acid-free paper

This Springer imprint is published by the registered company Springer Science+Business Media B.V. part of Springer Nature.
The registered company address is: Van Godewijckstraat 30, 3311 GX Dordrecht, The Netherlands

To my late parents, Dr. Jageshwar Sharan Agrawal and Snehlata Agrawal

Foreword

One of the greatest challenges faced by the governments of all countries today lies in creating institutional convergence that integrates global goals emanating from the Sustainable Development Goals, the Sendai Framework for Disaster Risk Reduction, the Paris Agreement on Climate Change, and the World Humanitarian Summit. Disaster risk reduction and climate change adaptation are part of key agendas pursued in all these recent global agreements. The effective reduction of losses and risks stemming from natural hazards and climate extremes requires an integrated action at various levels of government and involves a wide range of stakeholders.

Canadians in all regions are exposed to risk from natural disasters. As Canada's natural and social environment changes, the complexity of managing the consequences of disasters also increases due to the ever-greater technological dependencies and interdependencies.

This is the context for this book, which puts the spotlight on the broad range of natural hazards that threaten Canada but also on the strategies and more practical actions that can challenge conventional perceptions of risk and inform decision-making to arrive at a more effective disaster risk reduction and resilience building.

In this extensive treatment of the subject, Dr. Nirupama Agrawal explores hazards from the small and medium scale such as erosion, landslides, and blizzards to the large scale such as epidemics, droughts, and earthquakes. The book systematically defines these key threats and explores their potential impacts on Canada.

After establishing the types and levels of threats to Canada, Dr. Agrawal proceeds to explore the many ways in which society must tackle these growing risks. Drawing upon a diverse body of literature that includes many disciplines and fields of study, the book investigates the interrelated concepts of disaster risk management and disaster resilience and considers the importance of risk perception by various groups. It argues that an all-hazard, risk-based, problem-solving, and results-oriented approach should be pursued in disaster risk reduction to address the multifactorial and interdependent nature of the disaster risk chain, identify relevant solutions, and optimize the use of resources.

Subsequent chapters describe a range of qualitative and quantitative methods that can provide a comprehensive approach to understanding disaster risk and vulnerabilities and inform decision-making to build more resilient communities. The book consistently draws upon examples and case studies from around the world, recognizing that both the challenge of disaster risk and the potential solutions can be found globally.

Although this book is entitled *An Introduction*, Dr. Agrawal must be commended for providing such comprehensive but also accessible insights into national disasters and risk management in Canada. It is our hope that this book will raise awareness and motivation across Canada and beyond, to find rational, balanced responses to these mounting threats to humankind.

University of Huddersfield
Huddersfield, UK

Professor Richard Haigh
Professor Dilanthi Amaratunga

Foreword

It gives me great pleasure to write the preface for this excellent monograph titled *Natural Disasters and Risk Management in Canada: An Introduction*, written by Professor Nirupama Agrawal of York University. I have known her for the past 12 years and have interacted with her professionally during this period. This is a first of its kind publication emphasizing disasters and disaster management in the Canadian context. The monograph is very relevant and timely in view of the ever-increasing vulnerability of the Canadian population to a variety of natural hazards. This monograph consists of eight chapters, and together they portray a comprehensive view and state of the art in disaster management in general and in Canada in particular.

Chapters 1 and 2 deals with a variety of natural hazards, such as earthquakes, landslides, hurricanes, ice storms, storm surges, tsunamis, floods, droughts, etc. Chapter 3 has the title "Disaster Risk Management" and discusses risk analysis, strategic planning, rehabilitation, and sustainable development. Chapter 4 is about disaster resilience and provides details about community resilience and decision support systems. Chapter 5 is about disaster perception and includes topics such as perceptions of risk, vulnerability, and emergency measures. Chapters 6 and 7 explain quantitative disaster risk evaluation methods practiced in Canada and around the world. Finally, Chap. 8 discusses qualitative methods to evaluate disaster risk. As one can see, this monograph provides very valuable information and data on a variety of natural hazards that can happen and happened in the past in Canada.

The bibliography is extensive and up to date, and the illustrations and tables add particular value to the contents of the monograph. I believe that this monograph will be very useful not only to the practitioners of disaster management but also to research scholars, graduate students, and the general public who are interested in natural disasters. I strongly feel that this should be in the personal collection of people interested in natural disasters and also should be in the library of scholarly institutions specializing in disaster research and mitigation.

Department of Civil Engineering
University of Ottawa
Ottawa, ON, Canada

Tad S. Murty

Preface

When York University launched a postsecondary program in the then-nascent discipline of disaster and emergency management in 2005, Canada had only one other undergraduate-level program in the field, at Brandon University. The discipline not only needed recognition, but given its complex and multifaceted nature, teaching and research resource material needed to be developed. Areas such as climate change, natural disasters, early warning systems, risk and vulnerability assessment, and humanitarian aid were essential to the curriculum. Although I found scholarly material developed by many government organizations in the USA and Canada and other well-known organizations (e.g., the World Health Organization, Munich Re, ReliefWeb, UN International Strategy for Disaster Reduction, Red Cross, etc.), I couldn't find everything I wanted for my classes in one place. So, this project was conceived out of the need to develop a course material for university-level degree programs, both undergraduate and graduate, on disasters, risks, vulnerability, and resilience for disaster risk reduction.

The book is divided into eight chapters. Chapters 1 and 2 deal with natural hazards, which are categorized by their size and impact as large-, medium-, or small-scale hazards. Chapter 1, "Defining Natural Hazards: Large-Scale Hazards," deals with droughts, earthquakes, extreme weather, floods, forest fires, ice storms, hurricanes, and biophysical hazards. Chapter 2, "Defining Natural Hazards: Medium- and Small-Scale Hazards," focuses on medium-scale hazards such as erosion, landslides, snowstorms/blizzards, subsidence, sinkholes, tornados, and windstorms and small-scale hazards, namely, extraterrestrial hazards, fog, geomagnetic storms, hailstorms, and lightning. Notably, tsunamis and volcanoes are excluded from these two chapters, as they are rarely a concern in Canada. Chapter 3, "Disaster Risk Management," discusses risk analysis strategic planning and sustainable development, drawing extensively on case studies and examples to do so. Decision support systems and the decision-making process are also integral parts of this chapter. Chapter 4, "Disaster Resilience," sheds light on the concept of resilience and related discussions on sustainable livelihood and community participation. Representative research studies are also included, supplementing these discussions. Chapter 5, "Disaster Perceptions," explores various aspects of perceptions such as disaster risk, vulnerability, and people, including influencing factors. In turn, Chapter 6, "Disaster Risk Evaluation: Quantitative Methods in Canada," highlights various methods and tools used by professionals to estimate the likelihood and consequences of hazards. Chapter 7, "Disaster Risk Evaluation: Other Quantitative Methods," describes tools and approaches developed in the USA (FEMA) and New Zealand and at the United Nations University. In closing, Chapter 8, "Disaster Risk Evaluation: Qualitative Methods in Canada," covers disaster models that account for a variety of aspects including the subjective and dynamic nature of risk and vulnerabilities over space and time, access to resources, and community perceptions of hazard-related risks.

The material compiled here is based on decades of my own research with colleagues and graduate students and includes examples and applications of various concepts and methodologies – existing and newly developed. Accordingly, numerous case studies constitute an integral part of this book. In addition, the Canadian federal government and provinces are increasingly taking a keen interest in hazard identification and risk assessment as a policy measure to mitigate disaster impact. This renewed focus on the part of the government is a testament to the importance of this field in light of the growing cost of disasters in Canada and around the globe.

Toronto, ON, Canada Nirupama Agrawal

Acknowledgements

This book could not have been written without the support and guidance of Dr. Tad Murty, my mentor, collaborator, and colleague. He encouraged, challenged, and believed in me throughout the research and writing process – for which I thank him sincerely. I wish to express my profound gratitude to Professor Richard Haigh and Professor Dilanthi Amaratunga, University of Huddersfield, for inspiring me with their research and for graciously agreeing to contribute the foreword; to Dr. Kumaraswamy Ponnambalam, University of Waterloo, for his careful review of the manuscript; and to Petra van Steenbergen, executive editor at Springer, for her incredible patience with me. I am also grateful to my numerous coauthors, collaborators, colleagues, and students over the years, as they have helped me tremendously through their knowledge and passion for the subject matters covered in this book, and I would especially like to thank Dr. Slobodan Simonovic, Western University, for his enduring support. Lastly, I am grateful to my family and friends for always being there for me when I need them most and to my two daughters, Nita and Nandita, whose unconditional love has always been my true strength.

Contents

1 Defining Natural Hazards – Large Scale Hazards 1
1.1 Definitions of Selected Large Scale Natural Hazards 1
1.2 Biophysical (health) Hazards 2
1.2.1 Epidemic 3
1.2.2 Pandemic 3
1.3 Drought 6
1.4 Earthquake 9
1.4.1 Earthquake Measurement and Monitoring 10
1.4.2 Earthquake Zones in Eastern Canada 11
1.4.3 Potential Impacts 12
1.5 Extreme Weather – Heat Wave and Cold Wave 14
1.5.1 Heat Wave 14
1.5.2 Cold Wave 16
1.6 Floods 17
1.6.1 Case Study – Toronto, Canada 18
1.6.2 Case Study – Flood Risk and Urbanization of London, Ontario 22
1.7 Forest Fire/Wildfire 23
1.8 Ice Storm 24
1.8.1 Case Study – Urban Impacts of Ice Storm of December 2013, Toronto, Canada 25
1.9 Hurricane 32
1.9.1 Hurricane Return Periods 33
1.9.2 Hurricane Intensity 33
1.9.3 Case Study – Hurricane Hazel – Toronto, Canada 37
1.9.4 Case Study – Hurricane Sandy, New York, USA 37
1.10 Exercise 38
References 38

2 Defining Natural Hazards – Medium and Small Scale Hazards 41
2.1 Medium Scale Hazards 41
2.1.1 Erosion 41
2.1.2 Landslide 42
2.1.3 Snowstorm/Blizzard 50
2.1.4 Subsidence and Sinkhole 53
2.1.5 Tornado 55
2.1.6 Windstorm 62
2.2 Small Scale Hazards 68
2.2.1 Extraterrestrial Hazard 68
2.2.2 Fog 73
2.2.3 Geomagnetic Storm 73
2.2.4 Hail Storm 74
2.2.5 Lightning 76
2.2.6 Exercise 78
References 78

3 Disaster Risk Management 81
3.1 Disasters 81
3.2 Risk 82
3.3 Disaster Risk Management – Key Elements 84
3.4 Threat Recognition 84
3.5 Risk Analysis and Assessment 85
3.5.1 Case Study – A Multi-tier Hazard in Northern India in 2013 86
3.6 Risk Control Options 91
3.6.1 Case Study – London, Canada 92
3.6.2 Case Study – Red River Basin, Manitoba, Canada 95
3.7 Strategic Planning 105
3.7.1 Case Study – The 2004 Indian Ocean Tsunami 109
3.8 Response, Recovery, Reconstruction, and Rehabilitation 111
3.8.1 Case of Vertical Evacuation Shelters in India 112
3.8.2 Case Discussion – California 117
3.8.3 Case of Nepal – Post-earthquake Reconstruction 118
3.8.4 Case of Indian Ocean Tsunami 2004 – India 119
3.9 Knowledge Management and Sustainable Development 126
3.9.1 Case of Canadian Coasts – Arctic, Western, and Eastern 127
3.9.2 Case of Climate Change and Sunspots in Canada 131
3.9.3 Case Study – Propane Explosion in Toronto, Canada 133
3.10 Exercise 141
References 141

4 Disaster Resilience 147
4.1 Resilience 147
4.2 Community Resilience 152
4.2.1 Case Study Based on Events in Canada, Japan and New Zealand 154
4.2.2 Case Study Based on a Women's Group in Toronto, Canada 157
4.2.3 Case Study Based on Women's Training Group in Pakistan 160
4.3 Resilience of the Built Environment 165
4.4 Decision Support Tool For Estimating Resilience 167
4.4.1 Data Requirement 168
4.5 A New and Comprehensive Approach to Evaluate Resilience . . . 173
4.5.1 A Canadian Case Study to Demonstrate the Approach . . . 174
4.5.2 Findings – Resilience Maps 184
4.6 Exercise 186
References 187

5 Disaster Perceptions 193
5.1 Perception of Risk 193
5.1.1 Media's Influence on Risk Perception 200
5.2 Perception of Vulnerability 200
5.3 Perception of People 202
5.3.1 Case Studies 205
5.4 Perspectives of Emergency Managers 208
5.4.1 Highlights of the Findings 210
5.5 Exercise 214
References 214

6 Disaster Risk Evaluation – Quantitative Methods in Canada 219
6.1 Hazard Identification and Risk Assessment Method 219
6.1.1 Purpose 220
6.1.2 Scope 220
6.1.3 Structure of the HIRA Process (Fig. 6.1) 221
6.1.4 Application of HIRA for Hazards in Ontario, Canada 224
6.2 Hazard Risk and Vulnerability Assessment Tool 226
6.2.1 Purpose 228
6.2.2 Objective 230
6.2.3 Steps Required for HRVA 230
6.2.4 Understanding Hazards 233
6.2.5 Identification of Vulnerabilities 233
6.2.6 Impact Assessment and Ranking 236
6.2.7 Case Studies on the Application of the HRVA 238
6.3 All Hazards Risk Assessment Method 253
6.3.1 Purpose 253

6.3.2 Overview of the AHRA Process 254
6.3.3 AHRA Business Cycle (Fig. 6.19) 254
6.4 Exercise 268
References 268

7 Disaster Risk Evaluation – Other Quantitative Methods 271
7.1 Federal Emergency Management Agency Method 271
7.1.1 Timing and Duration of Disruption 273
7.1.2 Conducting the BIA 274
7.1.3 Resources Required to Supporting Recovery Strategies 276
7.1.4 Testing & Exercises 277
7.2 SMUG Model 278
7.2.1 Seriousness 278
7.2.2 Manageability 279
7.2.3 Urgency 279
7.2.4 Growth 279
7.2.5 Application of the SMUG Model 279
7.2.6 Challenges in Prioritizing Hazards 281
7.2.7 Case Study: Risk Profile for Chatham Islands 282
7.3 World Risk Index Tool 284
7.4 Role and Assessment of Return Period 285
7.4.1 Working Exercise to Calculate Return Period 288
7.5 Exercise 291
References 292

8 Disaster Risk Evaluation – Qualitative Methods 295
8.1 Pressure and Release (PAR) Model 295
8.1.1 Application of Pressure and Release (PAR) Model to the Case of Hurricane Katrina 296
8.1.2 Application of PAR Model to the 2008 Sichuan, China Earthquake 297
8.2 Access to Resources (ATR) Model 299
8.2.1 Application of Access to Resources Model to the Case of Hurricane Katrina 305
8.3 Community Perception Model 306
8.3.1 Application of Community Perception Model 309
8.3.2 Comparison of Community Perception Model with Conventional Approach 312
8.4 Risk Aversion Concept 313
8.5 Exercise 314
References 314

Appendices 317

Glossary and Definitions of Terms 345

Learning Objectives 355

Index 357

Chapter 1
Defining Natural Hazards – Large Scale Hazards

The natural forces at work on planet Earth have been an integral part of life since the dawn of mankind. The impacts of hazards of natural origin can range from affecting infrastructure, personal possessions, and ecosystems to negatively affecting individuals' psychosocial wellbeing. Disasters are the aftermath of hazards caused by natural phenomena, set off by shifts in tectonic plates or atmospheric interactions in populated areas. The extant literature offers a variety of ways to classify natural hazards. For example, they can be categorized by their origin – geological, hydro-meteorological or biophysical; by their nature and speed – permanent, ephemeral or episodic; or on the basis of their size or scale – large, medium or small. Adopting the last of the three classification schemes, this chapter presents large scale hazards, which are more likely to occur on the North American continent, in alphabetical order. The list of hazards includes biophysical hazards, droughts, earthquakes, extreme weather, floods, forest fires, ice storms and hurricanes. To help readers follow the material, the chapter draws heavily on recent examples.

1.1 Definitions of Selected Large Scale Natural Hazards

Natural hazards are caused by natural forces of nature and the environment. Human activities may enhance the impact of a natural hazard but they do not trigger it. For example, excessive deforestation may cause landslides in the event of heavy rainfall.

A summary of the history of natural hazards in Canada since the 1900s is given in Table 1.1.

N. Agrawal, *Natural Disasters and Risk Management in Canada*, Advances in Natural and Technological Hazards Research 49,
https://doi.org/10.1007/978-94-024-1283-3_1

Table 1.1 A summary of the history of natural hazards in Canada since the 1900s (Public Safety Canada 2015)

Disaster type	Disaster subtype	Events count	Total deaths	Total affected	Total damage ('000 US$)
Drought	Drought	5	0	55,000	4,810,000
Earthquake	Tsunami	1	27	0	0
Epidemic	Viral disease	4	50,526	2,008,347	0
Epidemic	Bacterial disease	2	35	171	0
Epidemic	Parasitic disease	1	1	399	0
Extreme temperature	Heat wave	1	500	0	0
Extreme temperature	Cold wave	3	0	200	2,000,000
Extreme temperature	Severe winter conditions	1	10	0	0
Flood	Coastal flood	1	0	0	58,000
Flood	Riverine flood	25	43	167,770	7,815,100
Flood	Flash flood	3	0	7304	339,000
Mass movement (dry)	Rockfall	2	94	41	0
Mass movement (dry)	Landslide	3	67	3506	0
Mass movement (dry)	Avalanche	3	144	44	0
Storm	Convective storm	23	89	10,914	4,556,000
Storm		22	128	117,870	2,093,900
Storm	Tropical cyclone	8	88	203	310,100
Wildfire	Forest fire	19	119	67,600	6,462,500
Wildfire	Land fire (Brush, Bush, Pasture)	1	0	5000	0

1.2 Biophysical (health) Hazards

Health and biophysical hazards have affected society for centuries. Many cases in the past have been documented such as the black plague where a sizable proportion of the population was affected with many succumbing to their ailments. Malthus (1826) indicated that there are three major factors that control populations where disease and poor health being one of the control factors (the other two were famine and war).

1.2.1 *Epidemic*

An epidemic is defined by either an unusual increase in the number of cases of an infectious disease, which already exists in the region or population concerned; or the appearance of an infection previously absent from a region. EMDAT (2015b).

Ebola: Ebola virus disease (EVD) is a severe disease that causes hemorrhagic fever in humans and animals. Diseases that cause hemorrhagic fevers, such as Ebola, are often fatal as they affect the body's vascular system (how blood moves through the body). This can lead to significant internal bleeding and organ failure.

1.2.1.1 Case Study – Ebola Outbreak in 2014 – West Africa

A "mysterious" disease began silently spreading in a small village in Guinea on 26 December 2013 but was not identified as Ebola until 21 March 2014 (WHO 2015a). The average EVD case fatality rate is around 50%. Case fatality rates have varied from 25% to 90% in past outbreaks. Primarily, about 4 West African countries suffered from the Ebola outbreak at an unprecedented level (Fig. 1.1). Sierra Leone recorded a staggering 20,171 cases of EVD, 7890 deaths from it, with 660 case and 375 deaths reported for health care workers (Fig. 1.2) (WHO 2015b; Dumbuya and Nirupama 2016).

1.2.2 *Pandemic*

A pandemic is a worldwide outbreak of a specific disease which affects a large proportion of the population. The federal, provincial, and territorial governments in Canada are working on pandemic preparedness, and many municipalities, companies, and health care facilities also have plans in place (Health Canada 2015a).

1.2.2.1 Examples

Avian influenza (AI) is a contagious viral infection caused by the influenza virus Type "A", which can affect several species of food producing birds (chickens, turkeys, quails, guinea fowl, etc.), as well as pet birds and wild birds (Health Canada 2015a).

An outbreak of human infections with a new avian influenza A (H7N9) virus was first reported in China by the World Health Organization (WHO) in March 2013. Cases have been reported mostly in China. In addition, travel-related cases have been

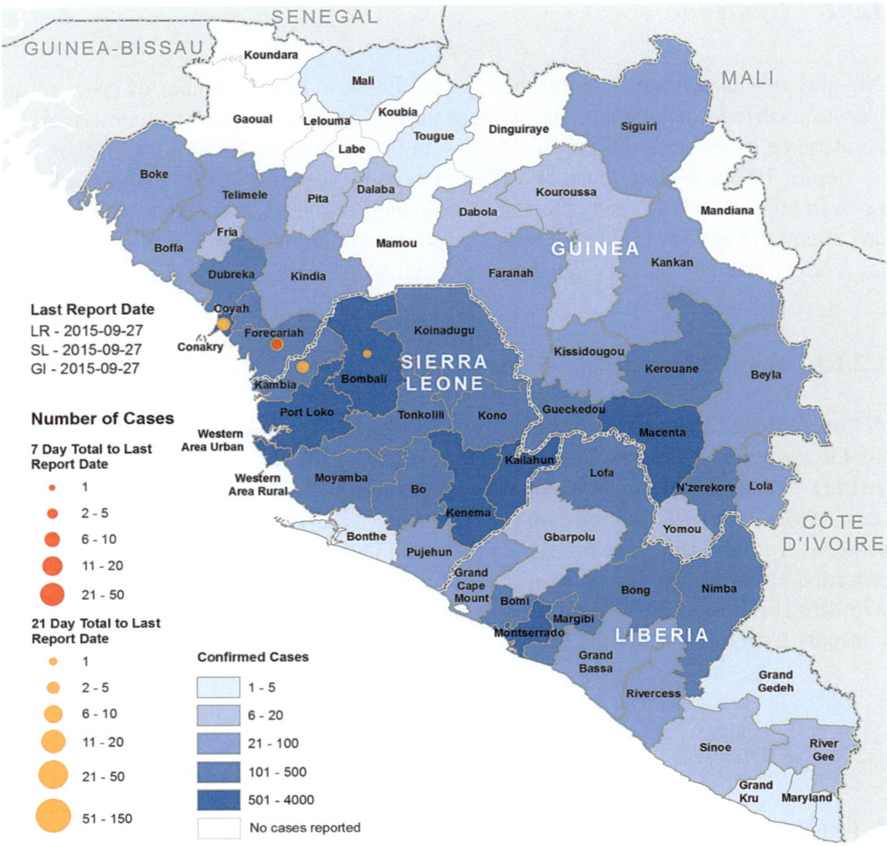

Fig. 1.1 Ebola outbreak showing geographical distribution of total confirmed cases as of February 2015 (WHO 2015a)

reported in Malaysia, Taiwan, Hong Kong, and most recently in Canada. The first case of avian influenza A (H7N9) in a human in North America was confirmed in January 2015. This individual lives in British Columbia and recently returned home from a trip to China. The person in question did not have symptoms while travelling, but became sick after returning to Canada. The individual was not sick enough to be hospitalized and recovered. As precautionary measures, all persons in contact with the infected individual were monitored by public health authorities. Generally, the risk of Canadians getting sick with avian influenza A H7N9 remains very low as evidence suggests that it does not transmit easily from person-to-person. The majority of people in China infected with avian influenza A (H7N9) had previously been exposed to the live birds, mostly chicken. This particular strain of avian influenza A (H7N9) virus has not been detected in birds in Canada.

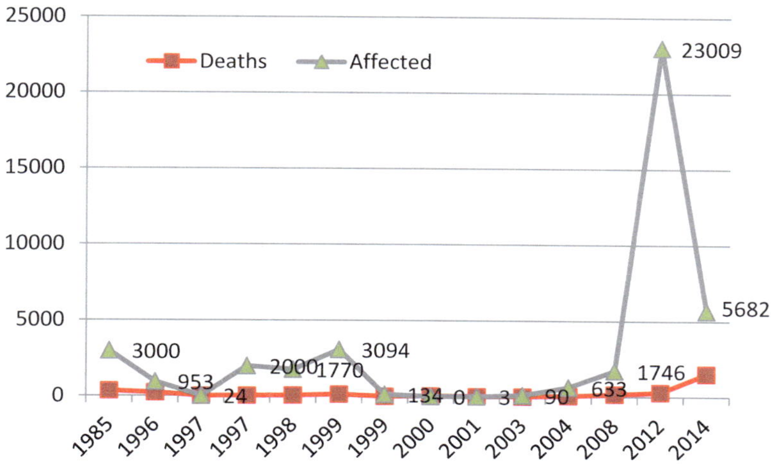

Fig. 1.2 Epidemics caused by, both viral and bacterial diseases in Sierra Leone since 1985 (based on OECD data)

1.2.2.2 H1N1 Pandemic

From 2009 to 2010, Ontario and many other parts of the world experienced its first pandemic in more than 40 years. In April 2009, the first cases were being reported in Mexico and the United States. The rapid spread of the virus and its appearance in other countries resulted in the WHO declaring a global pandemic on 11 June 2009. The pandemic was caused by influenza A virus subtype H1N1, which is colloquially known as 'swine flu' or simply referred to as H1N1. Similarly, the elderly were at first warned that they may be especially vulnerable. However, it was later found that people born prior to 1957 already had some immunity towards the virus due to exposure to a similar strain.

There were at least 8633 confirmed cases of H1N1 in Ontario, however, many other cases went unconfirmed and many people were able to recover without medical intervention. A total of 128 people died during this outbreak due direct to the virus or associated complications related to being infected by the virus in Ontario (Ministry of Health and Long-Term Care 2010).

1.2.2.3 Severe Acute Respiratory Syndrome (SARS)

Severe acute respiratory syndrome (SARS) was first reported in Asia in February 2003. it is a viral respiratory illness caused by a coronavirus, called SARS-associated coronavirus (SARS-CoV). The illness spread to more than two dozen countries in North America, South America, Europe, and Asia before the global outbreak of 2003 was contained. Since 2004, no known cases of SARS have been reported anywhere (CDC 2015).

SARS is spread by close contact with someone who is infected with the SARS coronavirus. Examples of close contact include living in the same household, providing care to someone with SARS, or having direct contact with respiratory secretions and body fluids of someone affected by SARS.

To date, it appears that people with SARS are not contagious until they develop symptoms. It may take up to ten days from the time they were in contact with someone who has SARS to show symptoms (Health Canada 2015b).

There are several ways in which diseases, viruses, illness, and health outbreaks can be introduced and spread throughout a community:

- *Direct contact*: a person can become infected through close physical contact (e.g. kissing, touching) a person who is already infected.
- *Indirect contact*: a person can become infected by coming into contact with a surface that has been contaminated.
- *Droplet contact*: a person can become infected from exposure to droplets that have touched the surfaces of the eyes, mouth, or nose of an infected person. Sneezing and coughing are two methods in which this type of illness can be spread. This differs from airborne transmission since the droplets are too large to remain in the air for long periods.
- *Airborne transmission*: a person can become infected from exposure to droplet nuclei and contaminated dust particles which are capable of staying airborne. Few diseases are capable of surviving airborne transmission (e.g. influenza, pneumonia).
- *Vector-borne transmission*: a person can become infected through contact with an infected animal or insect. Mosquitoes are the most common vector for disease in humans (WHO 2015b).

1.3 Drought

According to NOAA's national Centers for Environmental Information (NOAA 2016), drought is a complex phenomenon which is difficult to monitor and define. Unlike hurricanes that have a definite beginning and end and can easily be seen as they develop and move, drought can develop slowly over time and impact many sectors of the economy in a systematic manner, thereby influencing many different space and time scales. However, drought can be viewed with respect to two general time periods. There can be acute and chronic drought events. An acute (or short-term) drought example can be seen through the Canadian prairie drought where the system is able to rebound when weather patterns return moisture. In terms of chronic (or long-term) drought, a dramatic example is the drying of the Aral Sea. In this case, drought has essentially caused a large water body to vanish, going extinct. In the case of long-term drought, there is typically irreparable damage and high unlikeliness of conditions to return to a pervious status.

There is an alternative perspective of defining drought types. The climatological community and specifically the work of Wilhite and Glantz (1985) have defined four types of drought. These are as follows:

1. meteorological drought,
2. agricultural drought
3. hydrological drought,
4. socioeconomic drought

Meteorological drought is measured based on dry weather patterns that dominate an area. These weather patterns can be the result of longer-term variations in the earth's atmospheric patterns or alternations of the earth's circulatory system due to human activity. Hydrological drought is observed when restricted water supply becomes evident, especially in streams, reservoirs, and groundwater levels. This type of drought usually happens after many months of meteorological drought. The result of this type of drought has immediate impact to local hydrology of impacted areas that are reflected in biological activity and in some cases the economy and society. Agricultural drought happens when crops become affected and cannot support normal biological growth (function). This type of drought is by nature connected to both meteorological and hydrological drought controls. Lastly, socioeconomic drought relates to the supply and demand of various commodities impacted by drought. From a general perspective it is usually the last drought type to be measured as the immediate and arguably the most severe drought types have already wreaked much of the havoc.

Meteorological drought can begin and end rapidly, while hydrological drought takes much longer to develop and recover from due to elapse times for proper hydrology to return, especially normalized base flow. Many different indices have been developed over the decades to measure drought in these various sectors, with those in dollar amounts being the index to resonate most easily with the general public. The U.S. Drought Monitor depicts drought integrated across all time scales and differentiates between agricultural and hydrological impacts as a means for clarity.

In Canada, Ontarians are not particularly vulnerable to drought or water shortage emergencies. In this respect, it is extremely rare in developed countries for people to die or be injured by drought. In fact, there have been no deaths or fatalities reported in any of the droughts recorded in Ontario. A caveat to consider is for people living in isolated areas and those who rely on wells for water, as they may require assistance during periods of drought. For these vulnerable people, effective management plans and communication is essential, which will be discussed in further detail later in this book.

Although Ontario and some other parts of Canada are not dramatically susceptible to severe drought events, the Canadian Prairies, comprised of Manitoba, Saskatchewan, and Alberta, periodically experience drought conditions that have been quantified in large economic losses. During 1988, Saskatchewan suffered severe drought, which created widespread hardships to local industries like agriculture and animal husbandry. In 2001, Southwestern and Eastern Ontario had an eight week dry

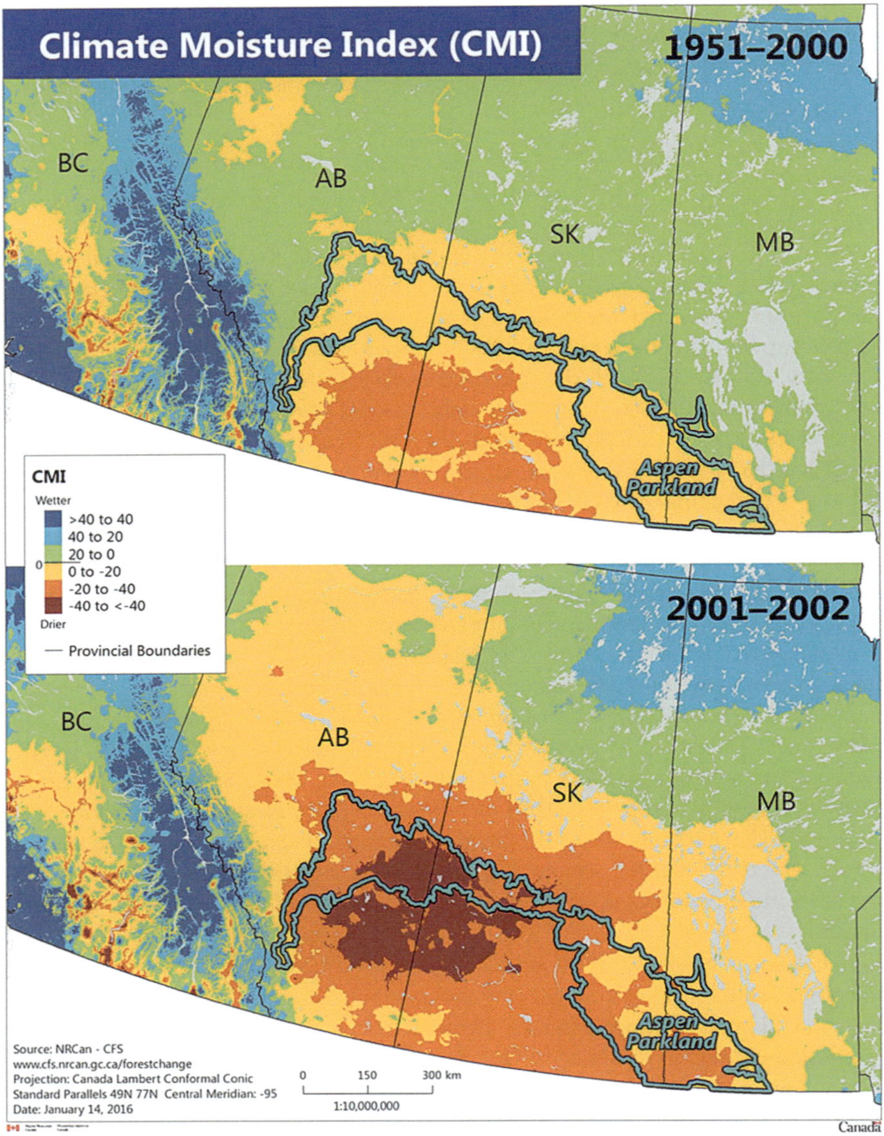

Fig. 1.3 A drainage ditch west of Osage, Saskatchewan, April 1988. Regina Leader-Post http://esask.uregina.ca/entry/drought.html Climate Moisture Index (CMI) in Canadian prairies during 1951-2002 that including the two severe drought years of 1988 and 2001 (Source: Natural Resource Canada)

Table 1.2 historical drought events occurred in Canada since 1900. Data extracted from EMDAT (www.emdat.be)

Year	Disaster type	Total deaths	Affected	Injured	Homeless	Total damage (CND)
1931	Drought	0	25,000	0	0	0
1961	Drought	0	0	0	0	0
1977	Drought	0	0	0	0	3,000,000
1984	Drought	0	30,000	0	0	1,000,000
1988	Drought	0	0	0	0	810,000

period in the middle of the growing season. During this period, agricultural drought became present and crops were severely damaged. Some areas received less than 15% of their normal rainfall during the 54-day time period the drought was measured. Figure 1.3 shows Climate Moisture Index, a measure of drought, during 1951 to 2002 in Canadian prairies. Over a stretch of 82 days, several communities in Southern Ontario had no significant rainfall. During the same period, some localities observed 21 days of temperatures above 30 °C. The Ottawa River came within 11 cm of its lowest level in 50 years on August 14 setting a new precedence for the river's hydrology and possible outlook of its environmental status (Environment Canada 2014).

In California, USA, Canada's neighbours to the south, drought concerns are growing exponentially in the twenty-first century. Many are worried what will happen in the near future, as many farmers have been forced to destroy their orange and almond orchards/trees because of the lack of economic feasibility to irrigate and maintain their vitality. In regions like California, groundwater well drilling is high in demand but come at a very expensive cost to initiate and maintain due to significant drops in the ground water table depth. Kristof reported in New York Time (Kristof 2015) that creeks, springs, and canals are going to dry up along the Pacific Crest Trail, as a likely result due to increased well drilling and extended water extraction. A CBC (2009) documentary 'Last call at the Oasis' makes fascinating discoveries and observations on the topic.

In Canadian natural disaster history, among the top ten worst natural disasters people were affected by, two are drought events that occurred in 1931 and 1984 (Table 1.2). This indicates that drought events are a real concern as a natural disaster to contend with and to be better understood for emergency preparedness and regional and national planning.

1.4 Earthquake

An earthquake is the physical interaction of two adjacent tectonic plates suddenly slipping past one another along the faultline. The location below the earth's surface where the earthquake starts is known as the hypocenter, whereas the location directly above it on the surface of the earth is termed the epicenter. Thus, the distinction between the hypocenter and the epicenter is their closeness to the earth's surface. In some cases, an earthquake may have foreshocks. Foreshocks are smaller earthquakes that happen at the same location as the larger earthquake that follows. Scientists are unable to determine that an earthquake is a foreshock until the larger earthquake happens, and need to wait in anticipation before sending formal reports on seismic activity. The mainshock is the largest, main earthquake of large tectonic plate slip. Mainshocks are always followed by aftershocks due to surficial geologic structure finding stability after the initial event. Aftershocks are smaller earthquakes that occur in the same place as the mainshock but only afterwards. It is not well understood but aftershocks can happen continuously for weeks to years after the mainshock, as dependent on the size of the mainshock.

Earthquake hazards include any physical phenomenon associated with an earthquake that may produce adverse effects on human activities and livelihoods. While they are often used as synonyms, it is useful to distinguish between "hazards" and "risk". Hazards are the natural phenomena that might impact a region, regardless of whether there is anyone around to experience them or not. Risk refers to what we stand to lose when the hazard occurs; it is what we have built that's threatened. Risk can usually be measured in dollars or fatalities. Hazard is generally measured in more physical units: energy, shaking strength, depth of water inundation, etc. (PNSN 2015).

1.4.1 Earthquake Measurement and Monitoring

A number of scales are used to measure the magnitude of earthquakes. The Richter scale is one of those measurement scales, which assigns a magnitude number to quantify the energy released by an earthquake. The Richter scale was developed in the 1935 by the seismologist Charles Francis Richter but was succeeded in the 1970s by the Moment Magnitude Scale (MMS), which is now the scale used by the United States Geological Survey to estimate magnitudes for all modern large earthquakes. Another earthquake measurement scale is the Modified Mercalli Intensity scale and it measures the effect of an earthquake. It consists of a series of specific response variables such as people awakening, movement of furniture, damage to chimneys, and total destruction.

Although numerous *intensity scales* have been developed over the last several hundred years to evaluate the effects of earthquakes, the one currently used in the United States is the Modified Mercalli (MM) Intensity Scale. It was developed in 1931 by the American seismologists Harry Wood and Frank Neumann. This scale, composed of increasing levels of intensity that range from imperceptible shaking to catastrophic destruction, is designated by Roman numerals. It does not have a mathematical basis thus being based on subjective and qualitative assessment. It is instead an arbitrary ranking based on observed effects. The lower numbers of the intensity scale generally deal with the manner in which the earthquake is felt by people. The **higher** numbers of the scale are based on observed structural damage. Structural engineers usually contribute information for assigning intensity values of VIII or above (USGS 2015a).

The following is an abbreviated description of the 12 levels (denoted in roman numerals) of Modified Mercalli intensity (USGS 2015b).

(I) Not felt except by a very few under especially favorable conditions.
(II) Felt only by a few persons at rest, especially on upper floors of buildings. Delicately suspended objects may swing.
(III) Felt quite noticeably by persons indoors, especially on upper floors of buildings. Many people do not recognize it as an earthquake. Standing motor cars may rock slightly. Vibration similar to the passing of a truck. Duration estimated.
(IV) Felt indoors by many, outdoors by few during the day. At night, some awakened. Dishes, windows, doors disturbed; walls make cracking sound. Sensation like heavy truck striking building. Standing motor cars rocked noticeably.
(V) Felt by nearly everyone; many awakened. some dishes, windows broken. Unstable objects overturned. Pendulum clocks may stop.
(VI) Felt by all, many frightened. Some heavy furniture moved; a few instances of fallen plaster. Damage slight.
(VII) Damage negligible in buildings of good design and construction; slight to moderate in well-built ordinary structures; considerable damage in poorly built or badly designed structures; some chimneys broken.
(VIII) Damage slight in specially designed structures; considerable damage in ordinary substantial buildings with partial collapse. Damage great in poorly built structures. Fall of chmineys, factory stacks, columns, monuments, walls. Heavy furniture overturned.
(IX) Damage considerable in specially designed structures; well-designed frame structures thrown out of plumb. Damage great in substantial buildings, with partial collapse. Buildings shifted off foundations.
(X) Some well-built wooden structures destroyed; most masonry and frame structures destroyed with foundations. Rail bent.
(XI) Few, if any (masonry) structures remain standing. Bridges destroyed. Rails bent greatly.
(XII) Damage total. Lines of sight and level are distorted. Objects thrown into the air.

1.4.2 Earthquake Zones in Eastern Canada

1.4.2.1 The Western Quebec Seismic Zone

The Western Quebec Zone includes the Ottawa Valley from Montreal to Temiscaming. This large zone also includes the Laurentians and Eastern Ontario. Large urban centres located in this zone include Montreal, Ottawa-Hull, and Cornwall. Two large earthquakes with magnitudes greater than five occurred here in 1935 and 1944 (HIRA 2005).

1.4.2.2 Southern Great Lakes

This zone is classified as having a low to moderate level of seismicity compared with the more active zones to the east, along the Ottawa River and in western Quebec. On average, only two or three earthquakes with a magnitude greater than 2.5 on the Richter scale are recorded annually. Only three magnitude five earthquakes have occurred in the past 250 years in 1929, 1986 and 1998. All of these have had epicenters across the border but were widely felt in Ontario (Natural Resources Canada 2009).

1.4.2.3 North-Eastern Ontario

Northern Ontario has historically experienced a very low level of seismic activity. This area has averaged only one or two earthquakes per year with a magnitude greater than 2.5. In 1905 and 1928 this zone experienced earthquakes each with a magnitude of five. Several studies have identified the Ottawa area as having the highest risk of an earthquake in Ontario. Although this area may have the highest risk within Ontario, it is classified as having a moderate risk compared to other parts of Canada. Since this is a heavily developed area, a significant earthquake close to this area has the potential to cause considerable damage to infrastructure, natural resources, and ecosystems. Several scientific studies, such as Ploeger et al. (2008), have noted that intraplate earthquakes are felt over a larger area than plate boundary earthquakes of the same magnitude. This is likely due to the relatively stable and un-fractured crust in the continental interior and the presence of soft and deep sediments in some areas. Soft and deep sediments can amplify seismic waves and this effect has resulted in areas of greater historical losses in eastern Canada.

Figure 1.4 shows earthquake zones in Canada.

1.4.3 Potential Impacts

While strong earthquakes are very rare in Ontario and a significant earthquake has never occurred in Ontario based on the historical record. The people, property, and infrastructure of Ontario would be very vulnerable to this hazard given that advanced planning to mitigate earthquakes is not management concerns, as reflected by the local disaster history. However, a powerful earthquake in the region could cause buildings and structures such as bridges to collapse, trapping people in the debris and reducing critical infrastructure and communication. The susceptibility of people being killed or injured by falling debris such as glass, chimneys, book cases, and roof tiles is of concern during earthquakes as their risk is higher. Further fatalities and injuries may occur during aftershocks if buildings compromised by the original earthquake are re-entered. Fires caused by ruptured gas mains etc., may also pose a significant risk.

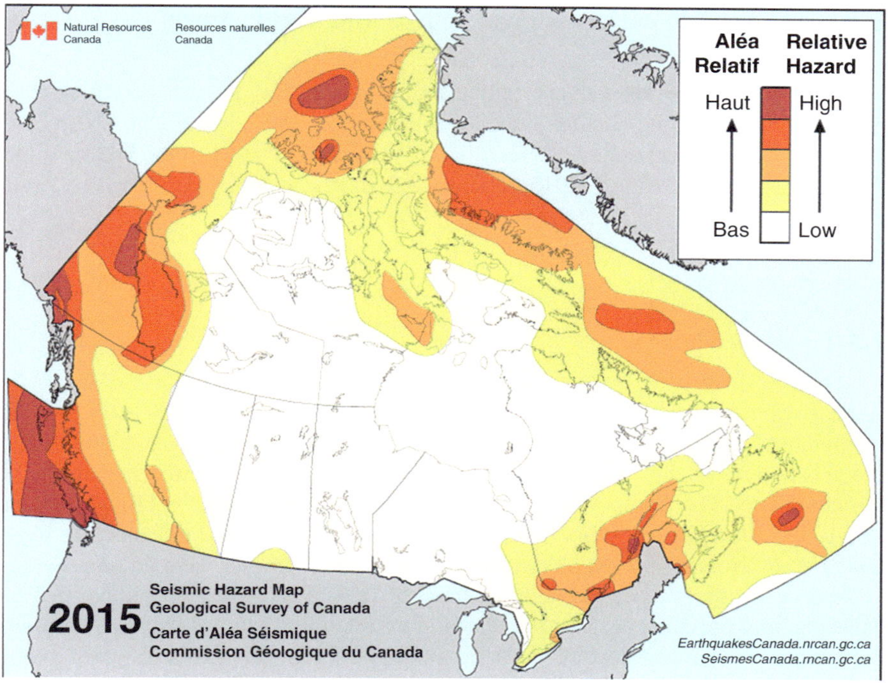

Fig. 1.4 The relative seismic hazard across Canada. (Natural Resources Canada 2009)

Buildings not constructed well and/or with high, unsupported roofs are more likely to be damaged during an earthquake than those that have been well built or properly enforced. Unanchored building materials and contents will further increase the amount of damage. The type of soil and rock the building is located on top can also influence the amount of damage. Buildings on a thick layer of loose sand, silty clays; soft and saturated granular soils; sand and gravel may experience more damage than a building built on deep, unbroken bedrock and stiff soils (Natural Resources Canada 2009). The movement from an earthquake can alter the soil characteristics from solid to liquid which can make the ground suddenly unable to support a building's foundation. This may result in the cracking or collapse of a building (Natural Resources Canada 2009).

The National Building Code of Canada has been revised to include a 2475 year return period (an earthquake with a 2% chance of exceedance in 50 years) for earthquakes. This is applied to the design of new buildings and the evaluation of existing buildings. Due to the infrequent nature of this hazard and the lifespan of buildings being approximately 50 years, a scientific examination of the revised building code concluded that this adjustment is perhaps redundant from an economic and engineering perspective; but that it is certainly sufficient from a public safety perspective (Searer 2007). Therefore, the outcome of a large earthquake in Ontario is not likely to be comparable in destruction as the outcome of the 2010 Haitian earthquake.

1.5 Extreme Weather – Heat Wave and Cold Wave

Extreme temperature poses a very unique and serious natural disaster that can disrupt normal human activities and debilitate infrastructure. In 1936, in Canada, 500 people died in the only severe heat wave recorded in the past 115 years of the international disaster database (EMDAT 2015a). Table 1.3 shows historical cold wave events occurred in Canada since 1900.

1.5.1 Heat Wave

A heat wave is usually defined as a period of 3 or more consecutive days with temperatures of 30 ° C or higher in North America. High humidity is not a requisite, yet most, but not the worst, heat waves are oppressively humid (Environment Canada 2015). The summer of 2012 was the year of the big heat – 16th warm year in a row. In the last ten years there have only been 4 out of 40 seasons that were colder than normal. In 2012 alone, winter, spring and summer were among the top 10 hottest for their respective seasons. Each of July, August and September tied or exceeded any previous year for the warmest on record. It follows that July through September was the warmest of any three-month period in Canada in 65 years. From January to November inclusive, 2012 was the fourth warmest since 1948 when record-keeping began on a nationwide basis. Every region felt the warmth, especially the millions of Canadians living in the Great Lakes/St. Lawrence Lowlands who experienced the warmest such period on record (Environment Canada). Figure 1.5 shows temperature anomalies for the 2012 Big Heat event (Canadian Environmental Health Atlas 2016).

The Government of Canada (GoC 2015) provides calculation of climatic normal for Canada on a 30 year period basis. For example, calculation of the 1981 to 2010 climate normal for Canada can be found at http://climate.weather.gc.ca/climate_normals/normals_documentation_e.html?docID=1981

Humidity plays a vital role in determining the severity of heat waves, as the heat felt with humidity taken into account may be much worse than the absolute air temperature. Figure 1.6 shows how to determine humidex values based on air temperatures. The humidex — short for humidity index — is a Canadian innovation first used in 1965, according to Environment Canada.

Table 1.3 Historical cold wave events occurred in Canada since 1900 (Environment Canada 2015)

year	disaster type	Total deaths	Affected	Injured	Homeless	Total damage
1982	Cold wave	0	200	0	0	0
1992	Cold wave	0	0	0	0	2,000,000
2013	Severe winter conditions	10	0	0	0	0
2014	Cold wave	0	0	0	0	0

Fig. 1.5 Record breaking temperature across Canada in March 2012 (Environment Canada 2014)

Humidex	Degree of Comfort
20 - 29	No discomfort
30 - 39	Some discomfort
40 - 45	Great discomfort; avoid exertion
46 and over	Dangerous; possible heat stroke

Humidex Temperature and Relative Humidity

Humidex for Relative Humidity from 100% to 65%. Refer to legend above.

Relative Humidity (%) / Temperature (°C)	100%	95%	90%	85%	80%	75%	70%	65%
21 °C	29	29	28	27	27	26	26	24
22 °C	31	29	29	28	28	27	26	26
23 °C	33	32	32	31	30	29	28	27
24 °C	35	34	33	33	32	31	30	29
25 °C	37	36	35	34	33	33	32	31
26 °C	39	38	37	36	35	34	33	32
27 °C	41	40	39	38	37	36	35	34
28 °C	43	42	41	41	39	38	37	36
29 °C	46	45	44	43	42	41	39	38
30 °C	48	47	46	44	43	42	41	40
31 °C	50	49	48	46	45	44	43	41
32 °C	52	51	50	49	47	46	45	43
33 °C	55	54	52	51	50	48	47	46
34 °C	58	57	55	53	52	51	49	48
35 °C		58	57	56	54	52	51	49
36 °C			58	57	56	54	53	51
37 °C					58	57	55	53
38 °C							57	56

Humidex for Relative Humidity from 60% to 20%

Humidex for Relative Humidity from 60% to 20%. Refer to legend above

Relative Humidity (%) / Temperature (°C)	60%	55%	50%	45%	40%	35%	30%	25%	20%
21 °C	24	23	23	22					
22 °C	24	24	23	23					
23 °C	27	26	25	24	23				
24 °C	28	28	27	26	26	25			
25 °C	30	29	28	27	27	26			
26 °C	31	31	29	28	28	27			
27 °C	33	32	31	30	29	28	28		
28 °C	35	34	33	32	31	29	28		
29 °C	37	36	34	33	32	31	30		
30 °C	38	37	36	35	34	33	31	31	
31 °C	40	39	38	36	35	34	33	31	
32 °C	42	41	39	38	37	36	34	33	
33 °C	44	43	42	40	38	37	36	34	
34 °C	47	45	43	42	41	39	37	36	
35 °C	48	47	45	43	42	41	38	37	
36 °C	50	48	47	45	43	42	40	38	
37 °C	51	50	49	47	45	43	42	40	
38 °C	54	52	51	49	47	46	43	42	40
39 °C	56	54	53	51	49	47	45	43	41
40 °C		57	54	52	51	49	47	44	43
41 °C			56	54	52	50	48	46	44
42 °C				56	54	52	50	48	46
43 °C					56	54	51	49	47

Fig. 1.6 Reference table for humidex estimation with reference legend (Environment and Climate Change Canada 2015)

If the forecast cites a humidex of 40, for example, it means that the temperature might be°C but, with the humidity, the discomfort feels like it would at a dry temperature of 40 °C. The index is based on a calculation of heat and humidity by using current air temperature and the dew point (the temperature and barometric pressure at which water vapour condenses into liquid). It matters because humidity can wreak havoc on a body's internal cooling systems (Hildebrandt 2013).

1.5.2 Cold Wave

A cold wave is defined as a period of abnormally cold weather. Typically, a cold wave lasts two or more days and may be exacerbated by high wind speed. The exact temperature criteria for what constitutes a cold wave vary by location. In Ontario, Canada, extreme cold warnings are issued in South central and south-western Ontario when minimum temperatures are expected to fall to −20 °C or less with maximum temperatures not expected to rise above −10 °C. For the rest of Ontario, they are issued when minimum temperatures are expected to fall to −30 °C or less with maximum temperatures not expected to rise above −20 °C. Figure 1.7 shows the 2008 deep freeze in central and western Canada. During this time, a strong Arctic ridge of high pressure ushered in teeth-chattering Siberian air and bone-chilling winds across the West. Temperatures tumbled to −40 °C in Prairie Provinces – and the "feel-like" temperatures were about 10 °C below actual air temperatures. On

Fig. 1.7 The cold wave of 2008 affected Prairie Provinces in Canada (Environment Canada)

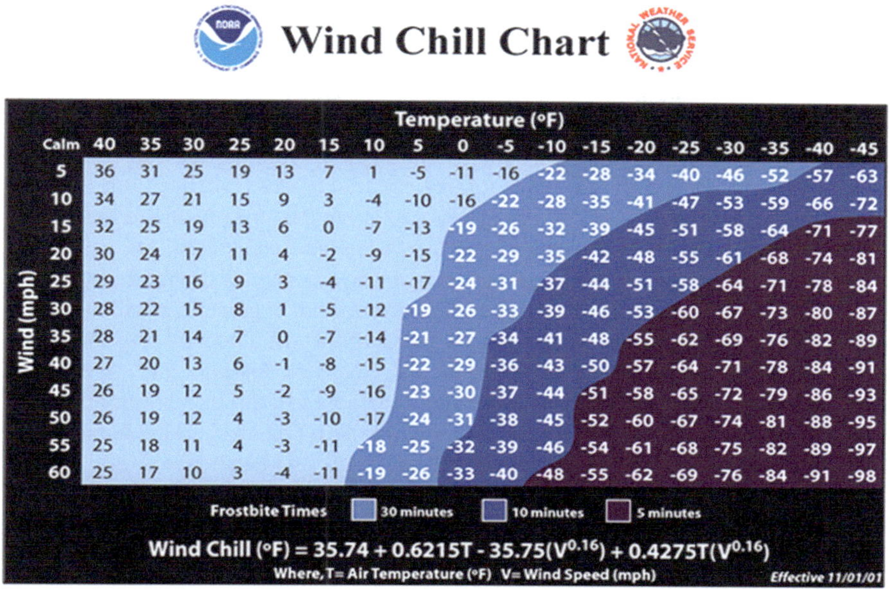

Wind Chill Chart

Wind (mph) \ Temperature (°F): Calm	40	35	30	25	20	15	10	5	0	-5	-10	-15	-20	-25	-30	-35	-40	-45
5	36	31	25	19	13	7	1	-5	-11	-16	-22	-28	-34	-40	-46	-52	-57	-63
10	34	27	21	15	9	3	-4	-10	-16	-22	-28	-35	-41	-47	-53	-59	-66	-72
15	32	25	19	13	6	0	-7	-13	-19	-26	-32	-39	-45	-51	-58	-64	-71	-77
20	30	24	17	11	4	-2	-9	-15	-22	-29	-35	-42	-48	-55	-61	-68	-74	-81
25	29	23	16	9	3	-4	-11	-17	-24	-31	-37	-44	-51	-58	-64	-71	-78	-84
30	28	22	15	8	1	-5	-12	-19	-26	-33	-39	-46	-53	-60	-67	-73	-80	-87
35	28	21	14	7	0	-7	-14	-21	-27	-34	-41	-48	-55	-62	-69	-76	-82	-89
40	27	20	13	6	-1	-8	-15	-22	-29	-36	-43	-50	-57	-64	-71	-78	-84	-91
45	26	19	12	5	-2	-9	-16	-23	-30	-37	-44	-51	-58	-65	-72	-79	-86	-93
50	26	19	12	4	-3	-10	-17	-24	-31	-38	-45	-52	-60	-67	-74	-81	-88	-95
55	25	18	11	4	-3	-11	-18	-25	-32	-39	-46	-54	-61	-68	-75	-82	-89	-97
60	25	17	10	3	-4	-11	-19	-26	-33	-40	-48	-55	-62	-69	-76	-84	-91	-98

Frostbite Times: 30 minutes, 10 minutes, 5 minutes

Wind Chill (°F) = $35.74 + 0.6215T - 35.75(V^{0.16}) + 0.4275T(V^{0.16})$

Where, T= Air Temperature (°F) V= Wind Speed (mph)

Effective 11/01/01

Fig. 1.8 Windchill Index (NOVA 2015)

January 29, wind chills dipped to a deadly −52 in Regina, the capital city of Saskatchewan Province (Environment Canada ||).

Internationally, severe weather warnings are provided by the World Meteorological Organization (WMO 2015). There are a number of websites (PhysLink.com; csgnetwork.com) that provide instant calculators to calculate Humidex or Windchill (a factor of air temperature and wind speed making it feel like colder than the actual temperatures) estimates for a given air temperature and the relative humidity or the wind speed respectively. Figure 1.8 shows Windchill Index (NOVA 2015).

1.6 Floods

Floods are a general term for the overflow of water from a stream channel onto normally dry land in the floodplain (riverine flooding), higher-than-normal levels along the coast and in lakes or reservoirs (coastal flooding) as well as ponding of water at or near the point where the rain fell (flash floods) (EMDAT 2015b).

In Canada, between 1900 and 2015, one major coastal flooding, 25 major riverine floods, and one event of flash flood have been reported in the international database EMDAT. Also, among the top ten most severe disasters ranked by people affected, four are flooding events – May 1950, June 2013, April 1997, and July 1996. The June 2013 flooding in the Alberta, Canada also ranks at number two in terms of economic damage of $5.7 million.

1.6.1 Case Study – Toronto, Canada

Toronto's population is about 2.5 million, concentrated in an area of 630 km^2. The Greater Toronto Area (GTA) covers 7100 km^2 with about 5.5 million people. Toronto is socially and geographically the most vulnerable city in Canada because it is the most populated city (6th in North America), it is located at the north shores of Lake Ontario of the Great Lakes. The Great Lakes are the largest surface fresh water system in the world, where numerous rivers, lakes and creeks that are part of the large watershed come together, and it is affected by air masses originating from the Gulf of Mexico, the Atlantic Ocean, and from the Arctic. In general, populations that surround the Great Lakes increase their susceptibility of flooding due to intense storm activity but it has been observed that Southern Ontario has experienced numerous tornadoes and impacts of passing hurricanes in the past several decades. Some notable storms that have impacted the Southern Ontario and the Great Lakes are Hurricane Hazel in 1954, Hurricane Fran in 1996, and Hurricane Sandy in 2012 causing immense damage and catastrophe to urban areas.

Toronto's topography is relatively smooth, starting at 75 m above sea level at Lake Ontario to 209 m elevation around the North York area located about a 25 km distance north from downtown Toronto. However, the area is characterized by deep ravines, such as the Don River valley, which is about 400 m wide but the river is only 15 m wide. The City of Toronto and GTA are located in the watersheds of the Don River in the east and the Humber River in the west, which drain into Lake Ontario (Fig. 1.9).

Due to the urbanized nature of the watershed, the Don River experiences low base flows and high volume floods. Even a small rainfall can cause the water level to rise very quickly. The average base flow for the Don River is about 4 m^3/s with peak flows occurring in late February and late September corresponding to seasonal variations.

The Don River has provided essential resources and opportunities by contributing and playing an important role in city's economic and social development. The Don River Valley has changed and been manipulated over the last two hundred years. For example, the Lower Don River was modified and the marsh area at the mouth area was filled, which changed the physical and ecological structure (Bonnell and Fortin 2009). This low lying land exposed the current dense population to high risk from flooding (TRC 2009).

The local weather phenomenon of the Toronto region consists of jet streams, high/low pressure systems, and other oceanic and atmospheric drivers. In July 2013, Toronto experienced a weather system where the polar jet stream was in an active phase, with many troughs and ridges occurring across the north Atlantic region in the upper-level circulation. Some of the slowly travelling troughs absorbed moisture from the Atlantic and Gulf of Mexico systems. Flash floods are usually caused by slow moving storms passing over an area or multiple thunderstorms gathering over the same region. This is exactly what took place over Toronto, where two thunderstorms merged right over the airport and downtown area in Toronto (Kimbell 2013).

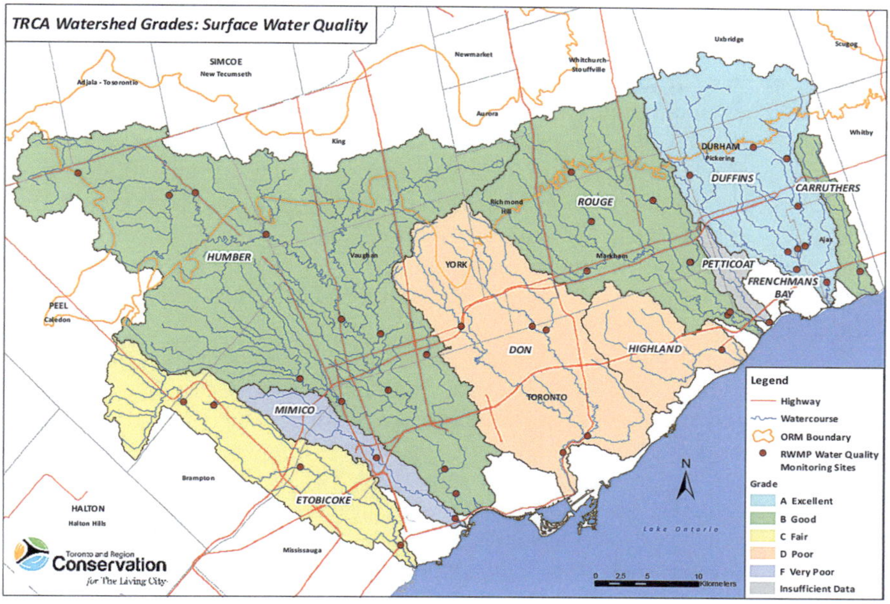

Fig. 1.9 Watersheds of the GTA (Source: wrapfordon, December 2, 2016 https://savethedonriver.wordpress.com/2016/12/02/we-can-help-the-our-watershed/)

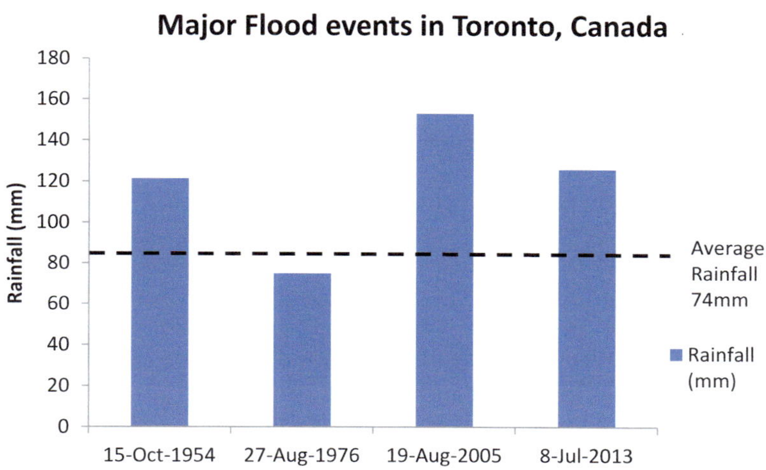

Fig. 1.10 Toronto rainfall during major flooding events

Over the last 100 years, the City of Toronto has experienced four major floods: the flood following hurricane Hazel in October 15, 1954, the August 27, 1976 floods, the August 19, 2005, and the recent flooding of July 8, 2013 (Fig. 1.10).

The flooding due to Hurricane Hazel in 1954 had significant impacts on the city. It left 81 dead and 7472 people were rendered homeless (GOC 2012). About 121 mm of rain fell in 12 h, where some areas experienced up to 210 mm over two days. The flooding impact was most severe in low lying areas of the Don and Humber Rivers, as well as the Etobicoke and Mimico Creeks. Hurricane Hazel was the most severe flooding in the Toronto area in 200 years that triggered the establishment of the then Metropolitan Toronto and Region Conservation Authority (MTRCA), now known as Toronto and Region Conservation Authority (TRCA). Infrastructural damage was unprecedented at the time, including 20 bridges being destroyed or damaged beyond repair and full blocks of homes being swept away (PSC 2013).

The next most significant flood event to impact the Toronto region was in August 1976 when the flooding event that lasted from approximately two days caused by 75 mm of rain by two large storms. The damage of the flooding was estimated to be over 1.3 million dollars. Critical infrastructure, such as bridges were destroyed in the GTA leaving the city incapacitated for some time (TRCA 1999).

On August 19, 2005, the City of Toronto rain gauge recorded 153 mm of rainfall over 3 h, which was only preceded by Hurricane Hazel in 1954 (D'Andrea 2010). This storm was a 100 year event north of the city. The unusually strong storm caused short term flooding in the Don Valley. Peak flow rates for that event were measured at 55.3 m^3/s. More than 10,000 Torontonians were left without power and City Hall received more than 1200 calls for flooded basements. This natural disaster resulted in $500 million in insured damage making the storm the province's most expensive natural disaster and the second costliest nationwide. Creeks, rivers, and ravines in the area were flooded causing bank erosion and damage to critical infrastructure and sewer backups.

The most recent flooding occurred late in the afternoon on July 8, 2013. Some parts of the GTA received over 90 mm of rain and in other rare cases the total exceeded 100 mm. At Pearson International Airport, more than 126 mm of rain was recorded; while the monthly average for Toronto is 74.4 mm. The power outages affected about 300,000 residents. Serious disruptions included flight cancellations, subway and other transportation closures, including the main train station of the city, Union station. Most of the public transit system was not available until the next day. This storm event was the most expensive disaster for Ontario thereby setting a new precedent of climatic and natural disaster events for Canada's most urban and populated city. According to the Insurance Bureau of Canada, the damage of insured properties exceeded $850 million (thestar 2013).

During these major events, the City of Toronto experienced significant impact on essential infrastructure and critical facilities, thus exposing physical, social, and economic vulnerabilities. Figures 1.11 and 1.12a, b are illustrative of damage that occurred during the recent major floods of 2005 and 2013 respectively. It took the emergency response team about seven hours to ferry all stuck on the train to dry ground using small inflatable boats. Almost all of Mississauga, the largest suburb of Toronto with 700,000 people, lost its power during the storm (Global News 2013; Armenakis and Nirupama 2014).

Fig. 1.11 A caved-in culvert at Finch and Sentinel in the north of the city during the 2005 flooding

Fig. 1.12 Impact of the 2013 flooding in GTA. (**a**) Stranded GO Train on flooded tracks with 1400 passengers on board (The Canadian Press/Winston Neutel). (**b**) Submerged roads and underpasses (www.blogto.com by Chris Bateman/July 10, 2013)

This case study was carried out by Nirupama et al. (2014). In the post-Hazel period, significant measures were taken to mitigate future impact due to flooding. The city has been able to manage so far. However, the flooding of 2013 renewed debate on a number of issues, such as failure prone and deteriorating infrastructure, insufficient flood management, and inadequate design codes and standards. Several factors contribute to flooding in this area, including low-lying road and rail network crossing the Don River. The very nature of the valley is wide but not deep making it a non-confining valley, with sharp angles in many places, varying seasonal water levels in Lake Ontario, sedimentation concern, and potential of ice and debris jams (TRCA 2006). Additionally, natural creeks have been buried by sewer pipes, thus altering the natural waterways towards Lake Ontario and forcing existing rivers and creeks to overflow their banks (Young 2013).

The weaknesses addressed here were exposed during the most recently recorded flooding events and led to failures of critical infrastructure, specifically transportation and power networks. In light of this, serious consideration must be made regarding infrastructural renewal, risk communication plans, early warning systems, and education and awareness. Being prepared for extreme storm events, such as a 500-year flood, is highly desirable because it would potentially allow adequate emergency preparedness but a cost benefit analysis is a must to determine feasibility of planning scenarios and resource allocation. For prevention strategies, a combination of adaptive designs and preparedness/ response plans maybe the methodological approach to clear understanding of the regions evolving disaster response capacity. For example, using technological advances and social media, crowd sourcing information and implementing rapid (near real time) two-way risk communication will support efficient and effective response and recovery that allows a type of self-organizing in disaster scenarios. Furthermore, flood damage mitigation strategies should include insurance coverage and/or tax break on retrofitting homes and businesses for those living in flood prone areas.

1.6.2 Case Study – Flood Risk and Urbanization of London, Ontario

This study is focused on whether or not increasing urbanization is leading to increased risk of flood for the City of London in Ontario, Canada. From 1974 to 2000 there has been an elevated risk from floods due to heavy urbanisation in the Upper Thames River watershed in London, Ontario. Databases were prepared using satellite remote sensing technology on landuse classification. This information is integrated with meteorological and hydrological data records and analysed to obtain quantitative estimates of the potential risk from river floods to London.

The goal of the study is to show that progressive urbanization considerably increases the risk of flooding using the City of London, Ontario, Canada as an example. The Upper Thames River (UTR) Watershed has been experiencing net population migration trends that are quite similar to a very large metropolitan area, namely the City of Toronto, which is already facing increased risk of flooding due to urbanization.

This study will illustrate in chapter three the process of establishing a relationship between an impervious area and river flows making use of remote sensing techniques and simultaneously analyzing the relevant meteorological and hydrological data. Results of this study have a direct application in the formulation of policies on land use planning and future balancing of urbanization through conservation means. Once the influence of urbanization on river flows is quantified, it becomes possible to predict the future trends of flooding so that measures can be taken to cope up with increasing demand for residential, commercial areas without risking the increased intensity and extent of storm water in rainy periods.

1.7 Forest Fire/Wildfire

Forest fires or wildfires are any uncontrolled and non-prescribed combustion or burning of plants in a natural setting such as a forest, grassland, brush land, or tundra, which consumes the natural fuels and spreads based on environmental conditions (e.g., wind, topography). Wildfires can be triggered by lightning or human actions (EMDAT). The zone where these fires occurs at the fringes of forest and where urban development has taken place is typically known as the zone of interface. Wildfire in a wooded area is called a forest fire and can cause great damage as the interface zone encroaches on properties or land of economic value and areas settled by populations.

In the past 25 years, Canadian wildfires have consumed an average of 2.3 million hectares a year. These fires occur in forests, shrub lands, and grasslands and can rage out of control for extended periods of time. Some uncontrolled wildfires are ignited by lightning or human carelessness (NRC 2015). Canada has a wildland fire information system that is managed by the NRC, which also monitors peatland fires and carbon emissions. To protect life and property and to minimize area and assets lost to forest fires, fire managers must make decisions every day about where to direct Canada's firefighting resources (Fig. 1.13). Fire managers take on the responsibility to assess and determine the fires that pose a threat to human safety, property and public assets

Fig. 1.13 Fire retardant being sprayed by planes (US Department of Agriculture, Public domain via Wikimedia Commons)

Table 1.4 Forest fires occurred in Canada during 1900–2016 (EMDAT 2015a)

Year	Disaster type	Deaths	Affected	Damage
1911	Forest fire	73	200	0
1922	Forest fire	43	11,000	8000
1980	Bush fire	0	5000	0
1985	Forest fire	0	0	0
1986	Forest fire	0	2000	0
1989	Forest fire	1	25,000	4,200,000
1992	Forest fire	0	0	120,000
1994	Forest fire	0	3000	0
1995	Forest fire	0	6500	89,500
1997	Forest fire	0	1600	0
1998	Forest fire	0	8000	0
1999	Forest fire	0	1500	0
2001	Forest fire	0	1200	0
2002	Forest fire	0	600	0
2003	Forest fire	1	0	545,000
2005	Forest fire	0	0	0
2011	Forest fire	1	7000	1,500,000
2016	Forest fire	2 (indirect)	90,000	7 Billion

(including homes, businesses, utility corridors, wildlife and merchantable timber) and then decide what fire-fighting resources are needed and where. In making these decisions, managers use their experience and information provided by the Canadian Forest Fire Danger Rating System (CFFDRS). According to EMDAT (2015a, b), the international disaster database maintained by the Center for Research on the Epidemiology of Disasters (CRED), Canada experienced a number of severe forest fires since the 1900s (Table 1.4).

1.8 Ice Storm

Freezing rain can result in heavy accumulation of ice on trees, powerlines, utility poles and communication towers. This can result in failure of these erect structures collapsing and causing widespread damage and loss of essential services like heat, electricity, and communication. In some instances, the accumulated ice can disrupt communications and power for days while utility companies attempt to repair extensive damage and restore normal function. Even small accumulations of ice can pose to be extremely dangerous to motorists and pedestrians as surfaces like roads and sidewalks become slippery reducing traction for safe travel. Bridges and overpasses are particularly dangerous during these natural events because they freeze before other surfaces (NWS 2015).

This type of ice accumulation usually occurs due to freezing rain which is due to precipitation initially falling as snow. This snow then encounters a layer of warm air

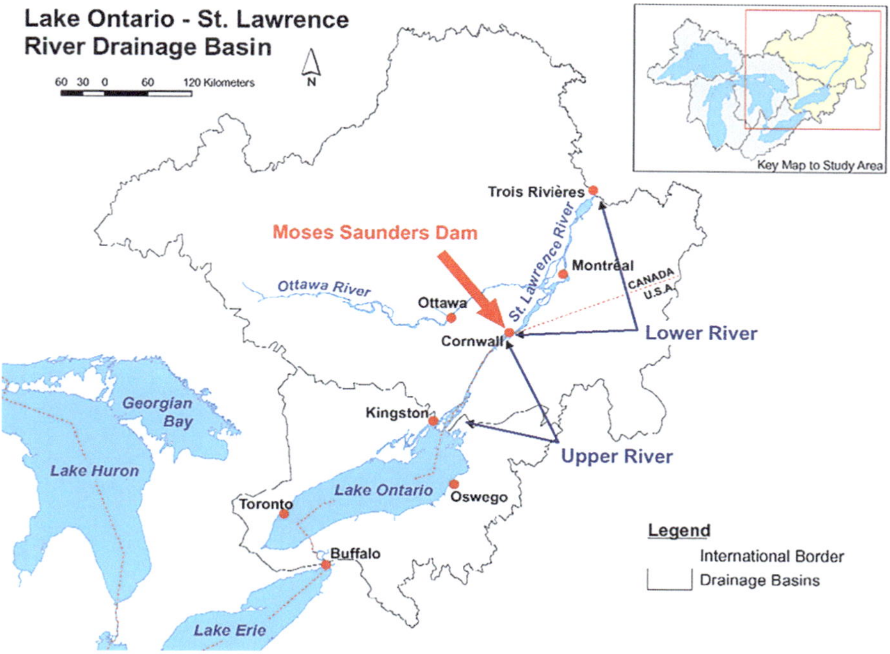

Fig. 1.14 St. Lawrence River Valley and Lake Ontario share jurisdiction with Canada and the United States (http://www.ijc.org/)

as it falls to the ground, which melts the snow changing it to rain. The rain then freezes as it encounters below freezing air at or near the surface creating a film of ice. To put the frequency of freezing rain events in perspective, most of the United States receives less than 10 h of freezing rain annually with the highest frequency in the Saint Lawrence River valley (Fig. 1.14) where over 40 h of freezing rain is observed annually. In recent history, the February 2015 ice storm in Northeast US was the worst in 20 years (http://www.srh.noaa.gov/). Urban impact of ice storms are discussed in the case study based on the 2013 ice storm in Toronto in the following section.

1.8.1 Case Study – Urban Impacts of Ice Storm of December 2013, Toronto, Canada

Toronto, Canada's largest urban centre, was among the hardest hit by the December 2013 ice storm that extended from North Eastern United States to Southern Ontario to Quebec to Maritimes in Canada (Fig. 1.15). The storm produced a significant layer of glaze ice on the ground that caused damage to plants, trees, vehicles, buildings,

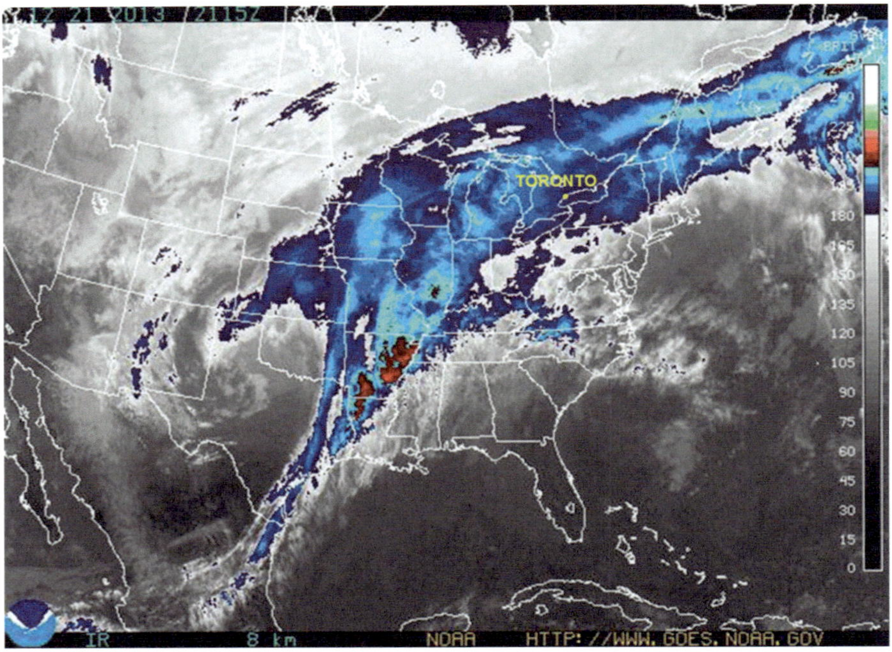

Fig. 1.15 Extent of the storm. GOES Eastern US Sector Infrared Image. Colours highlight colder regions (NOAA http://www.goes.noaa.gov/ECIR4.html) via Wikimedia Commons

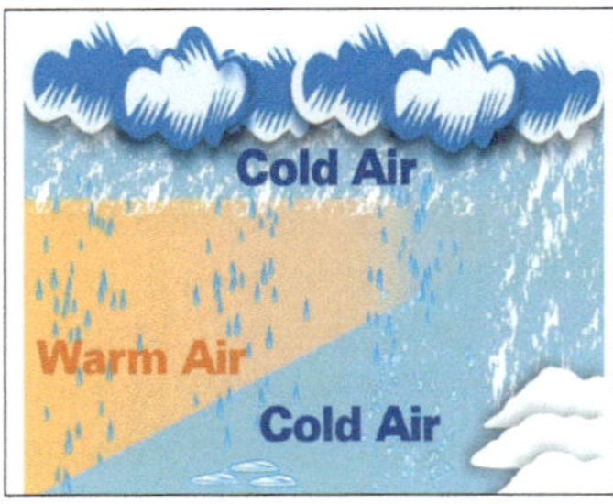

Rain	Freezing rain	Sleet	Snow
Frozen precipitation melts, reaches the ground as rain	Frozen precipitation melts in warm air. Rain falls and freezes on cold surfaces	Frozen precipitation melts in shallow warm air, refreezes into sleet before hitting ground	Snow fall through cold air and reaches the surface

Fig. 1.16 Formation of freezing rain that produces ice storms (Berger 2014)

and most importantly power lines. When ice accumulation is more than 6 mm, it is characterized as an ice storm. An ice storm is a type of winter storm that is defined by freezing rain that falls when ground temperatures are below zero because supercooled water droplets come in contact with the cold surface on the ground and freeze quickly on impact. Typically, freezing rain develops when a moist warm front moves on the top of a cold air mass. Rain falling from the warmer layer becomes supercooled (droplets do not freeze) going through the cold layer without freezing, but freezes as soon as it touches the surface on the ground (Fig. 1.16).

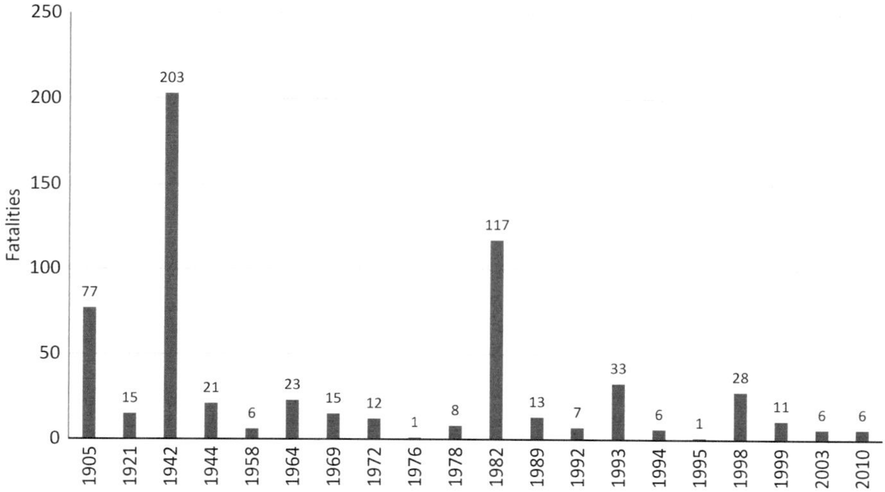

Fig. 1.17 Fatalities of historical winter storms in Eastern Canada during 1900–2014

Historically, between 1900 and 2014, eastern Canada has experienced 54 major winter storms and cold events, many of them with fatalities as shown in Fig. 1.17 and Table 1.5 (PSC 2014; Dolce 2014; Seguin 2013).

Severe weather warnings of freezing rain were forecasted on the eve of Thursday December 19th 2013 for the next day. True to the warning, freezing rain started late on Friday December 20th and left a significant coating of ice over the city. The day after a second wave of freezing rain hit even harder. The total freezing rain over this period amounted close to 40 mm causing ice accretion of up to 30 mm (Table 1.6). Considering that 1 linear meter of ice with 30 mm width and thickness respectively weighs about 0.8 kg, the combination of accumulated ice and strong winds snapped tree branches bringing down power lines, utility poles, and other structures of the distribution system in unprecedented proportions.

The freezing rain and ice accumulation occurred throughout southern Ontario's urban communities covering regions along the northern coast of Lake Ontario, up to Kingston along the HWY 401 through Oshawa and Whitby. In the north, it stretched within York Region affecting Markham, Richmond Hill, up to Aurora and Newmarket (Fig. 1.18). The Cities of Toronto and Brampton (northwest of Toronto, not shown in Fig. 1.18) were severely hit and Mississauga (west of Toronto, not shown in Fig. 1.18) and Hamilton to a lesser degree. A reduced impact was felt in the Niagara area to the west.

The damage suffered by urban forestry was unprecedented as the ice accretion caused major damages and losses to the tree canopy. Trees snapped as they were dormant and fragile during winter weather. In addition, tree limbs with branches heavily coated with ice brought down power lines (Fig. 1.19).

The Ice Storm hit the power distribution system predominately servicing urban communities and cities in southern Ontario. About 500 wires were down leaving

Table 1.5 Historical winter storms in Eastern Canada during 1900–2014

Date	Province	Date	Province	Date	Province	Date	Province	Date	Province	Date	Province
02/15/05	Nova Scotia (NS)	01/13/68	ON	01/26/78	ON, QC	12/24/86	ON, QC	12/08/94	NFL	02/18/04	NS
04/01/14	Labrador	02/02/69	NS	02/08/79	Yukon to ON	01/30/89	Yukon to ON	12/10/95	ON	03/30/05	NFL
02/08/21	Newfoundland (NFL)	12/28/69	Quebec (QC)	01/05/82	Canada	12/18/89	Canada	01/06/98	ON to New Brunswick (NB)	12/01/06	ON
12/01/33	Manitoba (MB) to East Coast	02/27/70	NFL	01/19/82	NFL	01/01/92	Prairies and ON	01/03/99	ON	12/04/07	Maritimes
01/01/42	NFL	03/04/71	QC	02/14/82	NFL	02/01/92	Maritimes	01/13/99	ON	01/09/08	QC
12/11/44	Ontario (ON)	02/19/72	QC	02/22/82	Prince Edward Island (PEI)	03/15/93	East Coast	02/13/99	ON	01/28/08	PEI
03/01/58	NFL	03/18/73	ON	01/01/83	East Coast	11/01/93	QC, ON	01/17/00	Maritimes	03/21/08	NB
02/16/59	NFL	01/02/76	QC	01/01/84	QC	01/05/94	QC	02/10/01	QC	12/13/10	ON
12/01/64	Maritimes	01/28/77	ON	04/13/84	NFL	01/23/94	Alberta (AB) to Maritimes	03/31/03	PEI	12/20/14	ON, QC

Table 1.6 Precipitation during the event of December 2013 in Toronto (Environment and Climate Change Canada 2014)

Date	Max Temp (°C)	Min Temp (°C)	Total Rain (mm)	Total Snow (cm)	Total Precip. (mm)	Snow on Ground (cm)
12/20/2013	0.6	−0.5	8.6	1	9.6	9
12/21/2013	0.2	−1.2	16.6	0	16.6	3
12/22/2013	1.9	−2.6	13.6	0.4	14	3

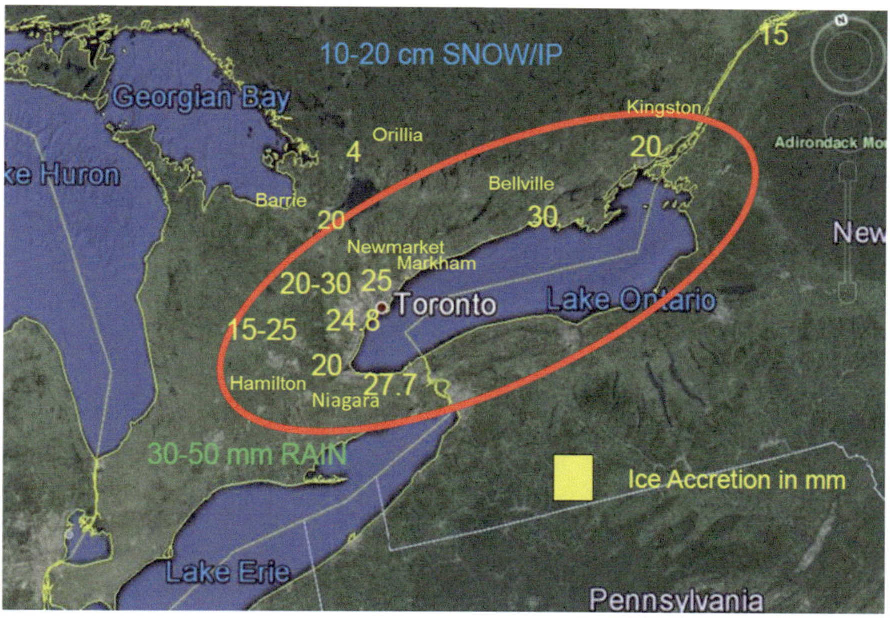

Fig. 1.18 The geographic extent of ice accumulation during the December 2013 storm (Coulson 2014)

more than 300,000 customers (over a million people; a customer corresponds to a household of 3–4 people) without power as well as live wires downed in many areas (Fig. 1.20). Outage of 800 traffic lights caused havoc in the city. Police were dispatched to about 160 locations to maintain order and provide assistance during the chaos. Roads were blocked by fallen tree branches creating traffic backup for kilometers on busy streets. Exposed power lines being knocked down by tree branches exposed vulnerability of the city. Ice on the road caused dangerous slippery conditions triggering dozens of collisions.

Telecommunications broke down during the storm as electricity transmitters, generators, and distributor operations are dependent on it. Mobile phones, social

Fig. 1.19 Image showing snapped tree limbs and knocked down power lines (By Ron Bulovs (Flickr: Crushed!) [CC BY 2.0 (http://creativecommons.org/licenses/by/2.0)], via Wikimedia Commons

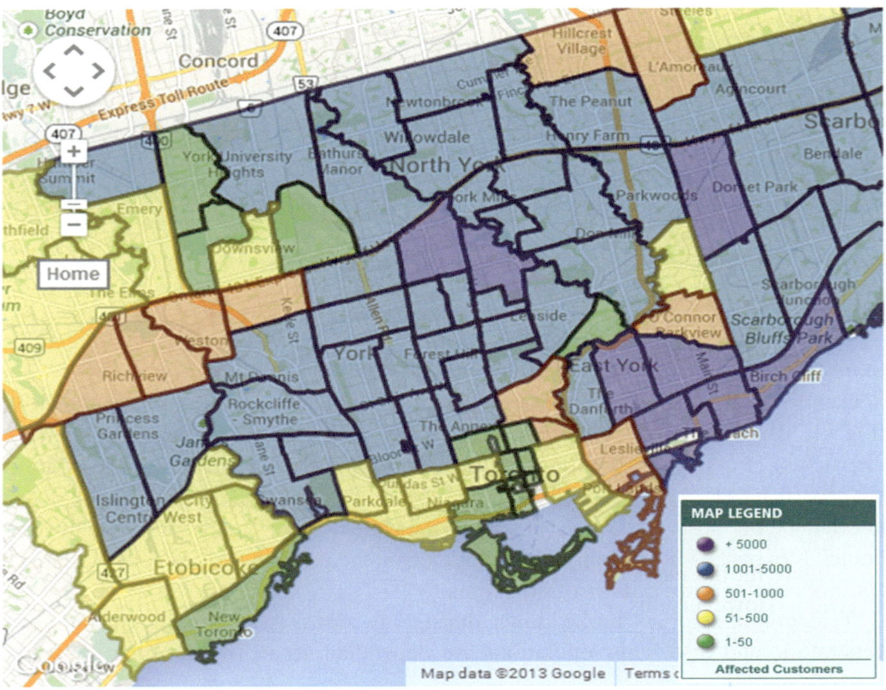

Fig. 1.20 Power outage map of the Greater Toronto Area on Dec 21, 2013 (Toronto Hydro 2013)

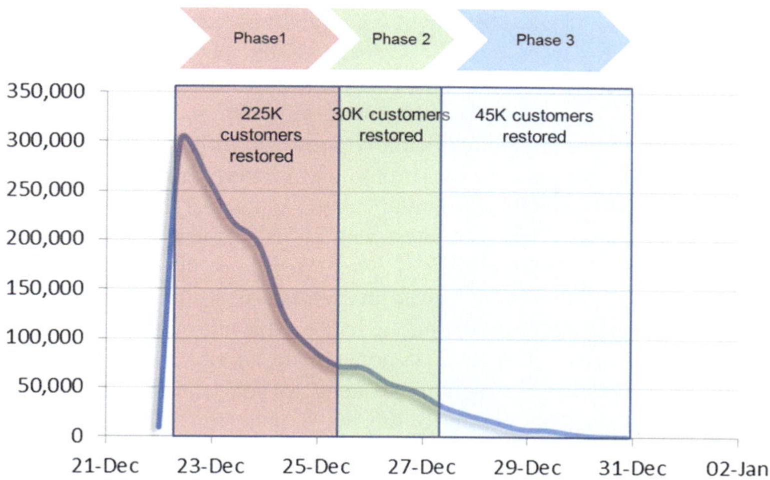

Fig. 1.21 Overview of the power restoration progress (Haines et al. 2014)

networks, personal computers, all became inoperable. By December 25th, there were still 69,800 customers without power across the city (Fig. 1.21).

Obviously, the power outage caused shutdowns and disruptions in public transit system, particularly, the subway and street car lines. Delays also occurred in the long distance train services. Cancellations and delays at Toronto's Pearson International Airport had bigger impact on stranded travellers in the holiday season. Two major hospitals in Toronto had to run on backup generators in the absence of regular power supply. Health impact was evident from lack of heat in the houses, falling ice, slippery sidewalks, consumption of spoiled food, possible electrocution, and carbon monoxide poisoning from operating generators and barbeques in enclosed spaces (Schwartz 2014). Issues were raised about reaching out to the vulnerable population regarding their access to health services and their accessibility by emergency response providers.

To summarize, emergency warming centers were opened up by the City and the Police for people to sleep and eat until their power was restored. Approximately 1000 people spent their Christmas Eve in the warming centres. Part of the immediate and longer term response was the management of tree canopies to address immediate threat posed by the large number of broken trees and branches (City Report 2014). It took more than a week to restore power in the city even though a number of power utility personnel from neighbouring cities contributed to the efforts of power restoration.

Due to the Christmas holidays, schools were closed and people were on holidays therefore, there was less demand for public transit and driving. As the aerial power wires were hit the most, there have been discussions about underground electricity networks. However, this is an expensive option, as the cost is approximately seven times greater than the overhead wires. The City of Toronto's 15,000 km network of overhead power lines would cost about $1.5 billion if transitioned underground, bumping up electivity rates by 300%. Additionally, an underground electrical system is also vulnerable to floods that Toronto has been experiencing recently, as well as repairs and maintenance would be more difficult.

The tree canopy management, particularly pruning operation, is an issue that will have to be addressed by the City of Toronto and property owners. Obviously, the lack of electricity affected public communications and social media as the residents of Toronto could not use their computers and mobile phones, although there were some pockets in the city with electricity where people could charge their mobile/ electronic devices. As Internet access was problematic, Toronto Hydro was severally inundated with phone calls from the public looking for information and emergency support during this crisis. A registry system for the vulnerable would be an idea worth exploring as a future mitigation strategy to help and support people with special needs. In subsequent sections of this book, elaborated discussions will focus on emergency preparedness and those who have the means to be resilient to specific disaster scenarios. While, according to the Insurance Bureau of Canada, the cost of insured losses due to the ice storm was in the range of $200 million, the City of Toronto reported its cost as over $106 million (City Report 2014).

Acknowledgements The case study benefited from valuable insights and information on the event by Professor D. Baumken of York University.

1.9 Hurricane

Hurricanes are known as tropical cyclones. A tropical cyclone is a rotating, organized system of clouds and thunderstorms that originates over tropical or subtropical waters and has a closed low-level circulation. Tropical cyclones rotate counterclockwise in the Northern Hemisphere. They are classified as follows:

- *Tropical Depression*: A tropical cyclone with maximum sustained winds of 61 km/h or less.
- *Tropical Storm*: A tropical cyclone with maximum sustained winds of 62–118 km/h.
- *Hurricane*: A tropical cyclone with maximum sustained winds of 119 km/h or higher. In the western North Pacific, hurricanes are called typhoons; similar storms in the Indian Ocean and South Pacific Ocean are called cyclones.
- *Major Hurricane*: A tropical cyclone with maximum sustained winds of 179 km/h or higher, corresponding to a Category 3, 4 or 5 on the Saffir-Simpson Hurricane Wind Scale (NHC 2015a).

The official hurricane season for the Atlantic Basin (the Atlantic Ocean, the Caribbean Sea, and the Gulf of Mexico) is from 1 June to 30 November. As seen in the Fig. 1.22, the peak of the season is from mid-August to late October. However, deadly hurricanes can occur anytime in the hurricane season.

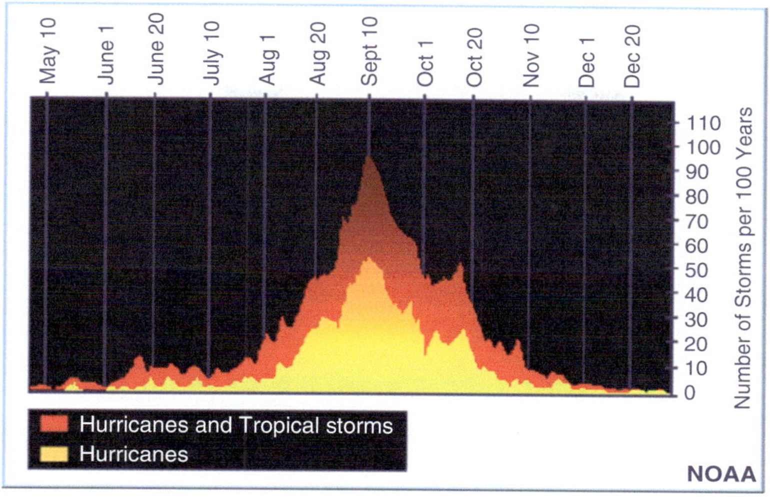

Fig. 1.22 Hurricane and tropical storm season (NHC 2015a)

1.9.1 Hurricane Return Periods

A hurricane return period explains the frequency at which a certain intensity hurricane can be expected within a certain geographic region (Fig. 1.23). For example, in the context of Fig. 1.23, a return period of 20 years would mean that on average during the previous 100 years, a Category 3 or greater hurricane passed within 92.6 km of that location about five times. Therefore, it would be expected that on average, an additional five Category 3 or greater hurricanes within that radius over the next 100 years can occur.

Atlantic hurricanes have been named since 1953 using lists originated by the National Hurricane Center at the beginning but now maintained and updated through a strict procedure by an international committee of the World Meteorological Organization. At any given time, a list for six years is made available, as shown in Table 1.7 (NHC 2015b).

1.9.2 Hurricane Intensity

A Saffir-Simpson Hurricane Wind Scale is a 1–5 rating based on a hurricane's sustained wind speed. This scale estimates potential property damage. Hurricanes reaching Category 3 and higher are considered major hurricanes because of their potential for significant loss of life and damage. Table 1.8 gives the description of the scale. A more detailed damage description is given in Appendix 1.

Often statements are made that a particular hurricane is ranked number one or two based on some impact but most often it is not clear what definition is used to rank the hurricane. Hurricanes can be ranked using different evaluation metric as shown in a study by Nirupama (2013) in which hurricanes during the period 1960–2012 along

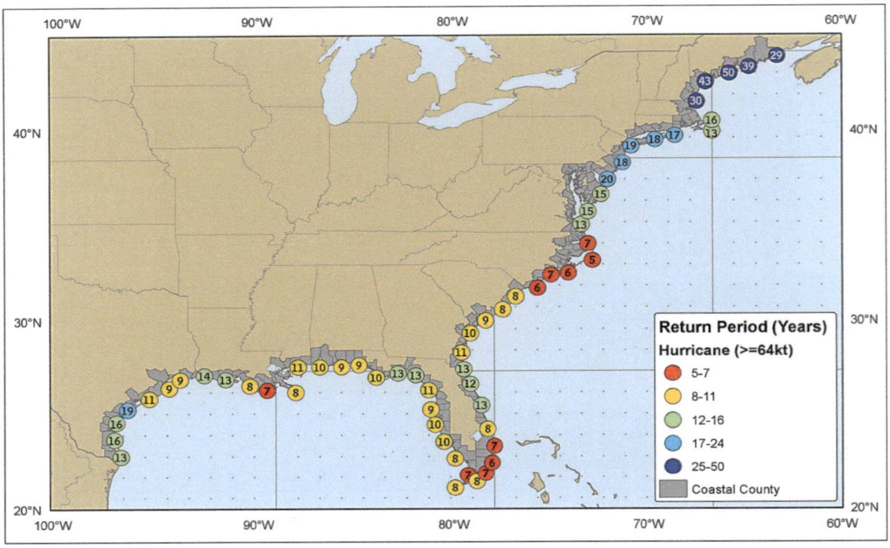

Fig. 1.23 Estimated return period in years for major hurricanes passing within 50 nautical miles (≅93 km) of various locations on the U.S. Coast (NHC 2015a)

Table 1.7 Atlantic hurricane names until year 2020 (NHC 2015b)

	2015	2016	2017	2018	2019	2020
1	Ana	Alex	Arlene	Alberto	Andrea	Arthur
2	Bill	Bonnie	Bret	Beryl	Barry	Bertha
3	Claudette	Colin	Cindy	Chris	Chantal	Cristobal
4	Danny	Danielle	Don	Debby	Dorian	Dolly
5	Erika	Earl	Emily	Ernesto	Erin	Edouard
6	Fred	Fiona	Franklin	Florence	Fernand	Fay
7	Grace	Gaston	Gert	Gordon	Gabrielle	Gonzalo
8	Henri	Hermine	Harvey	Helene	Humberto	Hanna
9	Ida	Ian	Irma	Isaac	Imelda	Isaias
10	Joaquin	Julia	Jose	Joyce	Jerry	Josephine
11	Kate	Karl	Katia	Kirk	Karen	Kyle
12	Larry	Lisa	Lee	Leslie	Lorenzo	Laura
13	Mindy	Matthew	Maria	Michael	Melissa	Marco
14	Nicholas	Nicole	Nate	Nadine	Nestor	Nana
15	Odette	Otto	Ophelia	Oscar	Olga	Omar
16	Peter	Paula	Philippe	Patty	Pablo	Paulette
17	Rose	Richard	Rina	Rafael	Rebekah	Rene
18	Sam	Shary	Sean	Sara	Sebastien	Sally
19	Teresa	Tobias	Tammy	Tony	Tanya	Teddy
20	Victor	Virginie	Vince	Valerie	Van	Vicky
21	Wanda	Walter	Whitney	William	Wendy	Wilfred

Table 1.8 The Saffir-Simpson hurricane wind scale (NHC 2015c)

Category	Sustained Winds	Types of Damage Due to Hurricane Winds
1	119–153 km/h	**Very dangerous winds will produce some damage:** Well-constructed frame homes could have damage to roof, shingles, vinyl siding and gutters. Large branches of trees will snap and shallowly rooted trees may be toppled. Extensive damage to power lines and poles likely will result in power outages that could last a few to several days.
2	154–177 km/h	**Extremely dangerous winds will cause extensive damage:** Well-constructed frame homes could sustain major roof and siding damage. Many shallowly rooted trees will be snapped or uprooted and block numerous roads. Near-total power loss is expected with outages that could last from several days to weeks.
3 (major)	178–208 km/h	**Devastating damage will occur:** Well-built framed homes may incur major damage or removal of roof decking and gable ends. Many trees will be snapped or uprooted, blocking numerous roads. Electricity and water will be unavailable for several days to weeks after the storm passes.
4 (major)	209–251 km/h	**Catastrophic damage will occur:** Well-built framed homes can sustain severe damage with loss of most of the roof structure and/or some exterior walls. Most trees will be snapped or uprooted and power poles downed. Fallen trees and power poles will isolate residential areas. Power outages will last weeks to possibly months. Most of the area will be uninhabitable for weeks or months.
5 (major)	252 km/h or higher	**Catastrophic damage will occur:** A high percentage of framed homes will be destroyed, with total roof failure and wall collapse. Fallen trees and power poles will isolate residential areas. Power outages will last for weeks to possibly months. Most of the area will be uninhabitable for weeks or months.

the US east coast and gulf coast (Fig. 1.24) have been taken into account to illustrate that hurricanes can be ranked according to at least ten different criteria. Therefore, it is not possible to declare a hurricane deadliest ever or the most severe hurricane in history, as can be understood from Table 1.9. In this table, a top ranking hurricane based on the evaluation criterion may or may not acquire top rank based on another evaluation criterion. Table 1.9 clearly proves this point as Hurricane Katrina tops in three categories, Hurricanes Donna and Camille make it to the top rank in two categories each.

1. Category according to the Saffir-Simpson scale
2. Lowest central pressure (which defines the intensity of the hurricane)
3. Maximum wind speed
4. Accumulated Cyclone Energy (ACE)
5. Maximum storm surge amplitude
6. Physical size of the hurricane
7. Loss of life
8. Population affected
9. Number of coastal counties affected
10. Economic damage

Fig. 1.24 Hurricanes that occurred along the US east coast and gulf coast during the period 1960–2012 (Nirupama 2013)

Table 1.9 Top ranking hurricanes based on the ten criteria taken into account during the period 1960–2012

Criteria	Value for the top ranking hurricane	Hurricane name	Year
1. Category according to the Saffir-Simpson scale	5	Camille	1969
2. Lowest central pressure	882 hPa	Wilma	2005
3. Maximum wind speed	305 km/h	Camille	1969
4. Accumulated Cyclone Energy (ACE)	$70 \times 10^4 kt^2$	Ivan	2004
5. Maximum storm surge amplitude	8.6 meters	Katrina	2005
6. Physical size of the hurricane	1600 km in diameter	Sandy	2012
7. Loss of life	1833 persons	Katrina	2005
8. Population affected	>24 million (based on 2008 population)	Donna	1960
9. Number of coastal counties affected	63 counties	Donna	1960
10. Economic damage	US$145 billion (2011 value)	Katrina	2005

1.9.3 Case Study – Hurricane Hazel – Toronto, Canada

In 1954, Hurricane Hazel was the result of when a weak hurricane system intercepted with another system and amplified into one of the most tragic hurricanes experienced in Canada. This hurricane is the only recorded tropical storm that has caused sustained hurricane force winds in Ontario, Canada. There was severe flooding in Southern Ontario due to Hurricane Hazel and the Greater Toronto Area experienced the worst flooding event ever witnessed in 200 years. Hurricane Hazel's destruction was felt everywhere with bridges washed out and homes and properties destroyed. There were 81 deaths because of Hurricane Hazel and approximately 7472 people needed evacuation because of the hazardous conditions left in the aftermath. One of the outcomes of Hurricane Hazel was the reimagination of Southern Ontario's emergency preparedness in case violent storm events were to occur again that could transfer large quantities of water and debris. Thus, following 1954's impact of Hurricane Hazel saw the formation of conservation authorities and initiation of emergency preparedness measures and mitigation strategies in Ontario as a means to damper the effects of future storms and hurricanes. Following sections provide in-depth insights into monitoring and warning systems and flood plain development regulations for watersheds (HIRA 2012).

1.9.4 Case Study – Hurricane Sandy, New York, USA

In 2012, Hurricane Sandy was the 18th tropical cyclone of the North American Atlantic hurricane season. The storm began developing in the central Caribbean region on October 22nd. Hurricane Sandy intensified into a hurricane as it reached Jamaica, Cuba and the Bahamas. As Sandy moved, the hurricane amplified as it moved up the northeast of the United States until turning west toward the mid-Atlantic coast on October 28th. As seen in previous years, Sandy showed classic late-season hurricane characteristics in the southwestern Caribbean Sea. However, Hurricane Sandy changed and took a complex evolution not seen before. Hurricane Sandy intensified in size and strength while over the Bahamas, despite weakening into a tropical storm north of those islands. Hurricane Sandy's system gather more strength as it turned into a formal hurricane as it moved northeast on the US coastline, parallel to the southeastern US coast. During this period of the hurricane, it reached a secondary peak intensity of over 157 km/h while it turned northwestward toward the Mid-Atlantic States. Initial estimates in the US were near $50 billion, making Sandy the second-costliest cyclone to hit the United States since 1901. Sandy caused at least 147 direct deaths in its path, with 72 being in the US, making it the greatest number of direct fatalities in the US since Hurricane Agnes in 1972.

1.10 Exercise

Find out recent natural hazards in a given calendar year in your area/region/country and analyze the information in light of historical similar events.

References

Armenakis C, Nirupama N (2014) Flood risk mapping for the City of Toronto. Proced Econ Financ, Elsevier 18:320–326

Berger E (2014) What's the difference between freezing rain, sleet and ice pellets? www.blog.chron.com. Accessed 7 Apr 2014

Bonnell J, Fortin M (2009) Don valley historical mapping project background. https://maps.library.utoronto.ca/dvhmp/. Accessed 21 Mar 2016

Canadian Environmental Health Atlas (2016) 2012 Heat anomaly – a winter heat wave. http://www.ehatlas.ca/heat-waves/trends/2012-heat-anomaly. Accessed 30 June 2016

CBC (2009) Doc Zone Documentaries, Last call at the oasis

CDC (2015) Centers for Disease Control and Prevention. http://www.cdc.gov/sars/index.html Accessed 21 Mar 2016

City Report (2014) Impacts from the December, 2013 extreme winter storm event, Staff Report, City of Toronto, 8 January 2014

Coulson G (2014) Recap of recent weather and a look ahead. Municipal flood management workshop, Black Creek Pioneer Village, 1 April 2014, Toronto, Canada

D'Andrea M (2010) Climate change – impacts on infrastructure and water management, climate science workshop: regional climate modelling capacity in Ontario, TRCA and York university workshop report, February 2010, Toronto, Canada

Dolce C (2014) Ice storms: why they're so dangerous and how to stay safe, Weather.com. http://www.weather.com/news/weather-winter/ice-storm-damage-impacts-20121123. Accessed 4 Apr 2014

Dumbuya B, Nirupama N (2016) Disasters and long term economic sustainability: a perspective on Sierra Leone, Special Issue: Investments in Disaster Resilience Int J Disast Resilien Built Environ 8(1)

EMDAT (2015a) D. GuhaSapir, R. Below, Ph. Hoyois EMDAT: The CRED/OFDA international disaster Brussels – Belgium. http://www.emdat.be/database. Accessed 21 Mar 2016

EMDAT (2015b) Glossary. http://www.emdat.be/glossary/9#letterc. Accessed 21 Mar 2016

Environment Canada (2014) Environment and climate change Canada, Past weather. http://climate.weather.gc.ca/. Accessed 4 Apr 2014

Environment Canada (2015) Environment and climate change Canada, Winter hazard. www.ec.gc.ca/meteo-weather/default.asp?lang=En&n=46FBA88B-1#Ice. Accessed 27 Jan 2016

Global News (2013) By the numbers: 2013 Toronto flood, By Irene Ogrodnik. http://globalnews.ca/news/704015/by-the-numbers-2013-toronto-summer-storm/. Accessed 21 Mar 2016

GOC (2012) Risk assessment 001 – 2012 Atlantic hurricane season, Government Operation Centre, Public Safety Canada

GoC (2015) Calculation of the 1981 to 2010 climate normals for Canada. http://climate.weather.gc.ca/climate_normals/normals_documentation_e.html?docID=1981. Accessed 21 March 2016

Haines A, LaPianta B, Tyrell C, Peberdy B (2014) Ice storm 2013 Technical briefing, Toronto Hydro Electric System Limited, 9 January 2014

Health Canada (2015a) Avian influenza. http://www.phac-aspc.gc.ca/influenza/avian-eng.php. Accessed 21 Mar 2016

Health Canada (2015b) SARS. http://www.phac-aspc.gc.ca/publicat/sars-sras/naylor/2-eng.php. Accessed 21 March 2016
Hildebrandt A (2013) The humidex, the flawed Canadian way to calculate summer discomfort. http://www.cbc.ca/news/canada/the-humidex-the-flawed-canadian-way-to-calculate-summer-discomfort-1.1394673. Accessed 21 Mar 2016
HIRA (2005) Hazard identification and risk assessment, Emergency Management Ontario. https://www.emergencymanagementontario.ca/english/emcommunity/ProvincialPrograms/hira/hira.html
HIRA (2012) Hazard identification risk assessment. https://www.emergencymanagementontario.ca/english/emcommunity/ProvincialPrograms/hira/hira_2012.html. Accessed 21 March 2016
Kimbell P (2013) Why Toronto was drenched by record rainfall. http://www.livescience.com/38052-toronto-floods-rainfall-record.html. Accessed 21 Mar 2016
Kristof N (2015) Our water-guzzling food factory, In New York Times, May 30, 2015
Malthus T (1826) An essay on the principle of population. J.M. Dent, London, pp 1–24
Ministry of Health and Long-Term Care (MOHLTC) (2010) Information for health care professionals: Pandemic (H1N1) 2009 update. http://www.health.gov.on.ca/en/. Accessed 14 Jan 2016
Natural Resources Canada (2009) Earthquakes Canada. http://earthquakescanada.nrcan.gc.ca. Accessed 14 Jan 2016
NHC (2015a) National Hurricane Center. http://www.nhc.noaa.gov/climo/. Accessed 21 Mar 2016
NHC (2015b) National Hurricane Center. http://www.nhc.noaa.gov/aboutnames.shtml. Accessed 21 Mar 2016
NHC (2015c) National Hurricane Center. http://www.nhc.noaa.gov/pdf/sshws_table.pdf. Accessed 21 Mar 2016
NRC (2015) Natural Resources Canada http://www.nrcan.gc.ca/forests/fire-insects-disturbances/fire/13143
Nirupama N (2013) Is it possible to rank hurricanes in a unique manner? Nat Hazards 67:963–968
Nirupama N, Armenakis C, Montpetit M (2014) Is flooding in Toronto a concern? Nat Hazards 72:1259–1264
NOAA (2016) Drought. http://www.ncdc.noaa.gov/sotc/drought/201602. Accessed 14 Jan 2016
NOVA (2015) http://www.novaweather.net/. Accessed 21 Mar 2016
NWS (2015) The varying types of winter weather. http://www.weather.gov/rlx/wintvar. Accessed 21 Mar 2016
Ploeger SK, O'Connell R, Atkinson GM, Samson C (2008) Earthquake loss estimation data collection and preparation for urban disaster management: a case study from the city of Ottawa, Canada. In: Proceedings of the 4th Canadian conference on geohazards Quebec City, Canada, 20–24 May 2008, pp 569–576
PNSN (2015) http://pnsn.org/outreach/earthquakehazards. Accessed 19 June 2015
PSC (2013/2014/2015) The Canadian disaster database, Public Safety Canada. http://cdd.publicsafety.gc.ca/srchpg-eng.aspx
Schwartz B (2014) Pandemics and SARS, Guest Lecture in DEMS 5040 Natural Disasters, York University
Searer G (2007) Does it make sense from engineering and economic perspectives to design for a 2475-year earthquake? In: Stein S, Mazzotti S (eds) Continental intraplate earthquakes: science, hazard, and policy issues. Geological Society of America Special Paper 425. https://doi.org/10.1130/2007.2425(23)
Seguin R (2013) At least six killed, thousands without power as ice storm hits Quebec. The Globe and Mail, December 22, 2013
thestar (2013) Toronto's July flood listed as Ontario's most costly natural disaster, by Carys Mills, August 14, 2013. http://www.thestar.com/business/2013/08/14/july_flood_ontarios_most_costly_natural_disaster.html
Toronto Hydro (2013) Power outages map. http://www.torontohydro.com. Accessed 21 Dec 2013

TRC (2009) Don River Watershed Plan – surface water hydrology/hydraulics and stormwater management, Toronto and Region Conservation Report, Toronto, Canada

TRCA (1999) State of the watershed report: highland creek watershed – Chapter 7 water. The Toronto and Region Conservation Authority, Toronto, Canada

TRCA (2006) Don Mouth naturalization and port lands flood protection project – revised terms of reference, The Toronto and region conservation authority, Toronto, Canada

USGS (2015a) The modified Mercalli Intensity scale. http://earthquake.usgs.gov/learn/topics/mercalli.php. Accessed 21 Mar 2016

USGS (2015b) The severity of an earthquake. http://pubs.usgs.gov/gip/earthq4/severitygip.html. Accessed 21 Mar 2016

WHO (2015a) Origins of the 2014 Ebola epidemic. http://who.int/csr/disease/ebola/one-year-report/virus-origin/en/. Accessed 21 Mar 2016

WHO (2015b) West Nile virus. http://www.who.int/mediacentre/factsheets/fs354/en/. Accessed 21 Mar 2016

Wilhite DA, Glantz MH (1985) Understanding the drought phenomenon: the role of definitions. Water Int 10(3):111–120

WMO (2015) Natural hazard. http://www.wmo.int/pages/themes/hazards/index_en.html. Accessed 21 Mar 2016

Young J (2013) The Toronto flood of 2013: actions from the past, a warning for the new normal? http://activehistory.ca/2013/07/the-toronto-flood-of-2013-decisions-of-the-past-a-warning-for-the-new-normal/. Accessed 21 Mar 2016.

Chapter 2
Defining Natural Hazards – Medium and Small Scale Hazards

Hazards are defined as medium or small scale hazards, based on their size and impact. In this chapter, the medium scale hazards examined in alphabetical order are erosion, landslides, snowstorms/blizzards, subsidence and sinkholes, tornados and windstorms. Small scale hazards include extraterrestrial hazards, fog, geomagnetic storms, hailstorms, and lightning. The chapter also highlights examples of recent hazards in North America and around the world.

2.1 Medium Scale Hazards

2.1.1 Erosion

Erosion is a natural environmental process that is mainly caused by removal of soil by running water, waves, currents, or wind. Coastal erosion is commonly known to threaten humans, their property, or the environment along the coastlines. Erosion can result in instability of structures found on once stable ground through the eroding of soil, soil and nutrient loss, decreased crop yields, damage to aquatic ecosystems, and dust storms. Some examples are loss of bridges, homes, road, river banks, trees, and hill sides. Below are two categories of erosions, those being natural and human-induced.

Natural causes of erosion include (HIRA 2012):

- Heavy and/or prolonged rainfall
- The effect of gravity on soils that rest on steep slopes
- Wind
- Flooding, wave action and/or currents
- Movement of glaciers
- Droughts, dry spells and/or high temperature

N. Agrawal, *Natural Disasters and Risk Management in Canada*, Advances in Natural and Technological Hazards Research 49,
https://doi.org/10.1007/978-94-024-1283-3_2

Fig. 2.1 An example of erosion and subsidence along the Lower Don River in Toronto, Canada (Photo by author)

Human causes of erosion include:

- The overgrazing of hoofed livestock
- Removal of vegetation
- Construction
- Poor agricultural practices

To visualize the impact of ground stability loss due to soil erosion, Fig. 2.1 illustrates a recent example of erosion and subsidence along the Lower Don River in Toronto, Canada. Similarly, Fig. 2.2 shows the composite design of wire mesh mixed with hay around the slope as an immediate mitigation measure to limit further erosion.

2.1.2 Landslide

The movement of soil or rock controlled by gravity with speed of the movement usually ranging between slow and rapid, but not very slow, is known as a landslide. It can be superficial or deep, but the materials have to make up a mass that is a portion of the slope or the slope itself. The movement has to be downward and outward with a free face.

2.1.2.1 Types of Movements and Materials

(Based on VARNES (Novotný 2013) Landslide Classification)

1. Falls– rock/debris/earth
2. Topples– rock/debris/earth
3. Rotational slide in different material types – rock/debris/earth
4. Translational slide in different material types – rock/debris/earth

Fig. 2.2 A quick fix to control erosion – a wire mesh mixed with hay around the slope (Photo by author)

5. Lateral spreads – rock/debris/earth
6. Flows – rock/debris/earth
7. Complex – combination of two or more principle types of movement

Although there are multiple types of causes of landslides, the three that cause most of the damaging landslides around the world are (USGS 2015):

- Slope saturation by water – this effect can occur in the form of intense rainfall, snowmelt, changes in ground-water levels, and water-level changes along coastlines, earth dams, and the banks of lakes, reservoirs, canals, and rivers.
- Seismic Activity – many mountainous areas experience at least moderate rates of earthquake occurrence in recorded times. The occurrence of earthquakes in steep landslide-prone areas greatly increases the likelihood that landslides will occur, due to ground shaking alone or shaking-caused dilation of soil materials, which allows rapid infiltration of water. Widespread rock falls also are caused by loosening of rocks as a result of ground shaking.
- Volcanic activity – volcanic lava may melt snow at a rapid rate, causing a deluge of rock, soil, ash, and water that accelerates rapidly on the steep slopes of volcanoes, devastating anything in its path. These volcanic debris flows (also known as lahars) reach great distances, once they leave the flanks of the volcano, and can damage structures in flat areas surrounding the volcanoes. The 1980 eruption of Mount St. Helens, in Washington triggered a massive landslide on the north flank of the volcano, the largest landslide in recorded times. Figure 2.3 shows the Mount Sinabung eruption in North Sumatra, Indonesia in 2014 that killed 17 people. Scientists were caught off guard because the volcano had been quiet for four centuries.

Fig. 2.3 Mount Sinabung eruption in North Sumatra, Indonesia in 2014 (By Rendy Cipta Muliawan [CC BY 2.0 (http://creativecommons.org/licenses/by/2.0)], via Wikimedia Commons)

2.1.2.2 Canadian History of Landslides

In Canada, a total of 39 landslides events have been recorded (Canadian Disaster Database 2015) since early twentieth century, out of which four have occurred in the twenty-first century. Figure 2.4 shows landslide prone areas, and Table 2.1 lists the 39 events.

2.1.2.3 Case Study – Quebec, Canada

In Quebec, Canada, a landslide occurred at a quarry in L'Epiphanie, as two workers went missing as shown in Fig. 2.5 (Globe and Mail 2013).

2.1.2.4 Case Study – Southern California, USA

A typical type of landslide occurred in La Conchita coastal area of southern California in the spring of 1995 and destroyed many houses and triggered large evacuation in the area (Fig. 2.6: Photo by R.L. Schuster, USGS 2004 http://pubs.usgs.gov/fs/2004/3072/fs-2004-3072.html).

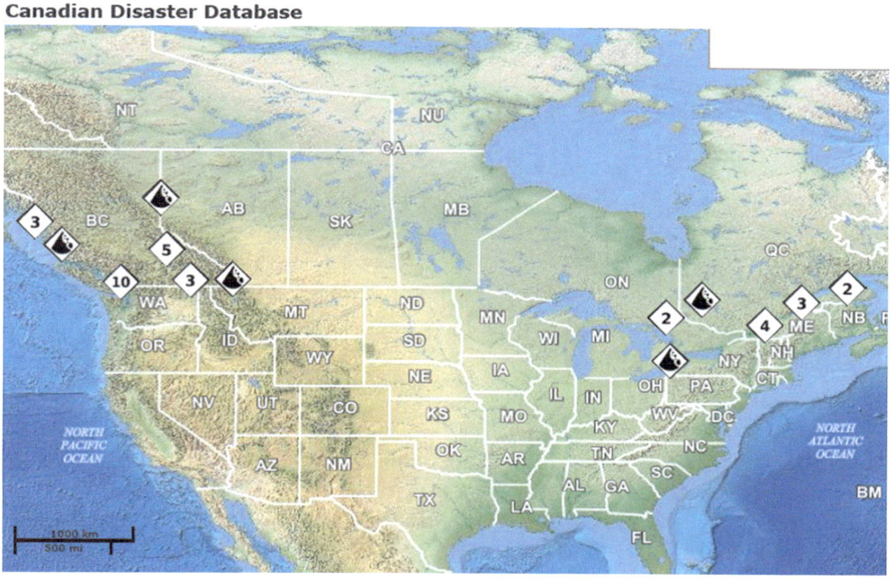

Fig. 2.4 Locations of the 39 landslide events recorded in Canadian Disaster Database (http://cdd.publicsafety.gc.ca/)

2.1.2.5 Case Study – Brazil

In southeastern Brazil, at least 64 people died in dual landslides on January 2, 2010. Hundreds of Brazilians die in mudslides each year, most of them slum dwellers living on precarious hillsides. But affluent tourists were struck in this event when they were vacationing in houses and a hotel at the foot of a jungle-shrouded cliff near Rio de Janeiro as shown in Fig. 2.7.

2.1.2.6 Case Study – India

In June 14–17, 2013, a rare weather system, last seen about 80 years ago, converged over northwestern India causing significant loss of life and substantial damage to property and critical infrastructure. One of the unique features of this disaster is its three tier nature, which is quite rare for natural disasters. First, owning to a synoptic situation of cloud formation, a cloud burst with precipitation higher than during a normal period occurred. Second, the carrying capacity of the rivers could not handle such excessive precipitation and river banks overflowed causing heavy flooding. Lastly, this area being in the Himalayan mountain range, and prone to geological hazards, landslides were triggered by the floods. The combined effects of these three hazards (cloudburst, river flooding, and landslides) was utterly devastating. According to the Indian Meteorological Department, the monsoon advancing

Table 2.1 List of the 39 landslides in Canada extracted from the Canadian Disaster Database (2015)

Location	Event Start Date	Fatalities	Injured	Evacuated	Estimated total cost
Daniels Harbour NL	15-Apr-07	0	0	0	$5,300,000
North Vancouver BC	19-Jan-05	1	1	300	$4,200,000
Khyex River BC	28-Nov-03	0	0	0	Unknown
Passmore BC	13-Apr-00	0	0	0	Unknown
Hummingbird Creek BC	11-Jul-97	0	Unknown	Unknown	Unknown
Notre-Dame de Pierreville QC	1-Apr-91	0	0	0	$304,756
Joe Rich and Southern BC	12-Jun-90	7	0	0	$18,208,847
Brantford ON	1-Jan-86	0	0	0	Unknown
Squamish BC	28-Oct-81	9	0	0	Unknown
Belmoral Mine, Val d'Or, QC	20-May-80	8	0	0	Unknown
Devastation Glacier BC	22-Jul-75	4	0	0	Unknown
Harbour Breton NL	1-Aug-73	4	0	0	Unknown
Fraser Canyon BC	4-May-71	3	0	0	Unknown
Saint-Jean-Vianney QC	4-May-71	31	0	1500	$1,250,000
Porteau BC	9-Feb-69	3	1	0	Unknown
Camp Creek BC	5-Jun-68	4	0	0	Unknown
Ocean Falls BC	14-Jan-65	7	0	0	Unknown
Hope BC	9-Jan-65	4	0	0	Unknown
Ramsey Arm BC	16-Sep-64	5	0	0	Unknown
Saint-Joachim-de-Tourelle QC	11-Dec-63	4	0	236	Unknown
Toulnustouc River QC	11-Dec-62	9	0	0	Unknown
McBride BC	7-Sep-60	3	0	0	Unknown
Revelstoke BC	27-Mar-59	4	0	0	Unknown
Prince Rupert BC	22-Nov-57	7	0	0	Unknown
Peace River BC	15-Oct-57	0	0	0	Unknown
Nicolet QC	12-Nov-55	3	6	2000	Unknown
St-Gregoire-de-Montmercy QC	1-Sep-38	4	12	0	Unknown
Crerar ON	27-Jun-30	8	0	0	Unknown
Capreol ON	26-Jun-30	4	0	0	Unknown
Echo Harbour BC	30-Sep-22	5	0	0	Unknown
Britannia Beach BC	28-Oct-21	37	0	0	Unknown
Cooper Mine, Jane Camp BC	22-Mar-15	56	22	0	Unknown
Coucoucache QC	18-Apr-10	6	0	0	Unknown
St-Alphonse-de-Bagotville QC	15-Apr-10	4	0	0	Unknown
Burnaby and New Westminster BC	28-Nov-09	22	15	0	Unknown

(continued)

Table 2.1 (continued)

Location	Event Start Date	Fatalities	Injured	Evacuated	Estimated total cost
Notre-Dame-de-la-Salette QC	26-Apr-08	33	0	0	Unknown
Spences Bridge BC	13-Aug-05	15	0	0	Unknown
Frank AB	29-Apr-03	70	23	0	Unknown
Upper Arrow Lake BC	28-Feb-03	0	0	0	Unknown

Fig. 2.5 Rescue workers, along with a dog looking for two missing workers at a quarry in L'Epiphanie, Quebec, East of Montreal (The Canadian Press/Graham Hughes)

towards the west of South Asia, combined with westerly winds resulted in several days of torrential rains. The weather interaction of this kind normally occurs between October and April, but extended to June. The intensity was also much higher than normally seen. Media networks like BBC and CNN reported more than 5700 people presumed dead, and over US$500 million damage estimates. There is a great need to develop fully integrated early warning systems for multi-faceted hazards, which is challenging because a cloud burst is a meteorological hazard, river flood is a hydrological hazard, and landslide is a geological hazard, which are handled by different government departments in most countries. Figure 2.8 shows the impact region in the Indian subcontinent.

Heavy rainfall occurred in the region at the time when there was still snow on the ground, therefore, the combination of heavy rainfall on melting snow created conditions for widespread landslides (NIDM 2013; Ramachandran 2013; Bagla 2013; Climate Himalaya 2013). Figure 2.9 shows the landslides with reference to the main towns in the impact region (Mukharji 2013).

Fig. 2.6 A typical type of landslide occurred in La Conchita coastal area of southern California in the spring of 1995 and destroyed many houses and triggered large evacuation in the area (USGS 2004)

The massive landslide hit the Hindu shrine in Kedarnath which lies just a short distance from the snout of two mountain glaciers. Figure 2.10 shows a satellite image of the impacted region, and Fig. 2.11 shows the devastation from the ground. The shrine, being an important pilgrimage destination, was packed with visitors celebrating a religious holiday. Figs. 2.12 and 2.13 are examples of mudslides and houses on the verge of collapsing in the background of severely flooded river.

According to Dobhal et al. 2013 and Ramachandran 2013, first the flow from the north east came down the margin of the glacier and spread out to strike the town. Next, the northwest flow descended from the other glacier to the town on its west side, and struck it directly. The debris flow from the north-east was triggered by a 75 m wide landslide, which then descended the steep slope about 500 m, gathering debris in its path. The flow was initially channeled into a narrow gully formed by the glacier and on exiting it the flow spread out in the floodplains before striking the town after traversing about 1200 m. The steepness of the slope would have given the debris enormous velocity.

Fig. 2.7 A rain soaked hillside collapsed on homes killing at least 12 people near Rio de Janeiro, Brazil, after New Year celebrations, Jan 2, 2010. AP Photo/Felipe Dana https://feww.wordpress.com/tag/brazil-mudslides/

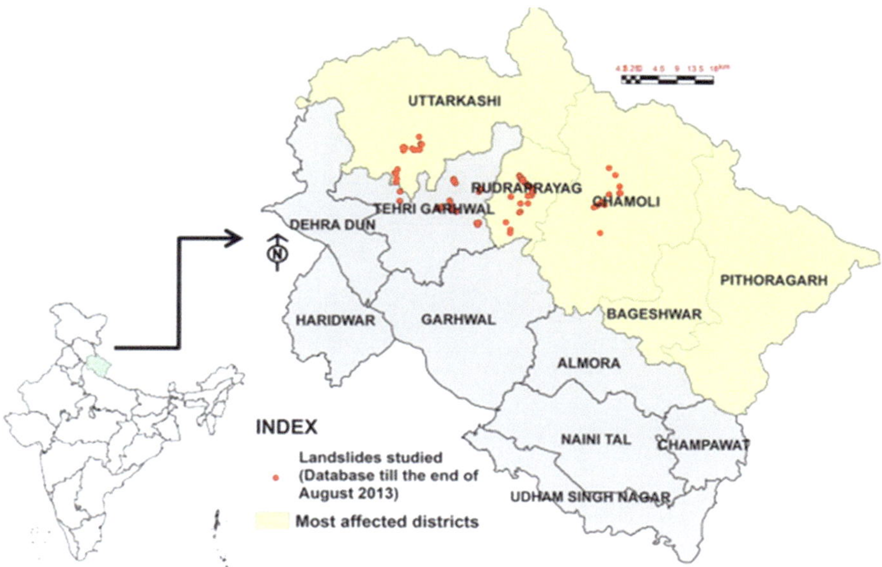

Fig. 2.8 Impact area in the Indian subcontinent (Sundaramoorthy 2013)

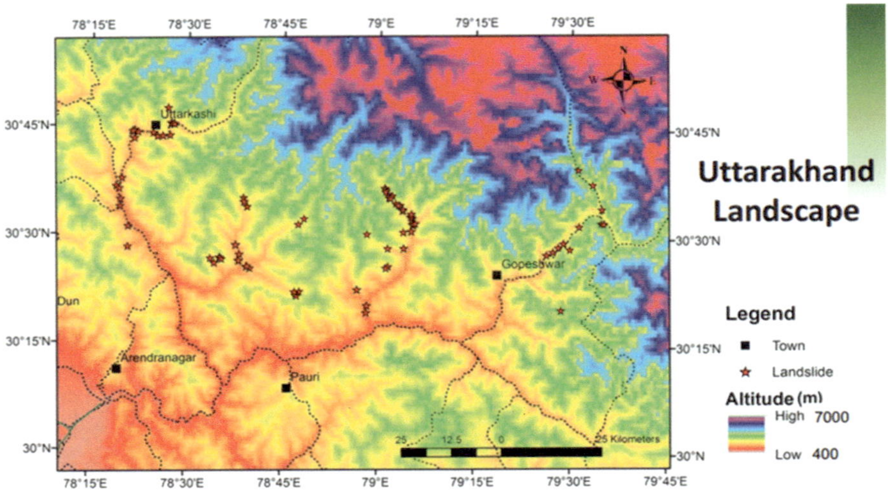

Fig. 2.9 Uttarakhand impact area showing towns and landslides (Mukharji 2013)

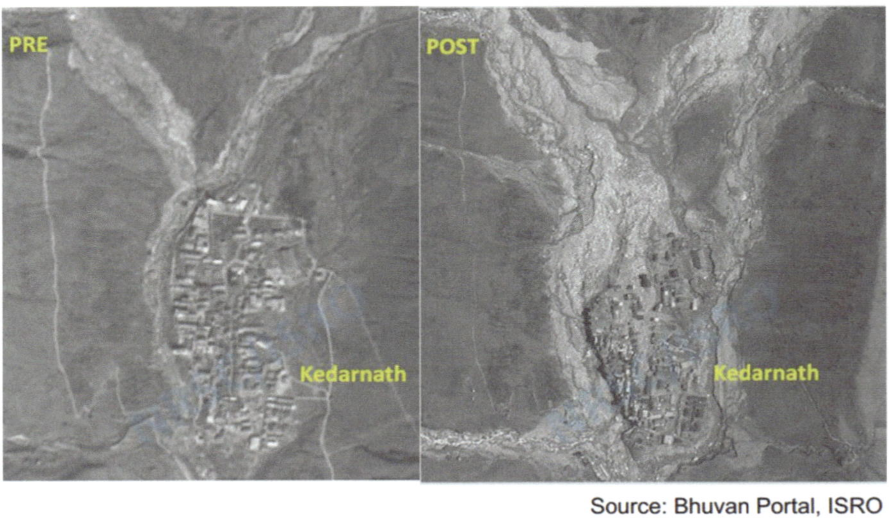

Fig. 2.10 The town of Hindu shrine, Kedarnath before and after the event. (Source: Bhuvan Portal, ISRO 2013)

2.1.3 *Snowstorm/Blizzard*

A snow storm is a low pressure system in winter months with significant accumulations of snow. A blizzard is a severe snow storm with winds exceeding 56 km/h for three or more hours, producing reduced visibility of less than 400 m.

Fig. 2.11 Landslide impact in Kedarnath before (left) and after (right). The geology of the area is still the same as in the 1880s (Uttarakhand For You June 15, 2014)

Fig. 2.12 Mudslide resulting in stranded vehicles and pilgrims (http://news.xinhuanet.com/english/photo/2013-06/19/c_132469016_4.htm)

In the international database (EMDAT 2015), snow storms are logged as winter/snow storm as a sub-sub-disaster type under the sub disaster type as convective storms, and disaster type as climatological disasters. The author suggests visiting the database to view hierarchical organization for clear understanding. Table 2.2 lists such events in North America since 1900. It is noteworthy that as high as up to seven occurrences have been recorded in a given year indicating the seriousness of this type of convective storms.

Within Ontario, Canada, to this date, the 1998 ice storm had been the most damaging at US$4.6 billion in estimated total cost and 28 fatalities (HIRA 2005). The Government of Canada provides regular updates, warnings, and watches on its weather website http://weather.gc.ca/canada_e.html . Unisys Weather (http://weather.unisys.com/) is another good weather resource for similar information.

Various warning categories include blizzard, blowing snow, flash freeze, freezing rain, snow fall, snow squall, wind, and wind chill (Table 2.3).

Fig. 2.13 Mighty river and collapsing houses (http://casa-india.org/pages/news/323-update-uttarakhand-cloudburst-floods-and-landslides.html)

2.1.3.1 Case Study – Ontario, Canada

The Province of Ontario in Canada suffered a severe winter storm that brought a major highway (HWY 402) to a standstill for hours, as can be seen in Fig. 2.14, in December 2010 (Chatham Daily News 2010). Numerous drivers were stranded overnight due to the whiteout conditions triggering a rare move of activating ground and air rescue mission. Two counties declared emergencies, but no injuries were reported despite the wait time of up to 30 h for some (QMI Agency 2010).

In February 2011, a furious winter storm struck Canada and the US as shown in Fig. 2.15 (Metro News 2011a). Eastern Canadian cities (Ottawa and Toronto) and major cities in the United States (Dallas, Oklahoma, Tulsa, and Chicago) were paralyzed by a severe snow storm that dumped 48 cm of snow in Chicago. People were advised not to go out at all during that time. Work places and schools were closed causing a total shut down of impacted areas. The National Weather Service, USA predicted that transportation will be impossible for a couple of days in these areas (Metro News 2011b).

Table 2.2 Snow storms in North America since 1900 (data extracted from EMDAT 2015)

Year	Occurrence	Total deaths	Total affected	Total damage ($US)
1959	1	6	0	0
1966	1	0	0	0
1971	1	0	0	0
1978	1	65	0	100,000
1979	1	7	50	13,000
1982	2	61	0	200,000
1987	3	115	0	131,000
1994	2	75	0	0
1996	3	211	0	0
1997	3	41	24	512,000
1998	3	65	1378	262,500
1998	1	7	0	0
1999	1	48	0	1,000,000
2000	1	0	0	0
2001	1	11	100	0
2002	4	92	0	450,500
2003	3	57	120	233,600
2004	1	0	0	0
2005	2	30	0	350,000
2006	3	3	0	660,000
2007	2	18	0	140,000
2008	1	5	0	360,000
2008	1	0	0	0
2009	1	19	0	0
2010	4	3	0	2,500,000
2011	5	95	0	4,900,000
2012	4	27	1	150,000
2013	5	51	4882	1,400,000
2013	2	29	0	0
2014	7	91	30	1,710,000
2015	3	67	0	100,000

2.1.4 *Subsidence and Sinkhole*

Subsidence refers to the sinking of the ground due to groundwater removal, mining, dissolution of limestone (e.g. karst, sinkholes), extraction of natural gas, and earthquakes. Sinkholes are caused by collapse of the land surface due to the dissolving of the subsurface rocks such as limestone or carbonate rock by water. Different regions in the world may have different reasons for being prone to subsidence depending on factors listed earlier.

It does not take long for a major subsidence to develop following a heavy rain event as shown in Fig. 2.16 where within a matter of hours a section of the road

Table 2.3 List of warning types with threshold values (compiled from http://weather.gc.ca)

Warning type	Winds	Visibility	Wind chill	Snow/ice accumulation
Blizzard (all the conditions to meet for $\geq$4 h)	$\geq$ 50 km/h	$\leq$ 1 km	−25 or lower	
Blowing snow		$\leq$ 1 km for $\geq$3 h		
Flash freeze (not applicable for black ice)			Quick temp fall from zero to well below freezing	Significant
Freezing rain				Moderate or heavy drizzle for $\geq$2 h
Snowfall				$\geq$ 15 cm within 12 h
Snow squall		Near zero for $\geq$4 h		$\geq$ 15 cm within 12 h
Wind	$\geq$ 60 km/h for $\geq$3 h; and/or gusting $\geq$90 km/h			
Wind chill	$\geq$ 15 km/h		$\leq - 30$ to $\leq - 50$ (regionally variable) to last for $\geq$3 h	

Fig. 2.14 Winter blast paralyses Highway 402 in Ontario www.chathamdailynews.ca

Fig. 2.15 Snow storm batters North America (Metro News 2011a)

caved in near Finch and Sentinel in Toronto on August 19, 2005. The Black Creek culvert under the road was corroding over time and caved in during the downpour.

In the State of Oregon, USA, a massive sinkhole opened up on the US 101 near the coastal town of Harbor in January 2016 (Fig. 2.17, left). The 5 m deep sinkholes formed after a series of heavy rainfall poured over the Oregon coast. Sinkholes and landslides frequently occur in the region because of the geology of the area. Another example of how damaging and unpredictable sinkholes can be is from Fukuoka City in Japan in a busy city road as shown in Fig. 2.17 (right). The sinkhole disrupted traffic, power transmission and banking systems as authorities scrambled to evacuate surrounding areas at risk of more cave-ins. A combination of many factors contributing to the event, such as, construction on a subway line extension was underway; overall deterioration of the sewer system built in the 70s; the coastal region has high levels of groundwater; and heavy rainfall in previous months (Japan Times 2016).

2.1.5 *Tornado*

A tornado is a violently rotating column of air that is in contact with the ground or open water (waterspout). During this natural event, wind is invisible and therefore, it is difficult to see a tornado unless it creates a condensation funnel made up of water particles, dust, and other debris. Tornadoes are the most violent of all atmospheric storms. Tornadoes are not an isolated natural phenomenon as many parts of the

Fig. 2.16 A section of the road caved in within a matter of hours near Finch and Sentinel in Toronto, near York University on August 19, 2005

world experience tornadoes, including tornadoes made of fire like reported in Australia. Outside of the U.S., Bangladesh and Argentina are known for high number of tornadoes. Since record keeping began in 1950, it is common knowledge that on average about 1200 tornadoes hit the U.S. annually. Significant progress has been made in the understanding, spotting, and reporting of tornadoes. Mitigation measures and emergency preparedness has also advanced over the decades to help deal with the outcome of these natural disasters. Figure 2.18 depicts Tornado Alley, an area of relatively high occurrences of tornadoes known to exist in the mid-southwestern United States. However, tornadoes can hit outside of this region. Figure 2.19 illustrates the mechanism of a traditional tornado formation, where warm moist air rises from the ground when weather and atmospheric conditions allow. A vortex is formed when warm air meets cold air from the north. This vortex rotates with high winds and touches the ground causing devastation in its path.

Tornado season varies in the Southern plains (May to early June), the northern plains, and upper Midwest (usually between June and July). However, they can also happen anytime of the year, anytime of the day or night – although majority of them occur in late afternoon-evenings.

Fig. 2.17 Left, massive sinkhole on a stretch of US 101 in Oregon, USA (Oregon Department of Transportation; licensed under the Creative Commons via wikimedia); Right, Hakataekimae Avenue near Hakata Station caved in on November 8, 2016, Fukuoka City, Japan. By Muyo (Own work) [CC BY-SA 4.0 (http://creativecommons.org/licenses/by-sa/4.0)], via Wikimedia Commons

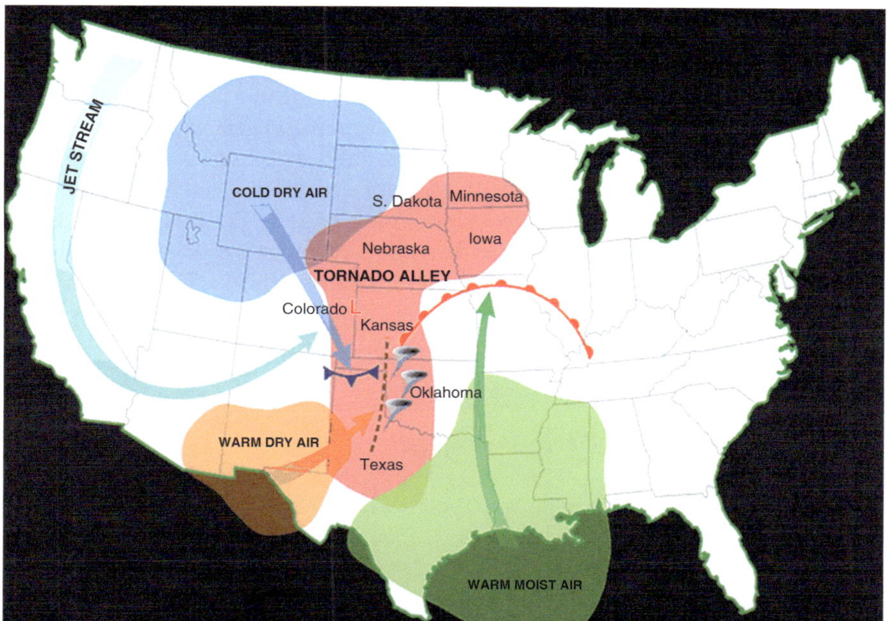

Fig. 2.18 Tornado formation conditions and Tornado Alley (NWS 2015a)

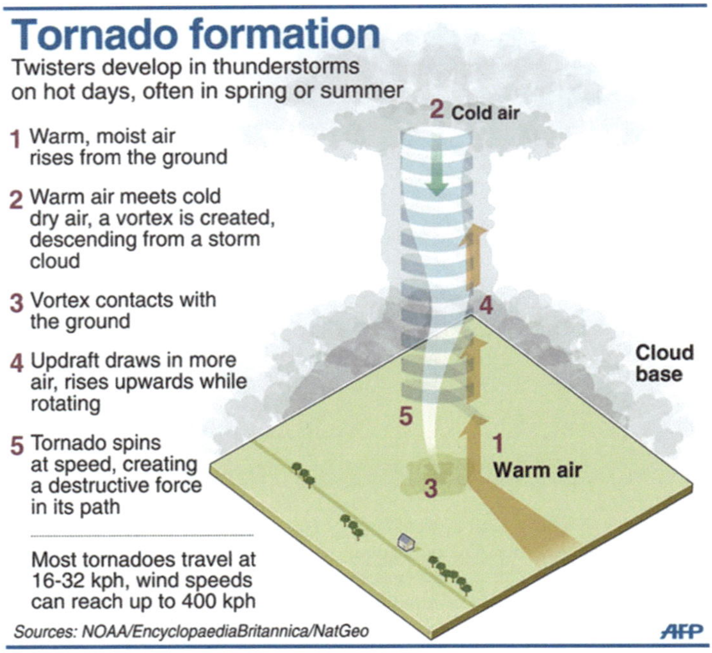

Fig. 2.19 Tornado formation (http://www.theborneopost.com/)

Tornado warnings are issued by local weather stations based on spotting or reporting of them. Radar data is effective in understanding of a possibility of tornado formations. Characteristics of a typical tornado thunderstorm are shown in Fig. 2.20. The most destructive and deadly tornadoes occur from supercells, which are rotating thunderstorms with a well-defined radar circulation called a mesocyclone.

Tornado formation is believed to be dictated mainly by the temperature differences across the edge of downdraft air wrapping around the mesocyclone. It should be noted here that during the most destructive tornadoes in history in Oklahoma also known as the Great Plains tornado outbreak during May 3–4, 1999, very little temperature variation was observed.

2.1.5.1 Measurement of Tornado Strength

Damage caused by a tornado is the most practical approach to determine the severity and strength of the tornado. This is because direct field measurements of tornadoes are dangerous and difficult to obtain. Table 2.4 gives the commonly used Enhanced Fujita (EF) Scale. The EF scale was implemented by the Weather Forecast Office of the National Weather Service (NWS 2015b) in 2007. The EF scale was the revised version of the original Fujita Scale which was solely relied on wind speed and did not take overall damage into account as a parameter variable. The use of EF scale

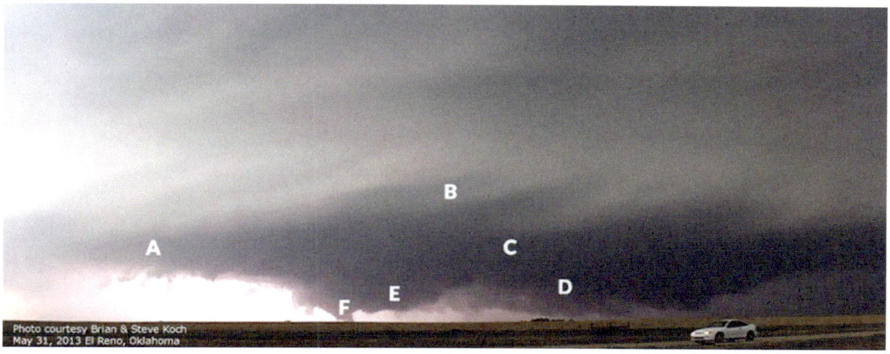

Fig. 2.20 Characteristics of a tornadic thunderstorm: A. Rear-flank, B. Striations indicating strong rotating updraft, C. Mesocyclone, D. Tail cloud, E. Wall cloud, F. Tornado. http://www.nssl.noaa.gov/projects/vortex2/

Table 2.4 The Enhanced Fujita (EF) Scale of tornado intensity (NWS 2015b)

	3 second gust	
EF scale	Miles/h	Km/h
0	65–85	105–137
1	86–110	138–177
2	111–135	179–217
3	136–165	219–266
4	166–200	267–322
5	over 200	322+

brought consistency in damage assessment that was much needed by the insurance industry allowing for damage indicators and degree of damage to be clearly defined. It is important to note that the EF scale is still a set of wind estimates, not measurement, based on damage. EF Scale damage indicators are given in Table 2.5. In order to maintain the continuity and consistency of the data, a correction formula has been worked out that allows conversion between the two Scales.

In Canada, Ontario, Alberta, Manitoba, and Saskatchewan are the four provinces that experience the most number of tornadoes per season. Quebec receives a bit less and New Brunswick and the interior of British Columbia are recognized tornado zones as well. However, other provinces and territories have significantly less threat from tornadoes. The peak time of year in Canada for tornadoes is in the summer months when competing air masses move north, as opposed to the spring season in the United States southern-central plains. However, tornadoes in Canada have occurred in spring, fall, and very rarely during winter (Wikipedia.org). Within the twenty-first century, 12 tornadoes have caused significant destruction in Canada, with the most being in Ontario. In 2010, Canada was impacted particularly badly with three tornado occurrences in (1) Kawacatoose First Nation, Saskatchewan,

Table 2.5 Damage indicators used in Enhanced Fujita Scale (NWS 2015b)

NUMBER (Details linked)	Damage indicator	Abbreviation
1	Small barns, farm outbuildings	SBO
2	One- or two-family residences	FR12
3	Single-wide mobile home (MHSW)	MHSW
4	Double-wide mobile home	MHDW
5	Apt, condo, townhouse (3 stories or less)	ACT
6	Motel	M
7	Masonry apt. Or motel	MAM
8	Small retail bldg. (fast food)	SRB
9	Small professional (doctor office, branch bank)	SPB
10	Strip mall	SM
11	Large shopping mall	LSM
12	Large, isolated ("big box") retail bldg.	LIRB
13	Automobile showroom	ASR
14	Automotive service building	ASB
15	School – 1-story elementary (interior or exterior halls)	ES
16	School – jr. or sr. high school	JHSH
17	Low-rise (1–4 story) bldg.	LRB
18	Mid-rise (5–20 story) bldg.	MRB
19	High-rise (over 20 stories)	HRB
20	Institutional bldg. (hospital, govt. or university)	IB
21	Metal building system	MBS
22	Service station canopy	SSC
23	Warehouse (tilt-up walls or heavy timber)	WHB
24	Transmission line tower	TLT
25	Free-standing tower	FST
26	Free standing pole (light, flag, luminary)	FSP
27	Tree – hardwood	TH
28	Tree – softwood	TS

(2) Midland, Ontario, and (3) Leamington, Ontario (Canadian Disaster Database 2015). Table 2.6 lists tornado events in North America since 1900.

2.1.5.2 Case Study – Leamington, Ontario June 6, 2010

A strong thunderstorm cell moved over the southern portions of Essex County. A series of four tornadoes had damaging wind gusts along an intermittent path of damage that measured about 40 km in length. The most significant damage was due to an F2 tornado which swept through the south end of the town of Leamington, with peak winds between 180 and 240 km/h. Approximately 4500 hydro customers were left without power, and 12 homes were deemed unsafe to reside in due to damages.

Table 2.6 An account of the occurrences of tornadoes in North America since 1900 (EMDAT)

Year	Disaster subtype	Disaster subsubtype	Occurrence	Deaths	Affected	Damage
1903	Convective storm	Tornado	1	98	0	0
1912	Convective storm	Tornado	1	28	2700	5000
1913	Convective storm	Tornado	1	732	0	200,000
1920	Convective storm	Tornado	1	224	0	0
1925	Convective storm	Tornado	1	739	0	18,000
1927	Convective storm	Tornado	1	87	0	22,000
1932	Convective storm	Tornado	1	332	0	0
1936	Convective storm	Tornado	1	455	0	21,000
1950	Convective storm	Tornado	1	0	0	0
1953	Convective storm	Tornado	1	92	0	52,000
1963	Convective storm	Tornado	1	203	0	0
1965	Convective storm	Tornado	1	271	1000	190,000
1966	Convective storm	Tornado	1	26	0	100,000
1970	Convective storm	Tornado	1	26	0	147,000
1974	Convective storm	Tornado	1	322	0	1,000,000
1979	Convective storm	Tornado	1	48	2110	550,000
1981	Convective storm	Tornado	1	20	0	400,000
1982	Convective storm	Tornado	2	93	0	600,000
1984	Convective storm	Tornado	2	680	600	1,005,600
1985	Convective storm	Tornado	2	116	1500	348,600
1986	Convective storm	Tornado	1	0	0	1500
1987	Convective storm	Tornado	3	62	3450	270,000
1989	Convective storm	Tornado	1	27	0	0
1991	Convective storm	Tornado	7	33	0	1,000,000
1993	Convective storm	Tornado	6	13	390	125,000
1994	Convective storm	Tornado	2	44	654	0
1995	Convective storm	Tornado	2	8	60	3,000,000
1996	Convective storm	Tornado	2	25	107	0
1997	Convective storm	Tornado	10	75	3668	1,260,000
1998	Convective storm	Tornado	3	65	1378	262,500
1999	Convective storm	Tornado	7	82	15,681	3,417,500
2000	Convective storm	Tornado	8	53	6764	799,600
2001	Convective storm	Tornado	13	17	7866	302,800
2002	Convective storm	Tornado	5	48	4983	2,920,000
2003	Convective storm	Tornado	4	51	17,226	9,000,000
2004	Convective storm	Tornado	3	5	2019	309,000
2005	Convective storm	Tornado	1	0	201	0
2006	Convective storm	Tornado	3	22	662	1,385,000
2007	Convective storm	Tornado	4	62	6940	1,080,000
2008	Convective storm	Tornado	7	97	747	5,860,000
2009	Convective storm	Tornado	4	33	874	7,050,000
2010	Convective storm	Tornado	3	25	2841	2,700,000
2011	Convective storm	Tornado	6	591	18,613	27,000,000
2012	Convective storm	Tornado	5	63	3897	9,025,000
2013	Convective storm	Tornado	5	76	172,592	7,100,000

Fig. 2.21 A 200 year old tree was uprooted during the storm in Essex County (CBC 2010a)

The Red Cross provided an emergency shelter to help the relief effort in the wake of the disaster. Estimated damage was $100,000 affecting 13,500 people (Canadian Disaster Database 2015). The town of about 30,000 residents on Lake Erie was sanctioned under a state of emergency due to severe damage, fallen trees (Fig. 2.21), and also to facilitate effective response to the disaster (CBC 2010a).

2.1.5.3 Case Study – Midland, Ontario June 23, 2010

An F2 tornado destroyed approximately 50 homes and caused damages estimated to be $15 million. The F2 classification means that wind speeds between 180 and 240 km/h were reached. 12 people were reported to be injured. Ontario provided immediate provincial assistance of up to $1 million to aid in cleanup and repairs (Canadian Disaster Database 2015). Figure 2.22 shows the damage (HydroOne http://www.hydroone.com).

2.1.5.4 Case Study – Kawacatoose First Nation, Saskatchewan July 2, 2010

An F3 tornado ripped through the Kawacatoose First Nation. No one lost their life, but 82 people were left homeless due to 18 homes being destroyed and uninhabitable. A total of $2.3 million in damages occurred during this disaster (Canada disaster database 2015; CBC 2010b). Figures 2.23 and 2.24 show the tornado as it occurred and the damage it caused.

2.1.6 Windstorm

Windstorms are winds that damage properties, infrastructure, and other structures that are produced by strong thunderstorms. Damaging winds are classified as those exceeding speed of 80–96 km/h. Damage from severe thunderstorm winds is

Fig. 2.22 Midland tornado damage and restoration efforts by Hydro One crew on June 24, 2010. http://www.hydroone.com/OurCompany/MediaCentre/PhotoGallery/Pages/default.aspx

Fig. 2.23 Errin Poorman took this photo of the tornado seconds before it struck Kawacatoose First Nation near Raymore, Saskatchewan (CBC 2010b)

experienced in North America much more than from tornadoes. Wind speeds can reach up to 161 km/h and can produce a path of destruction extending for hundreds of kilometers. Most thunderstorms produce straight-line winds as a result of outflow generated by the thunderstorm downdraft, and pose threat to people and properties located in thunderstorm-prone areas of the world. Particularly, mobile homes are at risk, including anchored mobile homes as shown in Fig. 2.25 when winds gust over 129 km/h (NOAA 2015a).

Fig. 2.24 About a dozen homes were destroyed when a tornado struck the Kawacatoose First Nation. (CBC 2010b)

Fig. 2.25 damage done by a strong windstorm (NOAA 2015a)

2.1.6.1 Types of Damaging Winds (NOAA 2015b)

Straight-line wind is a term used to define any thunderstorm wind that is not associated with rotation, and is used mainly to differentiate from tornadic winds.

A *downdraft* is a small-scale column of air that rapidly sinks toward the ground. It produces a *downburst* (Fig. 2.26) with horizontal dimensions of over 4 km resulting in an outward burst of damaging winds on or near the ground. Downburst winds may begin as a microburst and spread out over a wider area, sometimes producing damage similar to a strong tornado.

A *microburst* (Fig. 2.27) is a small concentrated downburst that produces an outward burst of damaging winds at the surface. Microbursts are generally small (less than 4 km across) and short-lived, lasting only 5–10 min, with maximum wind speeds up to 270 km/h. There are two kinds of microbursts: wet and dry. A wet microburst is accompanied by heavy precipitation at the surface. Dry microbursts, common in places like the high plains and the intermountain west, occur with little or no precipitation reaching the ground.

A **gust front**, the leading edge of rain-cooled air that clashes with warmer thunderstorm inflow, are characterized by a wind shift, temperature drop, and gusty winds out ahead of a thunderstorm producing damaging impact on their path.

A **derecho** (Fig. 2.28) is a widespread, long-lived wind storm that is associated with a band of rapidly moving showers or thunderstorms. A typical derecho consists

Fig. 2.26 A downburst is a strong downdraft that results in an outward burst of damaging winds on or near the ground. Downburst winds may begin as a microburst and spread out over a wider area

Fig. 2.27 A downburst is a strong downdraft that results in an outward burst of damaging winds on or near the ground. Downburst winds may begin as a microburst and spread out over a wider area (Federal Aviation Administration http://lessonslearned.faa.gov/)

of numerous microbursts, downbursts, and downburst clusters. By definition, if the wind damage swath extends more than 400 km and includes wind gusts of at least 93 km/h or greater along most of its length, then the event may be classified as a derecho (NOAA 2015c). The most severe derechos are called super derechos.

Fig. 2.28 Gust front on the leading edge of a derecho-producing convective system. The photo was taken on the evening of July 10, 2008 in Hampshire, Illinois as the storm neared the Chicago metropolitan area. The derecho had formed around noon in southern Minnesota. (Courtesy of Brittney Misialek on NOAA (2015c)

A **haboob** (Fig. 2.29) is a wall of dust that is pushed out along the ground from a thunderstorm downdraft at high speeds.

2.1.6.2 Case Study – Super Derecho across US States may 8, 2009

The derechos on May 8th, 2009 produced strong winds and caused vast devastation across U.S. States of Kansas and Kentucky (Fig. 2.30). Wind gusts of over 112 km/h were experienced along its path. Several clear bow and vortices were observed in this convective system that also caused widespread flash flooding in Missouri. Most significantly, an unusually large scale mesoscale convective vortex was experienced, accompanied by a band of intense surface winds and tornadoes that occurred independent of the severe weather directly associated with the large-scale bow.

The Super Derecho of May 2009 covered over 1600 km in 24 h causing havoc – fatalities, injuries, and tremendous economic loss. The event will be used for research purposes and better understanding of the phenomenon for a long time. The weather conditions at the time of the birth of this system were relatively mild causing no reason for alarm, and yet, the system proved to be shockingly devastating.

Fig. 2.29 A haboob is a wall of dust that is pushed out along the ground from a thunderstorm downdraft at high speeds NOAA

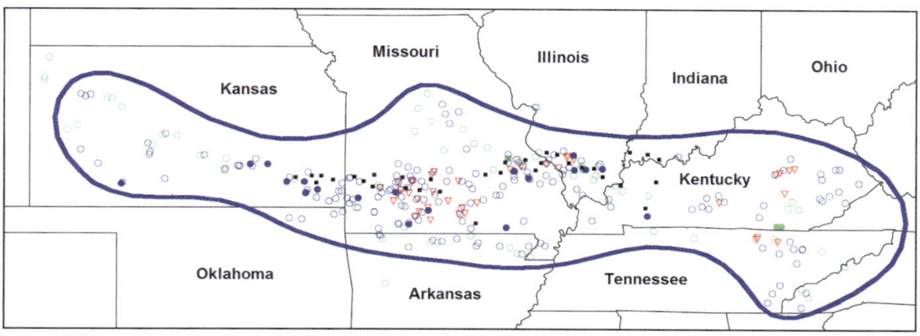

Fig. 2.30 Super Derecho of May 8, 2009 and associated affected area in blue (NOAA). Wind gusts of over 90 km/h are shown in open blue circles; gusts of over 119 km/h are shown in filled blue circles; hail of over 1.9 m as open green circles; hail of over 5 cm as filled green circles; tornadoes shown in red triangles. Flash flooding is denoted by black squares by county and intensely damaged area is shown by band of filled blue circles from Kansas to Missouri to Illinois

2.1.6.3 Case Study – Derecho in USA June 29, 2012

A super derecho across the Midwest and mid-Atlantic on June 29, 2012 was caused by a violent thunderstorm. The wind storm damaged more than 110 km long area on its path, killing thirteen people and causing power outage to millions. The derecho was triggered by a ripple in the jet stream and fueled by the record breaking intense heat in

Fig. 2.31 A tree toppled by severe storms sits atop a car in Washington's Dupont Circle neighborhood, June 30, 2012 in Washington, DC (AP Photo/Jessica Gresko: Accuweather.com)

Washington, D.C. area. The Storm Prediction Centre (http://www.spc.noaa.gov/) monitors and issues warnings and watches for potentially severe storms. The derecho lasted for about 12 h and covered northern Indiana to the southern mid-Atlantic coast. This event produced wind gusts equivalent to a category 1 hurricane. Downed trees (Fig. 2.31) dominated the damaging wind reports (Accuweather 2012). States of emergencies were declared in Virginia, West Virginia and Ohio. With 2.5 million without power or electricity, it was reported that Virginia experienced its largest non-hurricane power outage in history.

2.2 Small Scale Hazards

2.2.1 Extraterrestrial Hazard

Asteroids, meteoroids, and comets may cause hazards as they pass near-earth, enter Earth's atmosphere, and/or strike the Earth's surface thus causing changes in the interplanetary conditions that effect the Earth's magnetosphere, ionosphere, and thermosphere.

Meteoroids are tiny particles that can measure the size of a sand grain. They are usually the residue left from comets. If a meteoroid enters the Earth's upper atmosphere, it vaporizes and creates an event called a meteor due to rapid changes in structure. If the meteoroid is large enough, not completely vaporizing as it passes the upper atmosphere, and hits the ground, then it is referred to as a meteorite.

Comets can be characterized as more than just dirty ice-balls orbiting the Sun. As the Sun heats a comet, the ice vaporizes into gas. As the gas escapes, the dirt tags along. If one has ever seen a comet, or a photograph of a comet, you will notice the comet's tail pointing away from the Sun. This is the solar wind pushing the comet's gas and dust away from the Sun. The remaining residue of dust and small dirt particles remain close to the same orbit as the parent comet for years after, leaving a trail of dust. Meteor showers are produced when the Earth's orbit comes across streams of these very small particles, sometimes in spectacular fashion. For instance in 1965, some observers saw 40 meters per second from the Leonid meteor storm (NASA 2016a). The Leonid meteor storms of 1799 and 1833 were even more impressive. A famous artist conception of the 1833 Leonid meteor storm shows the brilliance of the meteors illuminating the night sky. The paint also highlights the fear and misunderstanding of the US public observing the spectacle.

A meteor shower's intensity depends on the size and density of a comet's trail. A comet that has a trail that is wide and loosely compact will result in a meteor shower that produces a few meteors seen per hour over a couple of weeks. However, if the dust trail is narrow and dense, the resulting meteor shower may result in hundreds, if not thousands, of meteors burning in periods of minutes to tens of seconds. This characteristic of a meteor shower is referred to as a meteor storm (NWS 2015c).

A risk-assessment system similar to the Richter scale for earthquakes is used to measure the impact of a meteor or comet. It is known as Torino scale (Fig. 2.32) and was adopted by a working group of the International Astronomical Union (IAU) in 1999 at a meeting in Torino, Italy. The author directs the readers to learn more about the Torino Scale at a recent publication by Morrison et al. (2004).

2.2.1.1 Cases-in-Point – Russian Meteor

A meteorite slammed into atmosphere above the city of Chelyabinsk, Russia on February 15, 2013 around 9:20 am local time (NASA 2016b), injuring 1200 people from flying debris, shattered glass, and lacerations (Fig. 2.33). The origin of the 20 m wide meteorite is still unknown. Originally, it was thought to be coming from a 2 km near-Earth asteroid called 1999 NC43. But a closer look at the asteroid's orbit and likely mineral composition, gained from spectroscopy, suggested that that was not the case (Howell 2015). Reddy et al. (2015) studied the event and demonstrated that it is difficult to make predictions about what particular asteroid could have shed pieces that slammed into Earth. Since most asteroids are so small and their orbits are chaotic, it is hard to make a firm link.

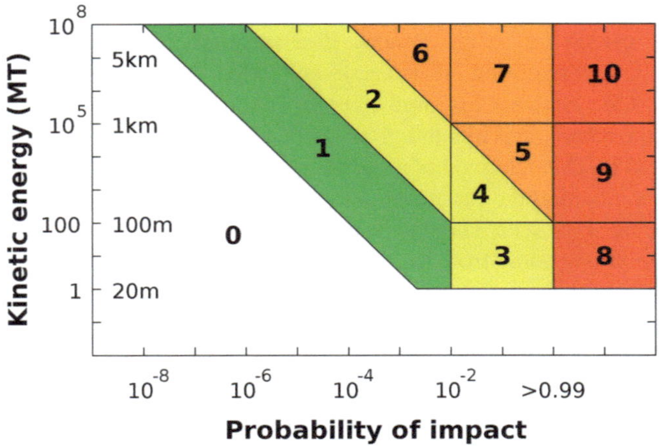

Fig. 2.32 Torino Scale (By Wrightbus at zh.wikipedia GFDL (http://www.gnu.org/copyleft/fdl.html from Wikimedia Commons)

Fig. 2.33 Trail of the asteroid that exploded over Chelyabinsk, Russia, on Feb. 15, 2013 (Credit: Alex Alishevskikh CC BY-SA 2.0 via http://www.flickr.com/photos/alexeya/)

2.2.1.2 Cases-in-Point – Lake Ontario, Canada

A small rimmed depression in Lake Ontario has been of intellectual curiosity among scientists. They believe the rimmed depression in Lake Ontario may have been an impact crater due to a meteorite. Detailed bathymetry of Lake Ontario reveals a small circular feature and adjoining SWtrending ridge associated with a small topographic

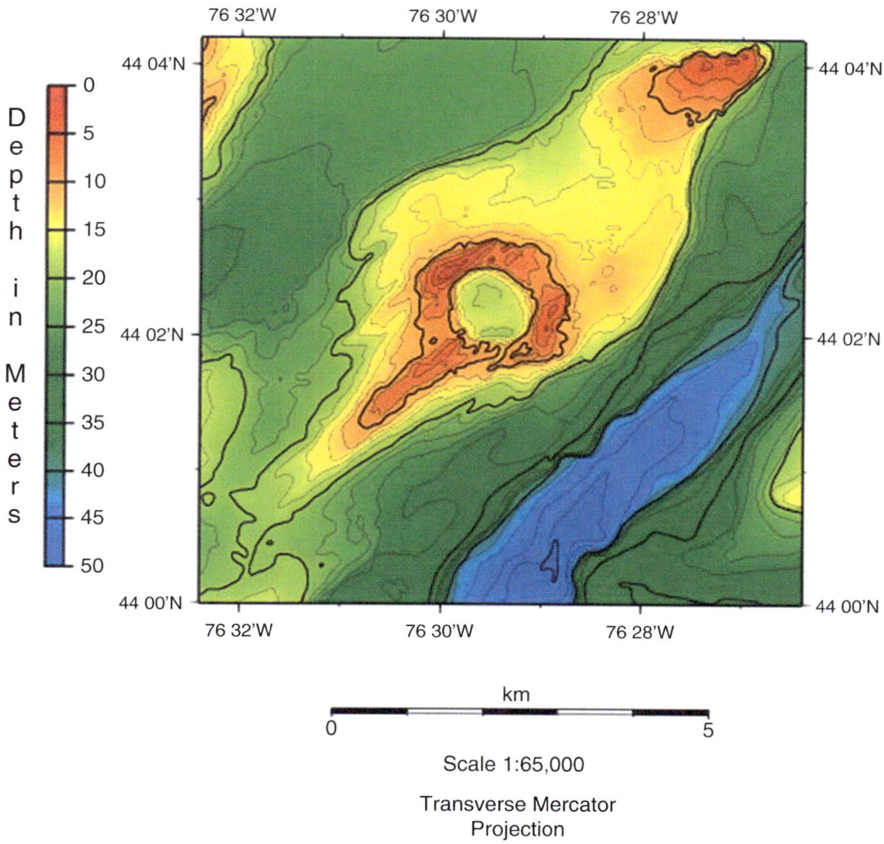

Fig. 2.34 Bathymetry of Lake Ontario, Canada, showing the Charity Shoal as a depression (https://www.ngdc.noaa.gov/mgg/greatlakes/lakeontario_cdrom/html/area12.htm)

high. The feature consists of a circular basin 1000 m in diameter and 19+ m deep, completely surrounded by a low-relief rim that rises to within 5 m of the water surface over much of its extent. Known as Charity Shoal (Fig. 2.34), the depression may be an extraterrestrial impact crater, but other origins such as sinkhole, volcanic cone, or kettle, are not ruled out. Time of formation is not known. A subtle negative magnetic anomaly coincides with the feature and is consistent with an impact origin, though not positively diagnostic. Relief of the feature is low compared to that typical of an impact crater of this size. Glaciation may have diminished relief by eroding the rim and filling the central basin with drift. Verification as an impact crater will require detailed geophysical surveys and collection and analyses of samples from in and around the structure (Holcombe et al. 2001).

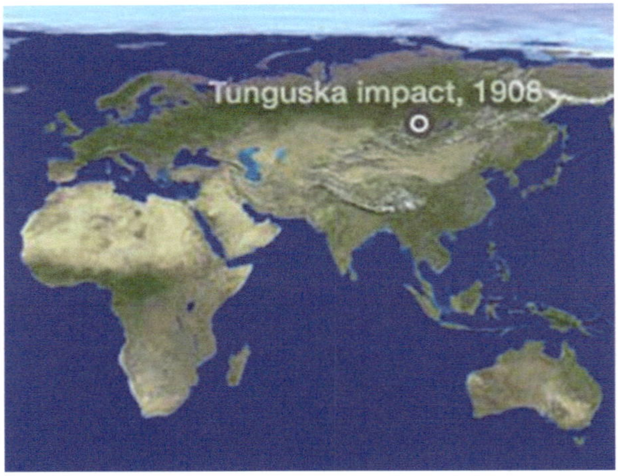

Fig. 2.35 Location of Tungushka where in 1908 a ferocious impact was witnessed near the Podkamennaya Tunguska River in remote Siberia (NASA 2008)

Fig. 2.36 Trees felled by the Tunguska explosion caused by a meteor. Credit: the Leonid Kulik Expedition (NASA 2008)

2.2.1.3 Tunguska, Siberia

The Tunguska impact is shown in Figs. 2.35 (location) and 2.36 (devastation), the place was flattened by a ferocious impact near Podkamennaya Tunguska River in Siberia in 1908. While the impact occurred in 1908, the first successful (one failed in 1921) scientific expedition to the area took place in 1927 led by Leonid Kulik, the chief curator for the meteorite collection of the St. Petersburg museum. Kulik found the locals to be reluctant to talk about the event because they thought that God Ogdy had visited the blast to curse the area by smashing trees and killing animals. The devastation was evident from 800 square miles of remote forest being ripped asunder and 80 million trees being on their sides, lying in a radial pattern (NASA 2008).

2.2.2 *Fog*

Fog can be characterized as a cloud that is in contact with the ground, where water droplets are suspended in the air near the Earth's surface (HIRA 2012). The United States National Weather Service monitors fog using satellite data, as observation platforms are sparse over the oceans. The Fog Monitor is a decision assistance tool engineered to continuously monitor satellite imagery for fog. It automatically alerts the World Food Organization (WFO) when probable fog is detected. The Fog Monitor provides a satellite-image-like display with pixels colored to indicate whether the location is probably fog-free, maybe fog-covered, or probably fog-covered. The Fog Monitor also provides the WFO the ability to customize the thresholds used to recognize and distinguish fog.

2.2.2.1 Ice Fog

Unlike the traditional fog events, which can severally reduce visibility due to suspended water droplets, ice fog reduces visibility due to suspended ice crystals. Ice fog occurs at very cold temperatures and is rare at temperatures warmer than -30° C. At higher latitudes ice fog can occur very quickly and has been known to develop immediately after a plane's liftoff (Environment and Climate Change Canada 2015).

2.2.3 *Geomagnetic Storm*

A geomagnetic storm is a significant displacement of Earth's magnetosphere that is due to the very efficient exchange of energy from solar winds released by the sun into the space environment surrounding Earth. This disturbance of the upper atmosphere is caused by magnetic activity of the Sun. According to the international disaster database EMDAT, geomagnetic storms can disrupt power grids, spacecraft operations, and satellite communications. Fig. 2.37 is an illustration of a geomagnetic storm captured by NOAA. These storms originate from erupting sunspots on the Sun's surface, releasing Coronal Mass Ejections (CMEs). CMEs are large clouds of ionized gas that carry superheated particles at a speed of approximately two million miles per hour (Public Safety Canada). If the CME is directed towards the Earth, it can impact the Earth's magnetosphere and cause a geomagnetic storm. This type of activity causes the aurora borealis and aurora Australis, allowing us to witness a spectacle in the night sky in the northern and southern pole regions (Fig. 2.38).

Fig. 2.37 Spectacular view of a geomagnetic storm (www.swpc.noaa.gov/phenomena/)

2.2.4 Hail Storm

Hail is a type of frozen precipitation that occurs within strong to severe thunderstorms, which can develop at any time of the year. Within these thunderstorms, very fast currents of air move upwards (updraft) and downwards (downdraft). Inside the updrafts, water vapor and rain are pushed extremely high into the cumulonimbus cloud. At a certain height, the air temperature drops below freezing, and the water condenses onto dust or other such particles, known as condensation nuclei. The supercooled water molecules condense onto the condensation nucleus and freeze, producing a tiny ice particle which serves as the center of the hailstone (Forbes 2015). A hail storm is a type of storm that is characterized by hail as the dominant part of its precipitation. The size of the hailstones can vary between pea size (6 mm) and softball size (112 mm) and therefore cause considerable damage.

2.2.4.1 Case Study – Calgary, Canada

Calgary, Canada experienced a monstrous hailstorm with hailstones measuring larger than golf balls late on August 12, 2012. In approximately 10 min, pounding hailstones dimpled vehicles and riddled house siding with millions of dents reported based on accounts of Environment and Climate Change Canada (2015) as shown in Fig. 2.39.

Southern regions of Canada (Fig. 2.40) are prone to hailstorms because of violent air masses and moisture interacting with each other. The conditions are set in these regions for masses of water in the air interacting with convective activity, which

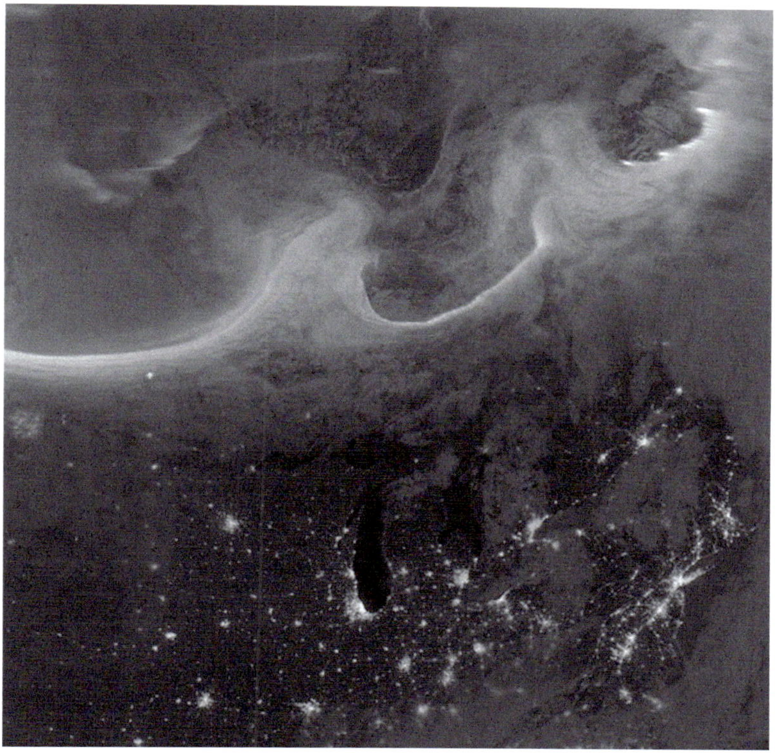

Fig. 2.38 Aurora borealis stretch across Quebec and Ontario, October 8, 2012 (Allen and Simmon 2015)

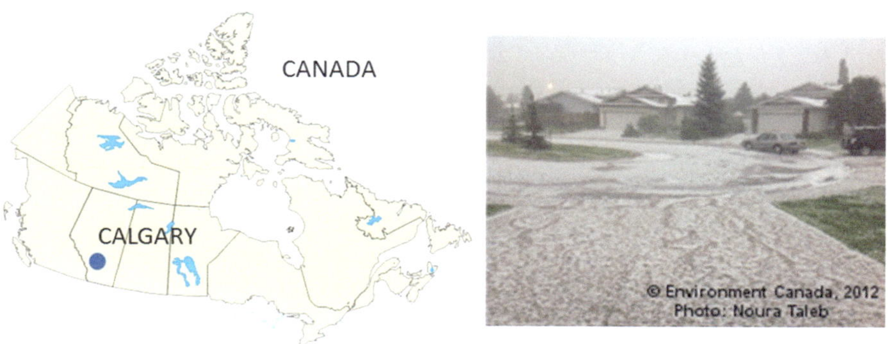

Fig. 2.39 The city of Calgary (left) and hailstones filled neighbourhood, Aug 12, 2012 (Environment Canada 2015)

Fig. 2.40 Southern Canadian region affected by severe hailstorm in 2008 http://ec.gc.ca/meteo-weather/default.asp?lang=En&n=D8F1A22A-1

produce this type of natural disaster. In 2010, another significant storm hammered Calgary causing $400 million dollars' worth of damage – the biggest and most damaging urban hailstorm in recorded Canadian history.

2.2.5 *Lightning*

Hazards and losses caused directly by lightning strikes are rare but do occur. Lightning is an atmospheric discharge of electricity, which typically occurs during thunderstorms, and sometimes during volcanic eruptions or dust storms. In the twenty-first century alone, 21 events of severe lightning have occurred, killing 63 people, affecting about 18,358, and causing economic damage of over $18 million as shown in Table 2.7 (extracted from EMDAT 2015). In comparison to other natural disasters, lightning impacts are not as damaging in terms of economic and social losses.

Lightning flashes occur in Canada about 2.34 million times a year, including about once every three seconds during the summer months. Lightning strike is estimated to kill up to ten people, seriously injure up to 164 others, and ignite some 4000 forest fires each year (Environment Canada 2015). Figure 2.41 shows a strong lightning strike on the CN Tower in Toronto.

Table 2.7 Lightning events in North America since the 1900s up until now (data extracted from EMDAT 2015)

Year	Occurrence	Total deaths	Total affected	Total damage
1997	2	21	0	150,000
1998	1	0	2200	0
1999	2	3	1097	300,500
2004	3	19	7624	715,000
2005	1	0	104	350,000
2006	1	1	600	450,000
2009	1	6	25	1,500,000
2010	1	20	100	0
2012	6	7	9005	7,610,000
2013	3	4	0	2,610,000
2014	5	6	900	5,444,000

Fig. 2.41 Lightning strikes the CN Tower during a thunderstorm in Toronto (By Raul Heinrich (Own work) https://upload.wikimedia.org/wikipedia/commons/b/b6/CN_Tower_struck_by_lightning.jpg

2.2.5.1 Types of Lightening

- Intra-Cloud: The most common type of lightning. It happens completely inside the cloud, jumping between different charge-regions in the cloud. Intra-cloud lightning is sometimes called sheet lightning because it lights up the sky with a 'sheet' of light.
- Cloud to Cloud: Lightning that occurs between two or more separate clouds.
- Cloud to Ground: Lightning that occurs between the cloud and the ground.
- Cloud to Air: Lightning that occurs when the air around a positively charged cloud top reaches out to the negatively charged air around it (NWS 2015d).

2.2.6 Exercise

Find out recent natural hazards of medium and small scale in a given calendar year in your area/region/country and analyze the information in light of historical similar events.

References

Accuweather (2012) Intense storms called a "Derecho" slam 700 miles of the US. http://www.accuweather.com/en/weather-news/deadly-super-derecho-strikes-m/67383. Accessed 21 Mar 2016

Allen J, Simmon R (2015) Auroras over North America, NASA Earth Observatory. https://commons.wikimedia.org/wiki/File:Auroras_over_North_America.jpg#/media/File:Auroras_over_North_America.jpg

Bagla P (2013) The making of Himalayan tsunami, NDTV media, July 18, 2013. Retrieved March 21, 2016. http://www.ndtv.com/article/india/uttarakhand-the-making-of-the-himalayan-tsunami-393901

Canadian Disaster Database (CDD) (2015) http://cdd.publicsafety.gc.ca/. Accessed 21 Mar 2016

CBC (2010a) Leamington tornado damage in the millions. http://www.cbc.ca/news/canada/windsor/leamington-tornado-damage-in-the-millions-1.910894. Accessed 21 Mar 2016

CBC (2010b) Tornado hits Sask. Reserve. http://www.cbc.ca/news/canada/saskatchewan/tornado-hits-sask-reserve-1.949847. Accessed 21 Mar 2016

Chatham Daily News (2010) Homepage. www.chathamdailynews.ca. Accessed 21 Mar 2016

Climate Himalaya (2013) Disaster in Uttarakhand, India: huge death toll. Climate Himalaya. http://chimalaya.org/2013/06/19/disaster-in-uttarakhand-india-huge-death-toll/

Dobhal DP, Gupta AK, Mehta M, Khandelwal DD (2013) Kedarnath disaster: facts and plausible causes. Wadia Institute of Himalayan Geology, Dehradun

EMDAT (2015) D. GuhaSapir, R. Below, Ph. Hoyois EMDAT: the CRED/OFDA international disaster Brussels – Belgium. http://www.emdat.be/database. Accessed 21 Mar 2016

Environment Canada (2015) Environment and climate change Canada, Canadian lightening fatalities and injury statistics. https://www.ec.gc.ca/foudre-lightning/default.asp?lang=En&n=5D5FB4F8-1

Forbes G (2015). The science behind hail formation, The Weather Channel. www.weather.com

Globe and Mail (2013) The Globe and Mail. http://www.theglobeandmail.com/news/national/in-pictures-rescuers-look-for-survivors-after-quebec-quarry-landslide/article7998305/. Accessed 21 Mar 2016

HIRA (2005) Hazard identification and risk assessment, Emergency Management Ontario. https://www.emergencymanagementontario.ca/english/emcommunity/ProvincialPrograms/hira/hira.html

HIRA (2012) Hazard identification risk assessment. https://www.emergencymanagementontario.ca/english/emcommunity/ProvincialPrograms/hira/hira_2012.html. Accessed 21 Mar 2016

Holcombe TL, Warren JS, Reid DF, Virden WT, Divins DL (2001) Small rimmed depression in Lake Ontario: an impact crater? J Great Lakes Res 27(4):510–517

Howell E (2015) Russian meteor's origin remains mysterious 2 years later. www.space.com. Accessed 30 June 2016

ISRO (2013) Bhuvan portal, gateway to Indian earth observation, National Remote Sensing Centre, Government of India. http://bhuvan.nrsc.gov.in/. Accessed 14 June 2014

Japan Times (2016) Huge street sinkhole disrupts services, forces evacuations near Fukuoka's Hakata Station. http://www.japantimes.co.jp/news/2016/11/08/national/huge-street-sinkhole-causes-turmoil-near-hakata-station/#.WLROcG8rJEY

Metro News (2011a) Snow storms batter North America. http://metro.co.uk/2011/02/02/snow-storms-batter-north-america-635821/. Accessed 21 Mar 2016

Metro News (2011b) Snow storms batter North America. http://metro.co.uk/2011/02/02/snow-storms-batter-north-america-635821/#ixzz3ekxwYU2w. Accessed 21 Mar 2016

Morrison D, Chapman CR, Steel D, Binzel RP (2004) Impacts and the public: communicating the nature of the impact hazard. In: Belton MJS, Morgan TH, Samarasinha NH, Yeomans DK (eds) Mitigation of hazardous comets and asteroids. Cambridge University Press, Cambridge

Mukharji B (2013) Monsoon speediest in over 50 years. Wall Street J, India, June 18, 2013

NASA (2008) The Tunguska impact – 100 years later. http://sciencenasagov/science-news/science-at-nasa/2008/30jun_tunguska/. Accessed 21 Mar 2016

NASA (2016a) Brief history of Leonid shower. http://leonid.arc.nasa.gov/history.html. Accessed 21 Mar 2016

NASA (2016b) Russian meteor plume. https://science.nasa.gov/science-news/science-at-nasa/2013/15aug_russianmeteorplume. Accessed 14 Jan 2016

NIDM (2013) National Institute of Disaster Management. http://nidm.gov.in/default.asp. Accessed 12 June 2014

NOAA (2015a) The National Severe Weather Laboratory. http://www.nssl.noaa.gov/education. Accessed 21 Mar 2016

NOAA (2015b) The National Severe Weather Laboratory. http://www.nssl.noaa.gov/education/svrwx101/wind/types/. Accessed 21 Mar 2016

NOAA (2015c) About Derechos. http://www.spc.noaa.gov/misc/AbtDerechos/derechofacts.htm. Accessed 21 Mar 2016

Novotný J (2013) VARNES landslide classification (1978). Presentation at Charles University, Czech Republic

NWS (2015a) The varying types of winter weather. http://www.weather.gov/rlx/wintvar. Accessed 21 Mar 2016

NWS (2015b) The enhanced Fujita scale (EF Scale). http://www.srh.noaa.gov/oun/?n=efscale. Accessed 21 Mar 2016

NWS (2015c) Meteors and Meteor showers. http://www.weather.gov/fsd/meteor. Accessed 21 Mar 2016

NWS (2015d) MWS Central Region Headquarters. http://www.crh.noaa.gov/. Accessed 21 Mar 2016

QMI Agency (2010) Winter blast paralyses Hwy 402 traffic. By O'Brien J, Richmond R, Miner J, Sher J. December 15, 2010. http://www.chathamdailynews.ca/2010/12/15/winter-blast-paralyses-hwy-402-traffic

Ramachandran R (2013) Why Kedarnath happened. Frontline, July 26, 2013. http://www.frontline.in/the-nation/why-kedarnath-happened/article4894867.ece?homepage=true
Reddy V, Vokrouhlichy D, Bottke WF, Pravec P, Sanchez JA, Gary BL, Klima R, Cloutis EA, Galad A, Guan TT, Hornoch K, Izawa MRM, Kušnirák P, Corre LL, Mann P, Moskovitz N, Skiff B, Vraštil J (2015) Link between the potentially hazardous Asteroid (86039) 2 1999 NC43 and the Chelyabinsk meteoroid tenuous. Planetary Science Institute, Tucson, AZ 85719, USA
Sundaramoorthy A (2013) Lessons learnt from landslides of Uttarakhand disaster. Geological Survey of India, Dehradun
USGS (2004) U.S. Geological Survey. http://pubs.usgs.gov/fs/2004/3072/pdf/fs2004-3072.pdf
USGS (2015) Landslide hazards program. http://landslides.usgs.gov/. Accessed 21 Mar 2016
UttarakhandForYou (2014) Kedarnath disaster 2013. http://uttarakhandforyou.com/2014/06/15/kedarnath-disaster-16-17th-june-2013/

Chapter 3
Disaster Risk Management

Disaster risk management is a comprehensive approach involving the identification of threats due to hazards; processing and analyzing threats; understanding people's vulnerability; assessing communities' resilience and coping capacity; developing strategies for future risk reduction; and building up capacities and operational skills to implement the proposed measures. Disaster risk cannot be eliminated completely, but it can be assessed and managed in order to mitigate the impact of disasters (Smith and Petley 2009). The management of disaster risks has attracted much attention since the 2005 initiative of the International Strategy for Disaster Reduction (ISDR 2004), which defined the Ten Essentials required to empower local governments and other agencies to implement the Hyogo Framework for Action by the year 2015.

3.1 Disasters

In the twenty-first century, our understanding of disasters that are caused by natural, technological, and/or human sources has improved significantly. Both the developing and the developed worlds have made considerable progress, within their capacity and limitations, towards the development of policies and mitigation measures to reduce future disasters. However, disasters continue to harm millions of people each year worldwide. A disaster can affect, or be affected by our natural environment, social processes, psychological elements, cultural issues, historical information, and political and economic ideologies. Certain risks are often inherent within a social system or physical location, but they can also be created due to certain natural or technological hazards (Alexander 1999). The consequences, however, can be similar in that they wreak havoc in communities and destroy social and economic systems. In order to effectively and efficiently manage disaster risks, attention should be on addressing

N. Agrawal, *Natural Disasters and Risk Management in Canada*, Advances in Natural and Technological Hazards Research 49,
https://doi.org/10.1007/978-94-024-1283-3_3

vulnerability[1] and improving the resilience and coping capacity of populations (Nirupama 2009, 2012; Twigg 2007; Canton 2007; Cutter 2001, 2003).

3.2 Risk

Risk identification is a measure of individual perception – how those perceptions are understood by society as a whole, as well as an objective assessment (Cardona 2006). The holistic approach of disaster risk management would involve risk identification and risk reduction components, a disaster management component, which is about response and recovery; and a financial protection piece that will account for institutional support, financial resources, and risk transfer tools. The shaping of risk identification, risk reduction, risk management strategies, policies, resource allocation, and operational plans should ideally engage all stakeholders in the process. A risk management team must have adequate information and understanding of high-probability/low-consequences versus low-probability/high-consequences events. A number of risk management strategies, such as education, awareness, economic incentives for individual mitigation measures, as well as legal, and legislative requirement can be considered. The process can be challenging as transfer of knowledge from science to politics is not easy (Schneider et al. 2006).

Risk is defined as a function of probability of occurrence of hazardous event, and potential loss to people, property, and/or the environment (Smith 2004; Wisner et al. 2004; ISDR 2004; HRVA 2004a, b) as shown in Eq. (3.1). Historical records of past disasters provide reasonable estimates of the probability of occurrence of hazards, hence risk is considered to be quantifiable using probabilities and consequences (Helm 1996; Green 2004; Smith and Petley 2009). Information on vulnerable populations and elements that are particularly exposed to risk can be assessed using a variety of indicators and criteria (Birkmann 2006; Armenakis and Nirupama 2013a,b; Nirupama 2012). Risk perception, demonstrated in Fig. 3.1, also plays a significant role in how disaster risk management is carried out in various societies and cultures (Slovic 2000). Therefore, perception becomes a noteworthy factor to be accounted for in risk management, and risks can vary with geographic location and local conditions. The standard risk formula is expressed as:

$$\mathrm{R} = \mathrm{H} \times \mathrm{V} \tag{3.1}$$

Here, R = risk; H = hazard, determined as a probability (or likelihood) of the occurrence of hazard; V = vulnerability (also loss, impact or consequences).

Several variations of standard risk formula have been proposed by experts (Table 3.1) and are as much practiced as the standard risk formula given in Eq. (3.1).

[1]The conditions determined by physical, social, economic and environmental factors or processes, which increase the susceptibility of a community to the impact of hazards (ISDR 2004).

Fig. 3.1 Various perceptions of hazard/disaster risk (Nirupama 2013)

Table 3.1 List of disaster risk assessment approaches that are similar to the conventional approach as given in Eq. (3.1)

Proposed risk evaluation equation	Variable other than probability and impact	Expert(s)
$R = p \cdot L^x$	x (>1) = people's perception	Whyte and Burton (1982)
$R = p \cdot S$	S = severity	Government of Michigan (2001)
$R = p \cdot V \cdot n$	n = social consequences	Ferrier and Haque (2003)
$R = \frac{H \cdot L}{preparedness\ (mitigation)}$	Preparedness or mitigation are measurable measures	Smith (2004)
$R = p \cdot L \cdot f(x)$	$f(x)$ = risk aversion factor	Schneider et al. (2006)
$R = H \cdot V \cdot M$	M = manageability or ability of humans	Noson (2009)
$R = H \cdot Elements\ at\ Risk \cdot V$	*Elements at Risk = physically exposed assets*	Smith and Petley (2009)
$R = H \cdot (V \cdot cp)$	cp = community perception	Nirupama (2012)

Here, commonly known variables are: *R* risk, *p (or H)* probability, *L* loss, *V* vulnerability, *I* impact

3.3 Disaster Risk Management – Key Elements

Key elements of a comprehensive disaster risk management system are shown in Fig. 3.2. Each key element is briefly explained in following sections.

3.4 Threat Recognition

This key element involves Identifying potential risks from natural, technological, or human-induced hazards; and recognizing vulnerable populations, such as very old, very young, single parents with young children, low income earners, unemployed, those facing language barriers, and physically and emotionally challenged persons and families.

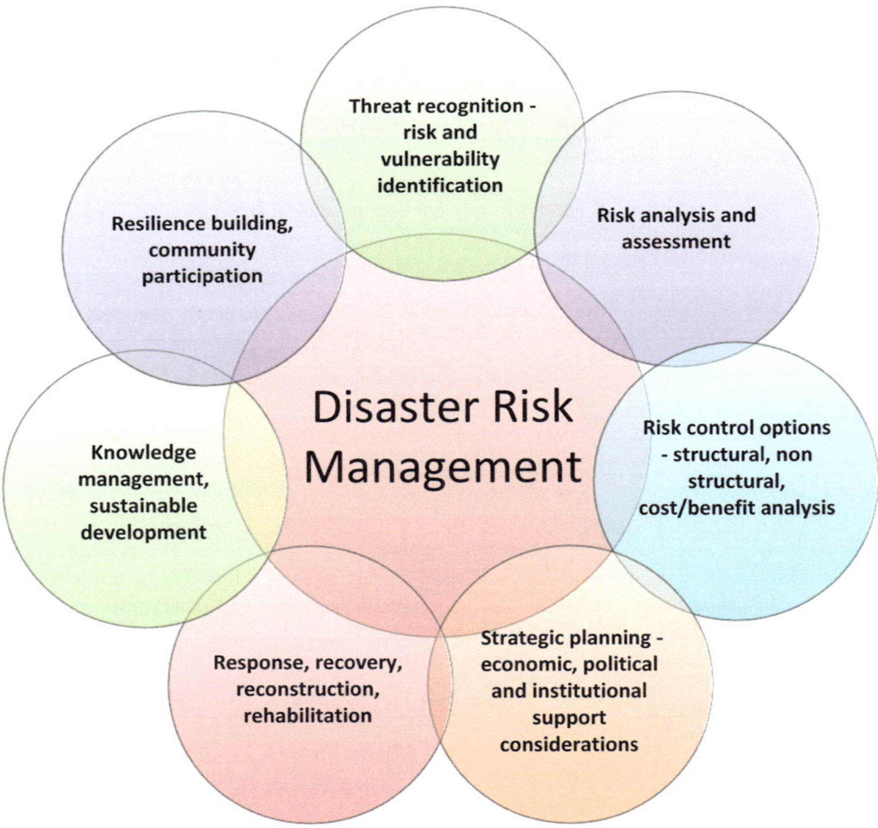

Fig. 3.2 Elements of comprehensive risk management (Nirupama 2013)

3.5 Risk Analysis and Assessment

This element is about understanding the magnitude, frequency of occurrence, and severity of consequences and prioritization of risks. The standard risk formula is given in Eq. (3.1). A few risk evaluation methods are discussed here in brief and elaborated in Chaps. 6, 7, and 8.

Qualitative and quantitative frameworks and methods have been developed to understand and evaluate disaster risk. Qualitatively speaking, all individual/institutional perceptions of risk carry equal weight as they choose to respond in a certain manner to a certain threat in certain circumstances (Nirupama and Etkin 2009). Among the qualitative models, *Pressure and Release (PR) and Access to Resources (AR)* models (Wisner et al. 2004) are widely used. The *PR* is a static model, founded on the concept of progression of vulnerability by looking at how underlying causes create an environment that allows for some dynamic pressures (e.g. lack of education, land degradation, population growth) to translate into unsafe conditions (e.g. exposure to risk, lack of social network etc.) in a given timeframe. Unlike the *PR*, the *AR* model is dynamic and community based. It focuses on access to income opportunities, and the development of coping strategies during and after a disaster.

In most quantitative risk assessment methods, two variables – probability of the occurrence of hazards and their potential impact – are commonly used. Although, detailed description of some of the methods is given in Chaps. 6, 7, and 8 Risk Evaluation Methods, a brief mention is given here.

The HRVA (Hazard, Risk and Vulnerability Analysis) method (HRVA 2004a, b) of British Columbia, Canada evaluates disaster risk based on event likelihood, assessment of vulnerability (social, physical, economic, and environmental), and severity of consequences (fatality, injury, damage and disruption of essential services – water, electricity, communication networks, physical and economic impact). Although the HIRA (Hazard Identification and Risk Assessment) (HIRA 2011) of Ontario, Canada follows similar steps (i.e. hazard identification, risk assessment, risk analysis and monitoring/review for future revisions), it accounts for psychosocial factors, such as panic and hoarding behavior, in assessing disaster impacts.

The FEMA (Federal Emergency Management Agency) model was developed in the USA to provide guidance to the nation for planning and decision making during disaster management through the use of mitigation. The model accounts for threat identification and rating, assessment of assets, vulnerability, and risk, and mitigation options. NOAA (National Oceanic and Atmospheric Administration)'s Geographic Information System (GIS) -based vulnerability assessment tool (NOAA) identifies opportunities beyond the existing built environment for reducing future hazard vulnerability, and identifies the large tracts of undeveloped land in communities that can be used for future land-use planning for sustainable growth.

The SMUG (Seriousness, Manageability, Urgency, and Growth) (CDEMG 2005a, b) model was developed by the Civil Defence Unit of Chatham Islands Council of New Zealand. The model describes the prioritization of potential hazard risks based on four criteria:

1. Seriousness – number of lives lost, potential for injury; physical, social and economic consequences
2. Manageability – ability to mitigate, both hazard and vulnerability
3. Urgency – measure of capability to address the hazard
4. Growth – rate at which hazard risk will increase through either an increase in the probability of occurrence, in the exposure of the community, or combination of the two); and four R's (Reduction, Readiness, Response, and Recovery)

In less developed regions such as Latin America and the Caribbean and Asian countries, national governments and NGOs usually play a pivotal role in managing disasters. The concept of risk evaluation, however, is similar to that of shown in Eq. (3.1) and risk assessment methodologies are similar to the ones used in developed world. In an ideal disaster risk management plan, a hazard and vulnerability analysis would be carried out and then appropriate action would be taken based upon the analysis (NDM 2012).

3.5.1 Case Study – A Multi-tier Hazard in Northern India in 2013

A rare weather system, last seen about 80 years ago, converged over northwest India in June 2013, leaving devastating impacts on communities in the region. The event involved three components to it. First, cloud burst with precipitation higher than normal occurred. Second, the rivers could not handle the excessive precipitation causing heavy flooding. Lastly, landslides were triggered by the flooding in this sensitive Himalayan region. The combined effects of these three hazards were extremely damaging. The weather data in the area suggested that the monsoon advancing towards the west of South Asia as well as westerly winds resulted in several days of torrential rains. Reports of more than 5700 people losing their lives and over US$500 million damage appeared in the media. This event has brought attention to a need to develop early warning systems for multi-faceted hazards. Development of an early warning system is challenging in this case because a cloud burst is a meteorological hazard, river flood is a hydrological hazard, and landslide is a geological hazard, which are handled by different government departments in most countries. This study is aimed at developing risk and vulnerability from such hazards through a critical analysis of the observed data (Nirupama et al. 2014; Sharma et al. 2014).

In order to carry out risk analysis on this three-tier disaster, hazards were examined individually, including cloud burst, flooding, and landslide, as well as historical perspective of these hazards in the region. A cloudburst is an extreme amount of precipitation, sometimes with hail and thunder, which normally lasts no longer than a few minutes but occasionally, can last much longer over an extended period, capable of creating flood conditions. Usually, rainfall rate is equal to or greater than 100 mm/h. The associated clouds can extend up to a height of 15 km

Fig. 3.3 A typical cumulonimbus cloud during a cloud burst event. The anvil shape at the top is due to the fact that the tropopause acts like a lid and does not allow the cloud to rise any further

above the ground. This rapid precipitation is mostly caused by the cumulonimbus clouds as shown in Fig. 3.3. Small scale intense vortices in the atmosphere generate strong convection currents which lift the moisture laden air with immaculate speed to generate cumulonimbus cloud (Aurora 2013; Climate Himalaya 2013; Khaladkar et al. 2009). Air currents rush upwards in a rainstorm that can hold a large amount of water. Figures 3.4 and 3.5 respectively show satellite coverage of rainfall concentration over Uttarakhand and the precise amounts of rainfall. Historically, somewhat similar events happened in July 1968 in Rajasthan and Gujarat States of India, with a death toll of 4892 (Dobhal et al. 2013; CWC 2013). While cloud burst is due to natural causes, the flooding and landslides are due to a combination of natural causes and human factors such as deforestation, urbanization, and poor land use management.

River flooding – disastrous floods resulting from monsoons are quite common in the Indian subcontinent causing heavy loss of life and significant damage. According to the EM-DAT database, 17 major flood disasters with each event killing more than 1000 people since flood records began in 1950 in India. According to the same database during 1950–2012, about 60,000 people were killed in India due to monsoon related floods (UWI 2013; CWC 2013; IMD 2013). The June 2013 monsoon rains in Uttarakhand were highly unusual, as the monsoon came to the region 2 weeks earlier than normal. The monsoon started in South India near the normal June 1st arrival date, but then advanced across India in an unusually rapid fashion, arriving in Pakistan along the western border of India on June 16, a full month earlier than normal. This was the fastest progression of the monsoon on record. The previous record for fastest monsoon progression occurred in 1961, when all of India was under monsoon conditions by June 21st.

On 17 June 2013 the state of Uttarakhand received more than 340 mm of rainfall, which is 375% above the normal benchmark of 65.9 mm rainfall during a normal monsoon. This cloudburst caused heavy floods in Uttarakhand as well as the neighbouring state of Himachal Pradesh. In the city of Dehra Dun, capital of

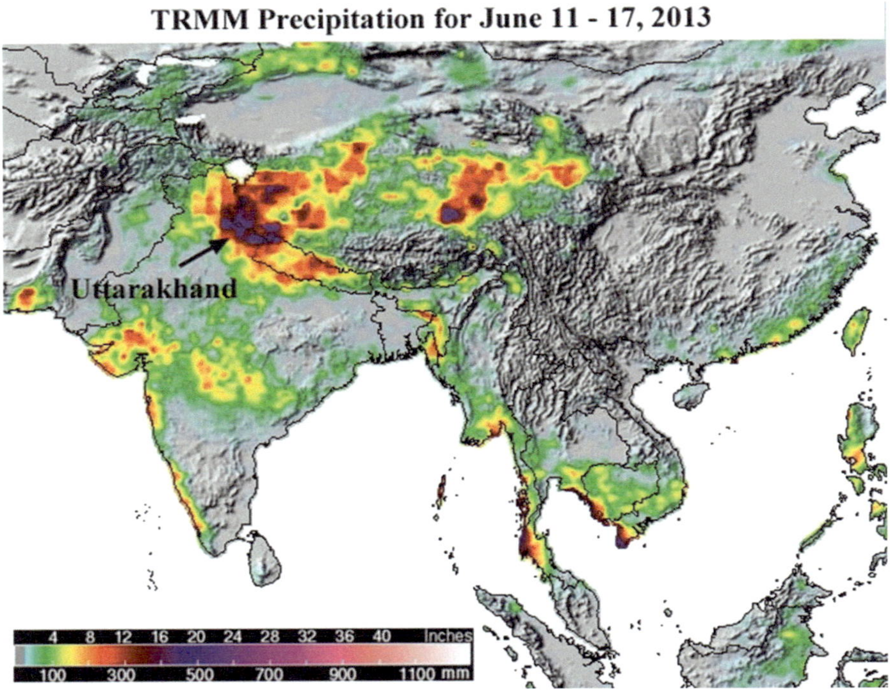

Fig. 3.4 Satellite-estimated rainfall for the 7-day period June 11–17, 2013, from NASA's TRMM satellite exceeded 508 mm (NASA)

Fig. 3.5 Rainfall amounts in mm during June 14–19, 2013 (Sundaramoorthy 2013)

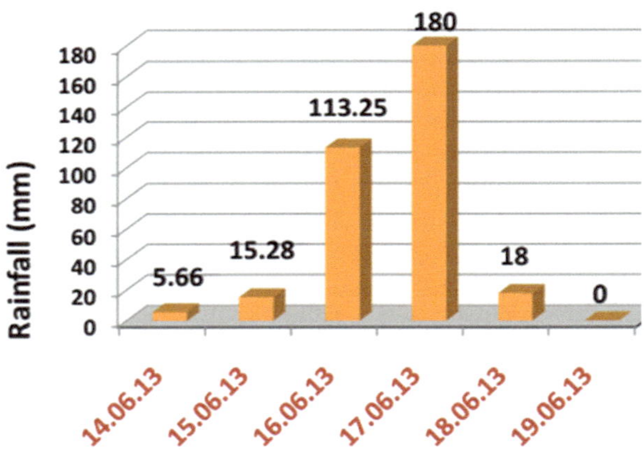

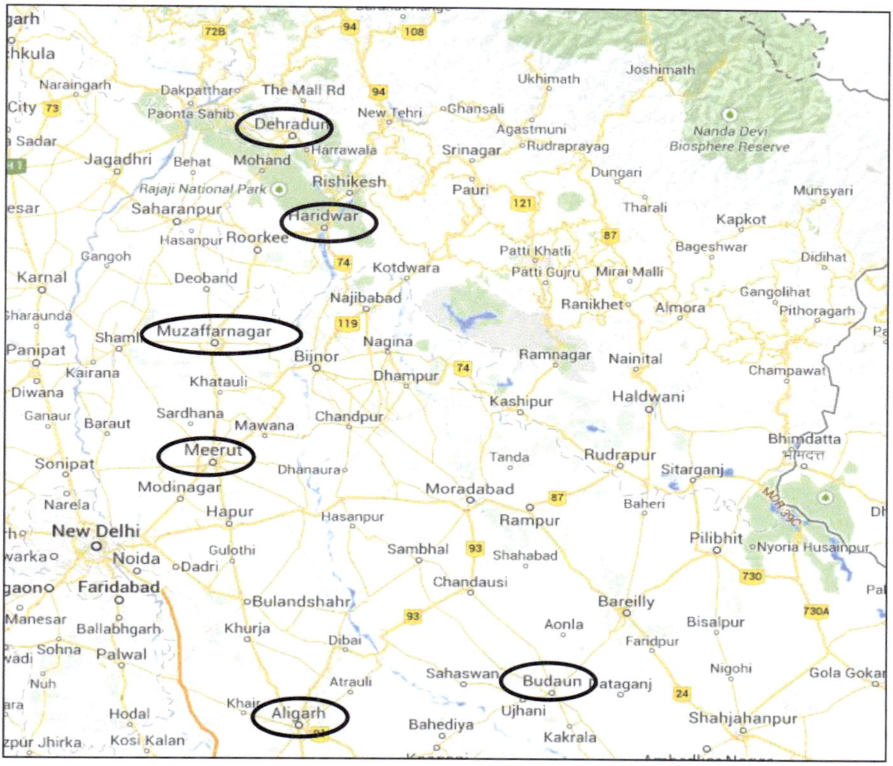

Fig. 3.6 Some of the districts affected by the monsoon floods of June 2013 in India (Google map)

Uttarakhand, this was the wettest day during the month of June in over five decades. Heavy rainfall for four consecutive days as well as melting snow during summer aggravated the floods further.

Mapping floods in the Northern Indian states of Uttarakhand and Uttara Pradesh using NASA's Satellite Images (as of 21 June 2013) shows that major flood affected districts in Uttarakhand include Haridwar and Dehradun, and in Uttar Pradesh, Meerut, Muzaffarnagar, Aligarh, and Badaun (Fig. 3.6). The heavy discharge in the Bhagirathi River valley of Uttarkashi that led to the major floods can be seen in Fig. 3.7.

Landslides – heavy rainfall occurred at the time when there was still snow on the ground, therefore, the combination of heavy rainfall on melting snow created conditions for widespread landslides (NIDM 2013; Ramachandran 2013; Bagla 2013; Climate Himalaya 2013). Figure 3.8 shows the landslides with reference to the main towns in the impact region (Mukherji 2013).

Severe flooding and massive landslide hit the Hindu shrine in Kedarnath (Fig. 3.9), which lies just a short distance from the snout of two mountain glaciers. The shrine being an important pilgrimage destination was packed with visitors

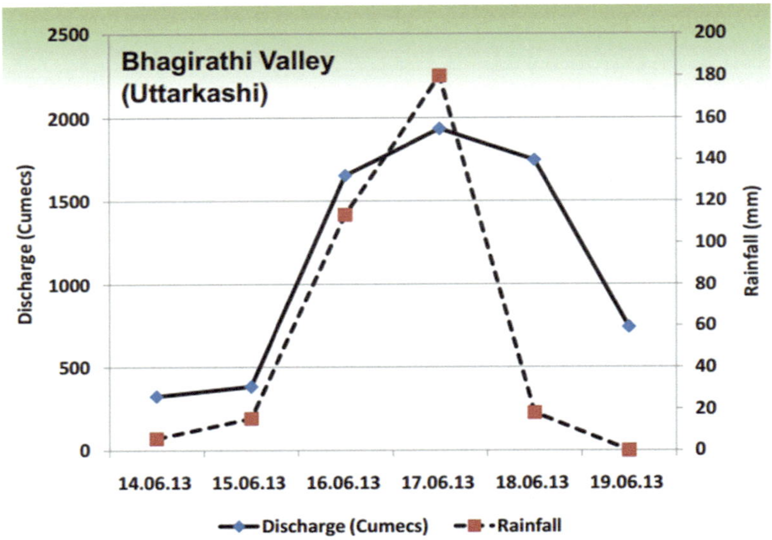

Fig. 3.7 Discharge in the Bhagirathi River valley of Uttarkashi (IMD 2013)

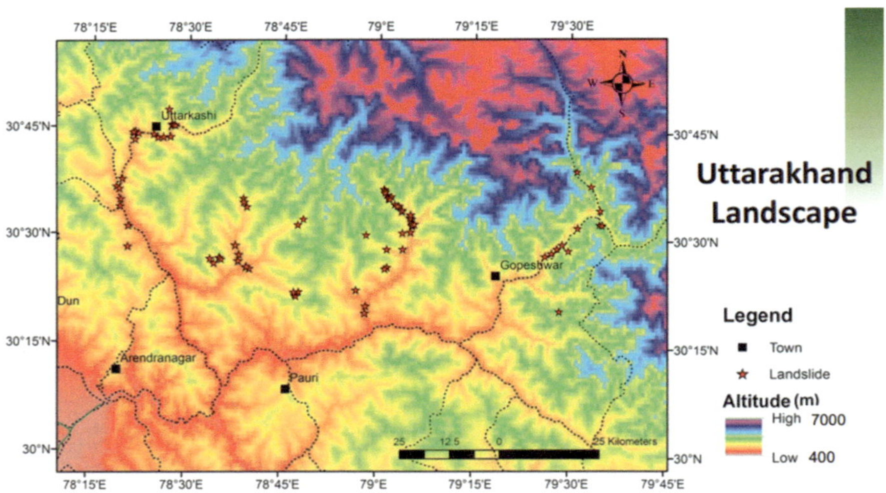

Fig. 3.8 Uttarakhand impact area showing towns and landslides (Mukherji 2013)

celebrating a religious holiday. According to Dobhal et al. (2013) and Ramachandran (2013), first the flow from the north east came down the margin of the glacier and spread out to strike the town. Next the northwest flow descended from the other glacier to the town on its west side, and struck it directly. The debris

Fig. 3.9 A bridge on the verge of collapse in Kedarnath Valley in Rudraprayag (Deccan Chronicle Oct 10, 2014 (left) and broken end of footbridge over the Mandakini River at Rudraprayag Sangam (By Mukerjee, via Wikimedia Commons (right))

flow from the north-east was triggered by a 75 m wide landslide, which then came down the steep slope about 500 m, gathering the debris in its path. The flow was initially channeled into a narrow gully formed by the glacier and on exiting it the flow spread out in the floodplains before striking the town after traversing about 1200 m. The steepness of the slope would have given the debris enormous velocity.

3.5.1.1 Risk Analysis

Disaster risk is a product of hazards and subsequent consequences. The three-tier hazard, as a combination of heavy rainfall, flooding, and landslides has been explained and analyzed in previous sections. In this section, the severity of impact is discussed (Theophilus 2013; Sphere India 2013; NDMA 2013; DMMC 2013). Table 3.2 presents a summary of social, economic, infrastructural, and environmental impacts.

To establish the return period of such an event, past cloud burst events in the same area were examined. According to the meteorological records (IMD 2013), a similar event had occurred in the same area on July 25, 1966, about 48 years ago. Therefore, the event of 2013 can be considered a 50 year return period event, which ties in well with the analysis of the event.

3.6 Risk Control Options

Structural, non-structural, cost/benefit analysis: these considerations are based on feasibility, effectiveness, and cost/benefit analysis. Structural measures may include the building of dykes, dams, and other protective structures. Non-structural measures

Table 3.2 Summary of the impacts of the three tier natural disaster in Uttarakhand, India

Social	5700 feared dead; Up to 11,000 missing; over a million tourists stranded; over 15 Primary Health Centres (PHC) facing shortage of medicine; drinking water shortage in 300 villages
Infrastructural	400 villages, covering 95,830 sq. km area; 1000 bridges damaged; 695 water distribution plants; 600 villages completely cut off from roads, 13,600 villages partially cut off; 70 hydroelectric stations and 505 dams badly affected
Economic	Over 1000 livestock killed; 4640 cases of illegal mining found; 50% of illegal housing found on riverbeds
Environmental	Water contamination concerns; over US$500 million – Trees, flora and fauna

may include land-use planning, hazard risk zoning, early warning systems, education and awareness campaigns, affordable disaster insurance, and legal and regulatory policy. Market-like tools, such as reinsured catastrophe funds (Mexico) and mitigation-focused insurance schemes (Barbados) have been implemented in a few countries (Freeman et al. 2002).

3.6.1 Case Study – London, Canada

The City of London, Ontario, Canada is situated on the Thames River. There was a growing concern in the Upper Thames River Watershed (Fig. 3.10) that the impact of rapid urbanization in the City may show as increased potential for flooding during increased receipt of rainfall. The government officials dealing with emergency preparedness and natural disasters felt that increasing urbanisation is enhancing the risk from river flooding in urban areas. Satellite imageries captured in 1974, 1990, and 2000 were examined for changes in landuse and land cover over the decades (Nirupama and Simonovic 2007). The images were classified for land areas of various landuse and landcover (shown in Figs. 3.11, 3.12, and 3.13) using remote sensing technology. The information was integrated with meteorological and hydrological historical data records after the classification scheme. It was evident from the study that over the past three decades the Upper Thames River watershed has been experiencing drastic landuse changes. Particularly, urbanization increased and now accounted for 22.25% coverage of the total watershed in 2000 compared to only 10.07% in 1974. Urbanization leads to increased impervious surfaces, which translates into precipitation being directed in to quick surface runoff, reducing the time to peak and producing higher peak-flows in the drainage channels. Surface water availability appears to be reducing slightly over the decades from 3.65% in 1974 to 2.73% in 2000 (Table 3.3). Forests are being cut down continuously at an alarming rate reducing the forest reserves to a mere 13.06% at present.

Figures 3.14 and 3.15 are examples of rainfall and runoff response pattern by plotting hydrographs using observed hydrologic and meteorological time series together on the same plot. This allows for a quick assessment of rainfall amount

Fig. 3.10 Upper Thames River Watershed (Upper Thames River Conservation Authority)

versus runoff in different years over the three decades covered in the study. Reviewing the hydrologic data, it can be seen that in 1970 (Fig. 3.14) total precipitation of nearly 400 mm resulted in 350 m^3/s of peak flow, whereas in 1997

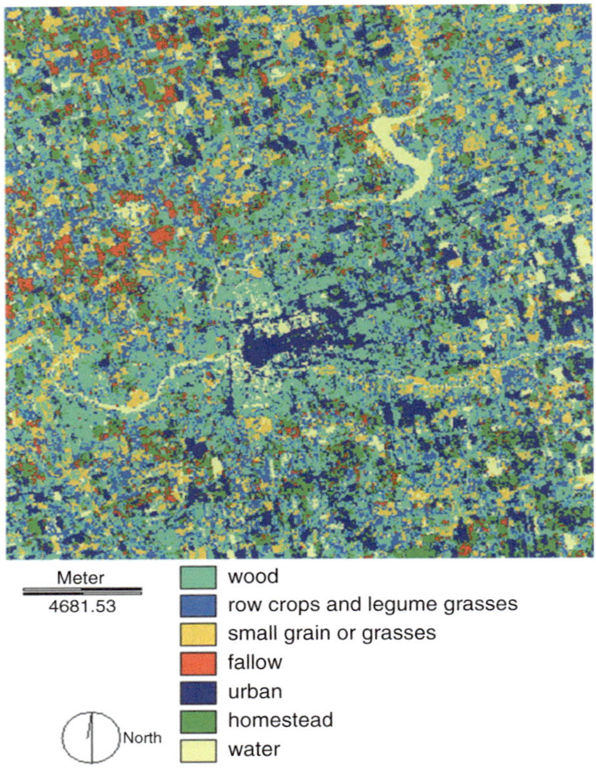

Fig. 3.11 LANDSAT 1 Imagery of July 1974 classified for land use

(Fig. 3.15) 200 mm of total precipitation brought about more than 800 m^3/s of river flows at Byron (outlet point of UTR watershed). Similar trends were observed in other years, except for dry years like in 1988 and 1989, according to a UTRCA report (thamesriver.on.ca). Inference of this is that increase of impervious areas enhances the river flows considerably.

This case study also demonstrates how remote sensing and GIS technology can help in better understanding the pattern of urbanization and its effect on the hydrology of the area, thereby assisting the authorities in flood control planning, mitigation, and adaptation.

Conventionally, flood emergency management, both public and private usually responds to crises rather than being concerned with the broader issues of vulnerability and its management (Shrubsole 2001). The present status quo of responding to environmental disaster events after the fact should be reviewed with emphasis on mitigation as high priority. Alternatives for mitigation of flood damages, landslides and erosion, such as, planned land use, should be explored, proposed and implemented to ensure a concerted emergency preparedness management structure is in place.

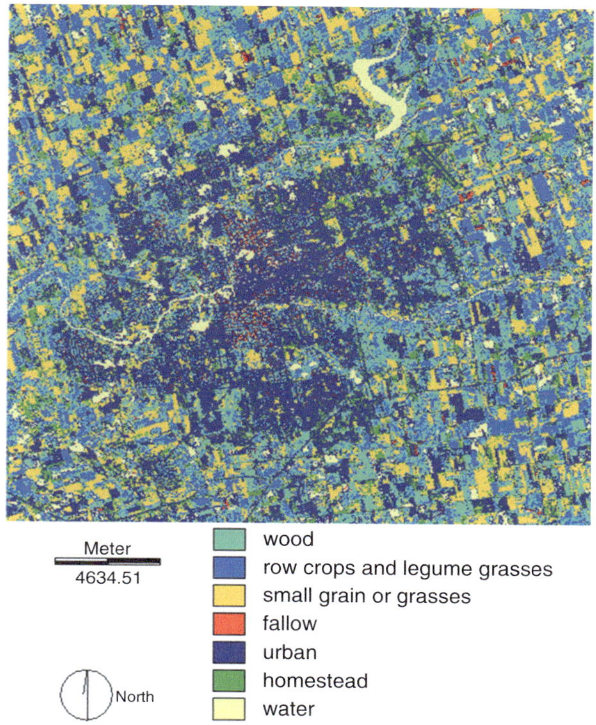

Fig. 3.12 LANDSAT 5 Imagery of July 1990 classified for land use

3.6.2 Case Study – Red River Basin, Manitoba, Canada

The nature of floods and their impact depend on both natural and human made conditions in the floodplain. Economic development and the installation of flood protection measures have political, economic, and social dimensions as well as engineering aspects. Hydrologic and hydraulic analysis of floods provides a sound technical basis for management decision making that must weigh numerous other factors such as, non-structural measures and virtual database-based decision support systems (Hoggan 1996; Simonovic 2002; Kundzewicz 2002).

A typical flood control problem requires selection and implementation of the best structural and/or nonstructural solution from the set of potential alternatives. Flood management problems include conflicting quantitative and qualitative evaluation objectives and multiple decision-makers. Multi-objective techniques help in evaluation and ranking of alternatives based on the objective values associated with each of the alternatives, and preferences of the various decision-makers. However, the flood management alternatives exhibit spatial variability. The Geographic Information System (GIS) is a useful computer-based tool to assist in flood risk management with spatially distributed variables (Carver 1991; Banai 1993; McKinney and Maidment 1993; Pereira and Duckstein 1993; Tim 1997; Wolfe 1997).

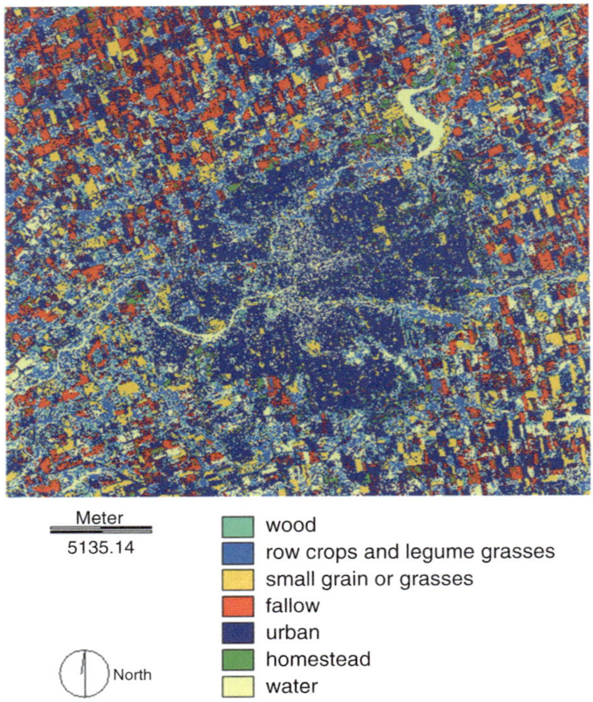

Fig. 3.13 LANDSAT 7 Imagery of October 2000 classified for land use

Table 3.3 Landuse classification corresponding to the three satellite images (Figs. 3.11, 3.12, and 3.13). The numbers shown are percentages

	Jul 7, 1974 (%)	Jul 23, 1990 (%)	Oct 30, 2000 (%)
Land use classes			
Woods	24.01	11.98	13.06
Row crops & legume grasses	22.78	29.18	13.20
Small grains or grass	31.56	34.91	16.84
Fallow land	4.79	2.34	30.06
Urban	10.07	16.72	22.25
Homestead	3.14	2.05	1.86
Water	3.65	2.82	2.73

Time and space play an important role in flood management and control. Therefore, there are uncertainties involved in flood prediction, in the evaluation of the inundated area and in the estimation of various physical, ecologic, economic, and social impacts. For better flood risk management it is essential to account for spatial variability and uncertainties involved in decision-making. A new technique combining these two objectives is developed and employed in a study (Simonovic and Nirupama 2005), in which the new technique is termed as Spatial Fuzzy Compromise Programming (SFCP). For illustrations purposes, SFCP was applied to the Red

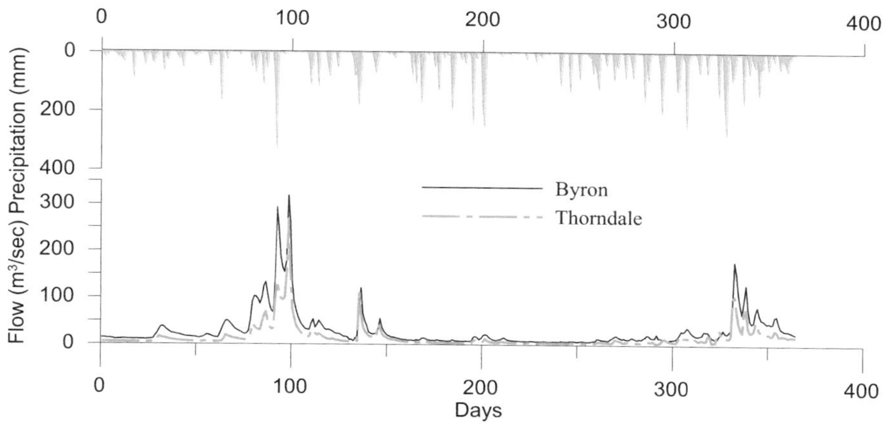

Fig. 3.14 1970 observed hydrographs at Byron and Thorndale and total precipitation at London, Ontario

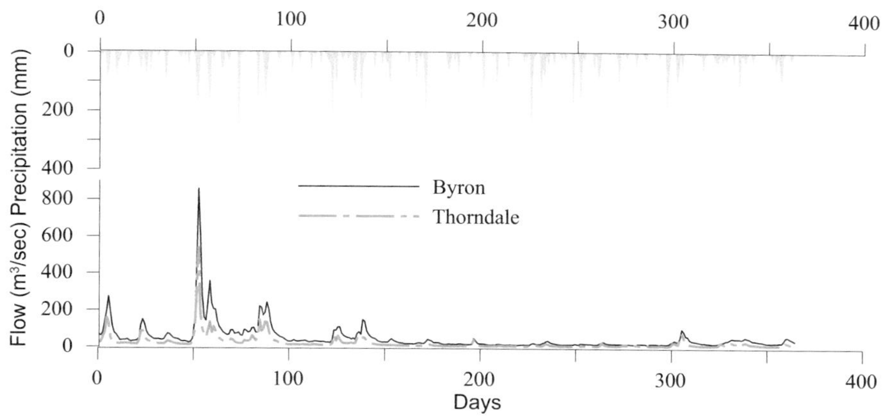

Fig. 3.15 1997 observed hydrographs at Byron and Thorndale and total precipitation at London, Ontario

River Basin, of Manitoba, Canada to demonstrate that this new approach can assist a decision-maker select the optimum flood management alternative for each location (5×5 m grid) in the entire study region.

For detailed analytical and numerical aspects of the Spatial Fuzzy Compromise Programming technique, readers are advised to refer to the original paper by Simonovic and Nirupama (2005). For the purpose of this textbook, a brief description is presented here.

The Red River Basin during the 1997 floods, known as the flood of the Century, is shown in Fig. 3.16. The floodplain consists of low-lying flat prairies predominantly used for agricultural purposes. The main population center in the region is the City of

Winnipeg, which is located in the downstream portion of the valley, at the confluence of the Red River and Assiniboine River (Fig. 3.17). Other communities of significant size further upstream in the Red River Valley include the towns of St. Adolphe, St. Agathe, Morris, and Emerson. The Red River Valley, which borders North Dakota and Minnesota in the US and expands north toward Lake Winnipeg in Manitoba, Canada, is prone to flooding and has historically (1826, 1950, 1979 and 1997), incurred extensive damage to both urban and agricultural areas from floodwaters. Figure 3.18 shows the flooded community of St. Adolphe during the 1997 floods.

Fig. 3.16 Red River Basin flooded during the 1997 Flood of the Century (in blue). Manitoba Centre for Remote Sensing website

Major floods are typically seasonal in nature and are the result of combined spring snowmelt and rainfall runoff along both the Red and Assiniboine Rivers (Krenz and Leitch 1993). For this study, the community of St. Adolphe, located 20 km south of Winnipeg, has been taken into consideration.

Flood Protection Alternatives a particular structural measure to alleviate flood risk to the City of Winnipeg, after the devastating flooding event in 1950, has been in place since 1968. It is a 47 km long channel, designed to carry up to 4000 cubic meters per second (cumecs), and expanded in the 2000s from its original capacity of 1700 cumecs. Since 1968, it has prevented more than $40 billion (in 2011 dollars) in flood damage in Winnipeg. As implied in Fig. 3.17, other flood mitigation measures in place include, dikes along both the Red and Assiniboine Rivers, flood pumping stations within the City of Winnipeg, and the Shellmouth Reservoir.

For the purpose of illustrating the SFCP method, two scenarios have been developed as flood protection measures for the community of St. Adolphe. These are: (a) a dike along the Red River bank and (b) modified operations of the Red River Floodway located immediately downstream from the town. The study focus is a 2.0 × 1.7 km region encompassing the community of St. Adolphe along the Red River. As St. Adolphe is the closest community upstream from the floodway inlet and gate structure, it is the one which is most heavily influenced by the floodway operation. In normal operations of the floodway, the backwater that it produces extends many kilometers upstream beyond St. Adolphe. As a result its operation is frequently responsible for heavy damage to the community and surrounding areas.

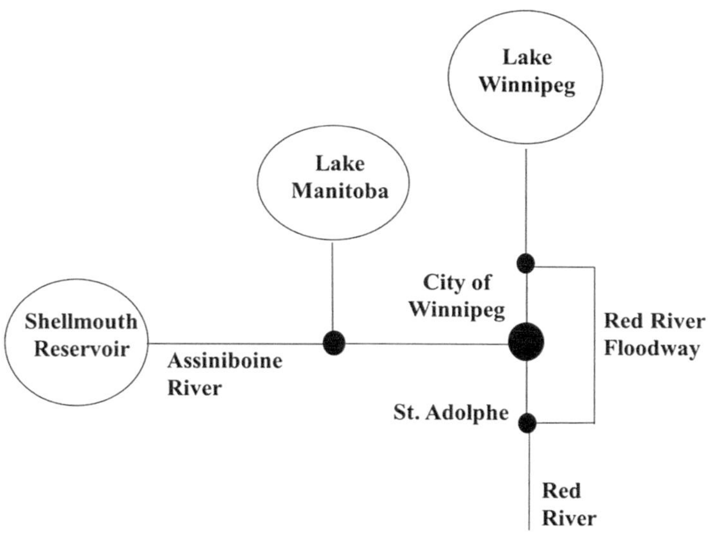

Fig. 3.17 Schematic diagram of the study area showing the Red River Floodway and the City of Winnipeg (Tkach and Simonovic 1997)

Fig. 3.18 The submerged community of St. Adolphe during the Red River flooding of 1997 http://www.gov.mb.ca/flooding/fighting/valley_ring_dike.html

For this reason, the largest conflict in the region is between St. Adolphe and the City of Winnipeg.

Computer simulation of three flood protection alternatives was developed and tested:

1. A dike around the community. This dike has been simulated only on the right bank of the river to protect the community of St. Adolphe.
2. Alteration of the controlled floodway operation so as to let more floodwater flow through the floodway in order to protect the larger city downstream. This is achieved by simulating a raise of the floodway gate height in such a way that the water surface elevation at the floodway entrance is increased by 1 m above the normal level. This alternative will be referred to as Floodway 1.
3. Alteration of the controlled floodway operation so as to let less floodwater flow through the floodway in order to protect a community upstream. This is achieved by simulating lowering of the floodway gate height in such a way that the water surface elevation at the floodway entrance is decreased by 1 m below the normal level. This alternative will be referred to as Floodway 2.

Data Used the basic spatial data set includes a digital elevation model (DEM) for the region under consideration. Figure 3.19 shows the DEM of the study region. The 5-meter resolution DEM was acquired from LIDAR (Light Detection and Ranging) remote sensing data. Feature image data sets were acquired for the purpose of damage assessment due to flooding. In Fig. 3.20, buildings in St. Adolphe are visible as small square-shaped clear yellow elements, roads can be seen as the straight lines around the buildings and also across the river and the agricultural fields are

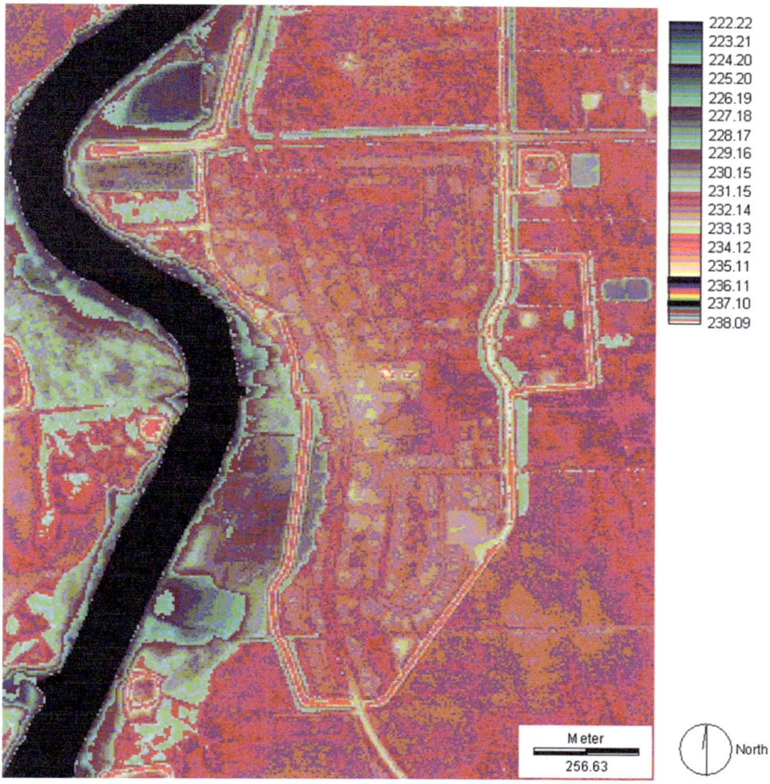

Fig. 3.19 Digital Elevation Model (DEM) of the study region derived from LIDAR

illustrated as polynomials. Red River is shown on the left side of the image in pink. The legend in the figure denotes the elevation of the features in meters.

Next, hydraulic data of the region was acquired which consisted of river cross section profiles, river flows and coefficients (Manning's n, contraction, and expansion). Using the HEC-RAS hydraulic model (Hydrologic Engineering Center 2001) the simulation of all three flood protection alternatives was performed, and results presented in Table 3.4. Figure 3.21 shows a screen capture of HEC-RAS simulation.

Criteria to Evaluate the Three Flood Protection Alternatives two criteria that exhibit a spatial variability were selected for evaluating the three alternatives described earlier: (a) water depth and (b) flood damage. The computational procedures necessary to produce the raster criteria images involve the use of GIS software and data on damage curves for buildings, agriculture, and roads. To implement the criteria, the simulated elevations of flooded areas were subtracted from the DEM. Raster cells in locations which were unaffected by floodwaters retained a value of zero. In this way an image containing the water depths for all

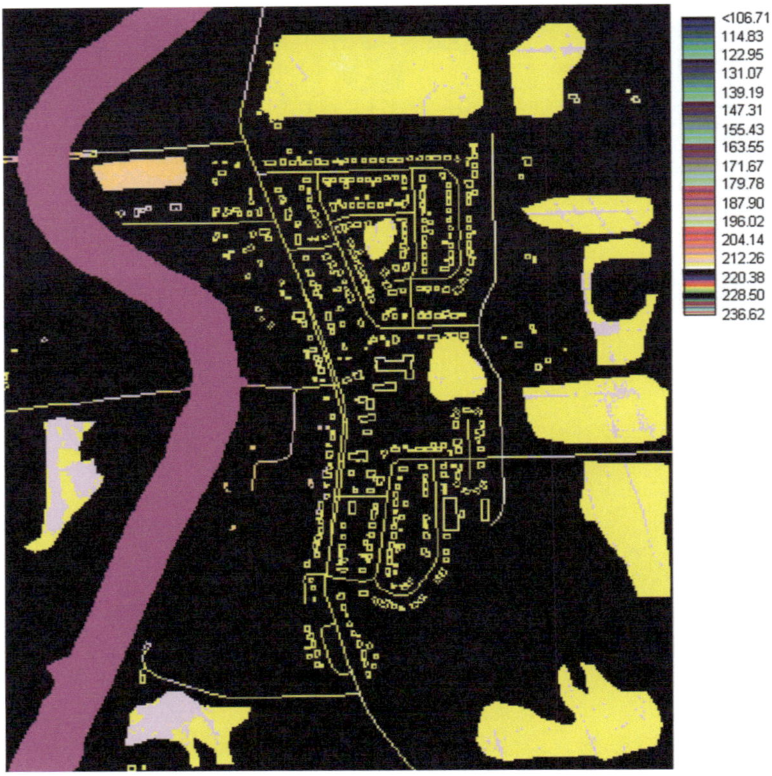

Fig. 3.20 Feature image comprising of buildings (clusters in yellow), roads (lines) and agricultural fields (yellow polynomials) in St. Adolphe region

Table 3.4 HEC-RAS simulation for the three alternatives (Simonovic and Nirupama 2005)

Alternative Name	Total discharge at floodway entry point (m^3/s)	Water surface elevation (m)
Dike	3650	232.89
Floodway 1	4730	233.83
Floodway 2	2900	231.71

flooded locations in the study region is produced for each alternative. The second criterion used in evaluation of the alternatives is the monetary value of damage to buildings, roads, and agricultural land within the region of interest. KGS Group (2000) recommendations, which are based on the 1997 flood event, were used to estimate the dollar value of damages associated with buildings, roads, and agriculture. Equations (3.2), (3.3), and (3.4) represent depth-damage relationship for the three categories – buildings, roads, agriculture respectively.

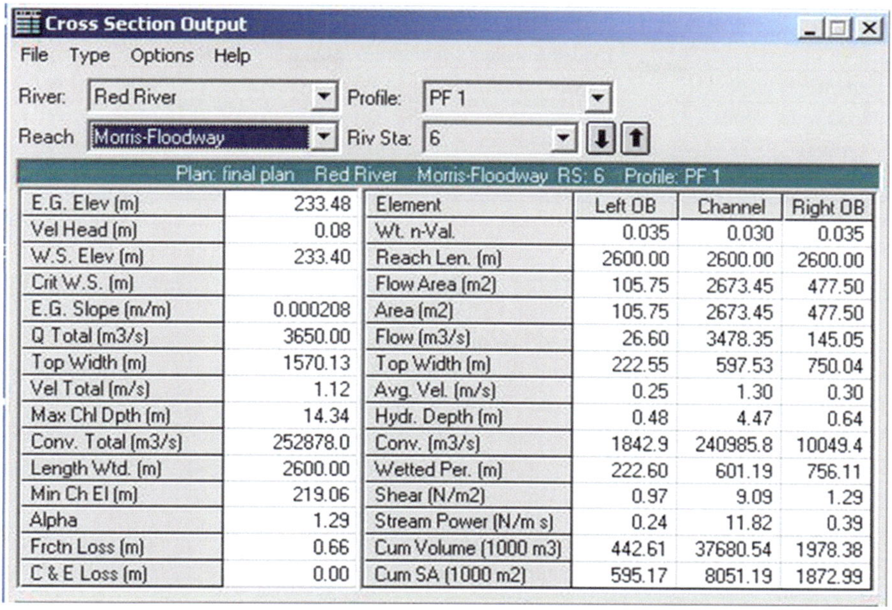

E.G. Elev (m)	233.48	Element	Left OB	Channel	Right OB
Vel Head (m)	0.08	Wt. n-Val.	0.035	0.030	0.035
W.S. Elev (m)	233.40	Reach Len. (m)	2600.00	2600.00	2600.00
Crit W.S. (m)		Flow Area (m2)	105.75	2673.45	477.50
E.G. Slope (m/m)	0.000208	Area (m2)	105.75	2673.45	477.50
Q Total (m3/s)	3650.00	Flow (m3/s)	26.60	3478.35	145.05
Top Width (m)	1570.13	Top Width (m)	222.55	597.53	750.04
Vel Total (m/s)	1.12	Avg. Vel. (m/s)	0.25	1.30	0.30
Max Chl Dpth (m)	14.34	Hydr. Depth (m)	0.48	4.47	0.64
Conv. Total (m3/s)	252878.0	Conv. (m3/s)	1842.9	240985.8	10049.4
Length Wtd. (m)	2600.00	Wetted Per. (m)	222.60	601.19	756.11
Min Ch El (m)	219.06	Shear (N/m2)	0.97	9.09	1.29
Alpha	1.29	Stream Power (N/m s)	0.24	11.82	0.39
Frctn Loss (m)	0.66	Cum Volume (1000 m3)	442.61	37680.54	1978.38
C & E Loss (m)	0.00	Cum SA (1000 m2)	595.17	8051.19	1872.99

Fig. 3.21 Water surface elevation computations in HEC-RAS

$$y = 76879x^3 - 344873x^2 + 470283x + 538659 \tag{3.2}$$

where y is the dollar value of damage to buildings; and x is the floodwater depth.

Damage to roads is expressed xas the relationship between the monetary value of damage and the total length of submerged roads:

$$rd = 18.889L^2 + 261.25L + 300000 \tag{3.3}$$

where rd is the dollar value of damage to roads and L is the total length of flooded roads.

Agricultural damage assessment depends on the time of year and the type of crop in the region of interest. Due to the lack of data availability, spatial variability in crop type could not be accounted for, and only one crop, R.S. Wheat, was assumed to be in the fields at the time of flooding. The following relationship is used to assess the agricultural damage in the region (KGS Group 2000):

$$ad = \sum [(1 - yield) * (cp) * A^* price] \tag{3.4}$$

Table 3.5 Weights w_i indicating decision-maker preferences

	Decision-maker's preferences (w_i)		
Criteria	Weight set # 1	Weight set # 2	Weight set # 3
Flood water depth	0.5	0.1	0.9
Damages	0.5	0.9	0.1

where *ad* is the dollar value of agricultural damage, *yield* is the expected yield (fraction of the optimum) as a function of seed date, *cp* is crop percentage of a typical distribution ($cp = 1$ in this case), *A* is the area of cropland (acres) and price is the three-year average price of the crop ($/bushel).

Next, weights were assigned to indicate various preferences of the decision makers involved in the process. Table 3.5 gives the three weight sets applied in this simulation. The first weight set is selected to give an equal level of importance to both of the criteria. The other two weight sets were chosen to represent the difference (to the order of extreme nature) in opinions and interests between various decision makers.

Numerical Analysis the deterministic and fuzzy analyses were conducted. Three flood protection alternatives were evaluated according to the two criteria. The following set of experiments was performed in the original study:

1. Deterministic spatial multi-objective analysis of flood management options with three different weight sets.
2. Fuzzy spatial multi-objective analysis of flood management options with the triangular membership function and three weight sets.
3. Fuzzy spatial multi-objective analysis of flood management options with the Z-shaped membership function and three weight sets.

In order to avoid an overly complex discussion of the above simulation, only the deterministic approach has been presented here. Illustrations are given in following sections. Figures 3.22, 3.23, 3.24, and 3.25 show flood depth simulations using the weight set #1 and flood control alternative 'Floodway 1' for: (i) the entire region, (ii) only buildings, (iii) agricultural fields, and (iv) roads.

To summarize, the Red River flood management case study suggested that appropriate computer techniques and required datasets are necessary to simulate various flood protection alternatives. Also tested in the study were multi decision makers' input into the decision making process, as well as multiple criteria to evaluate different options. All the results, not presented here for simplicity, but can be seen in Nirupama and Simonovic (2002) can be compared and the best alternative for effective flood risk reduction can be selected.

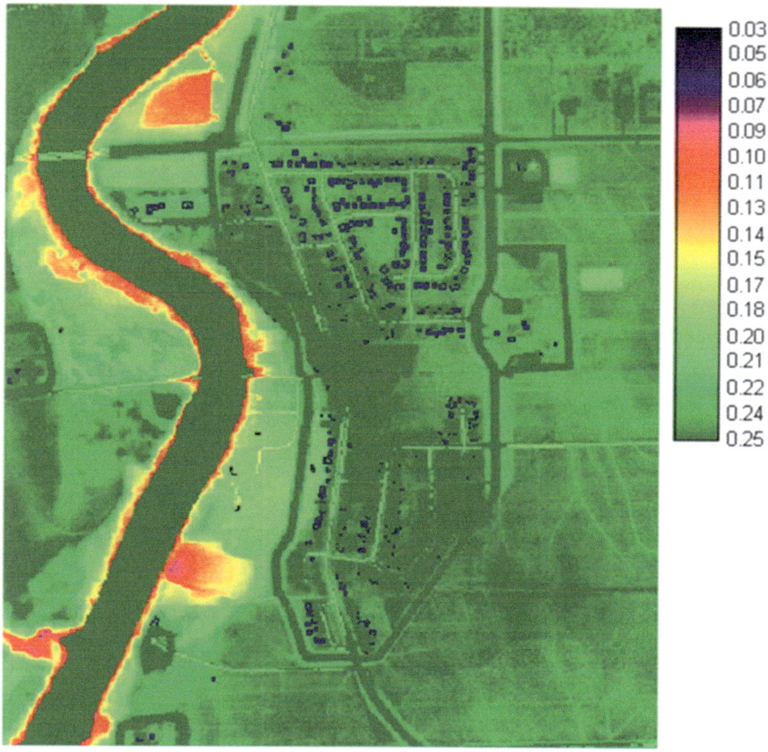

Fig. 3.22 Simulation of flood inundation (in meters) for 'Floodway 1' and weight set #1 using deterministic approach

3.7 Strategic Planning

This key element is about economic, political and institutional support considerations. Financial commitment and political will are fundamental to any successful disaster management program. The allocation of resources, the building of institutional support, the creation of social programs, and community based initiatives toward individual and collective protection measures are most important. In North America, Europe and other developed countries, disaster risk management programs are well established, structured, and fairly funded. These regions also have great early warning systems in place, remarkable disaster preparedness, and response and recovery capabilities. In the developing world, however, the focus has shifted to knowledge dissemination, disaster preparedness awareness, and community based programs. For example, in India, the authorities at the state level take the main

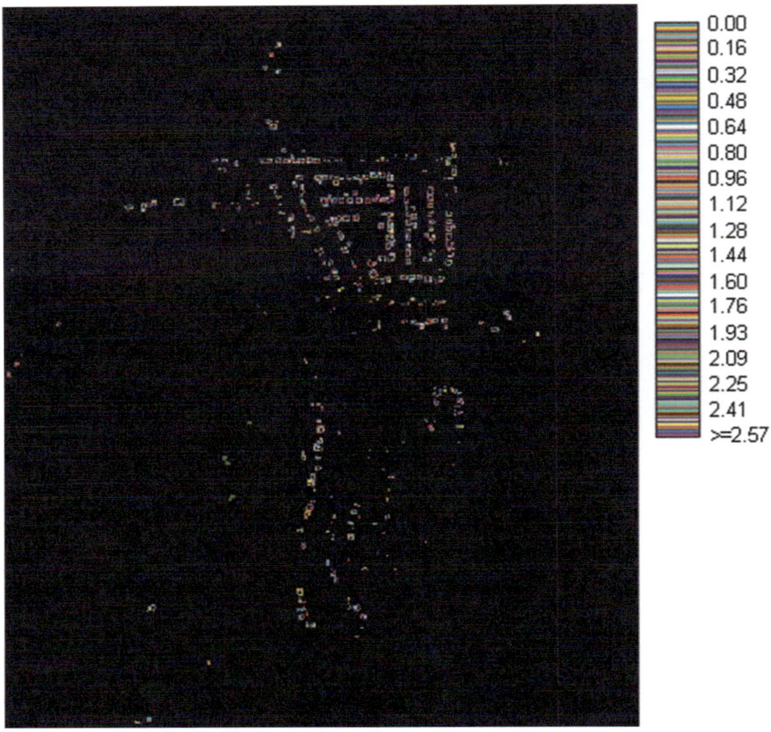

Fig. 3.23 Buildings protected against flooding by applying the alternative 'Floodway 1' and weight set #1 using deterministic approach

responsibility for disaster relief with financial assistance from the central government. A small Calamity Relief Fund, constituted with both state and central government contributions is managed by the Disaster Management Authority of India, under the Ministry of Home Affairs (Freeman et al. 2002). In case of a major disaster, the central government provides pre-determined reimbursement sums for loss of life, limb, and partial and total loss of housing and productive assets.

In Canada, the role of the Canadian Standards Association (CSA 2015) is very important in terms of policy and practices in emergency management. The organization is responsible for developing codes and standards for various sectors with an intent of creating a safe and sustainable environment for citizens and businesses. The CSA mandate includes certification marks and labels, listing certified products, developing safety and performance requirements, and developing consensus-based standards to support new technologies.

Recently, in response to emergency management professionals' calls for an integrated emergency management and business continuity planning standards, in

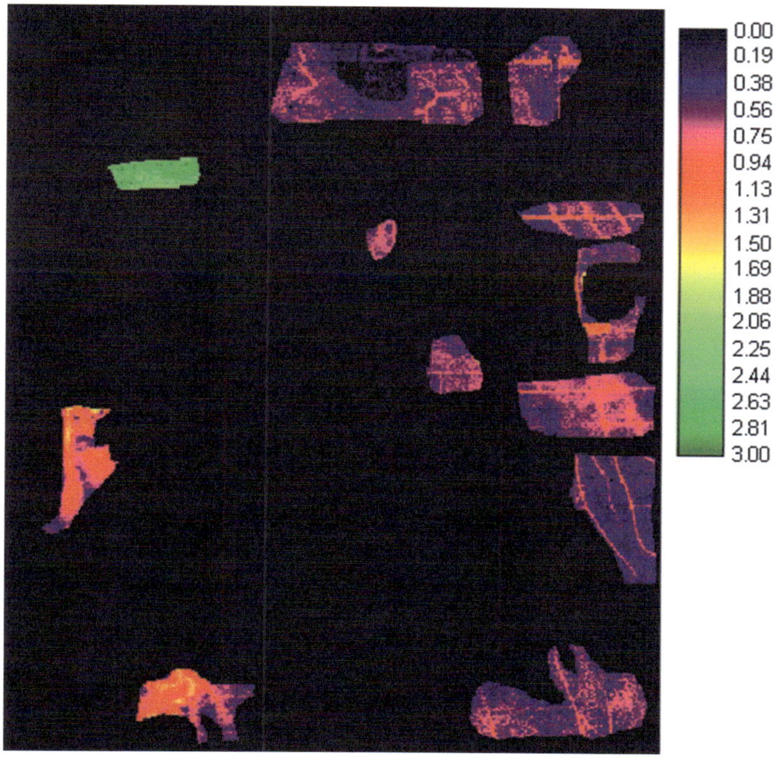

Fig. 3.24 Agricultural fields protected against flooding by applying the alternative 'Floodway 1' and weight set #1 using deterministic approach

August 2008, CSA developed and launched a single comprehensive document CSA Z1600-08 for the consumption of stakeholders. This new integrated approach also marks a paradigm shift toward a risk-mitigation approach from the previously popular response-focused mindset. The complete document is about 58 pages containing chapters on program management, planning, implementation, exercises and evaluations, and management review. In line with the focus of this book, Chap. 5 on Planning is deemed most relevant.

According to the CSA Z1600-08 (CSA 2015), after defining scope and overall management plan, planning is the most important aspect of setting standards. Planning begins with the identification of hazards, risks, and impact on businesses. Hazards have been classified in most universal manner as natural, human-caused, and technological. The risk assessment component comes next and includes probability of occurrence for each hazard based on their historical records and potential impact they may cause. The impact is considered on people, property, and the

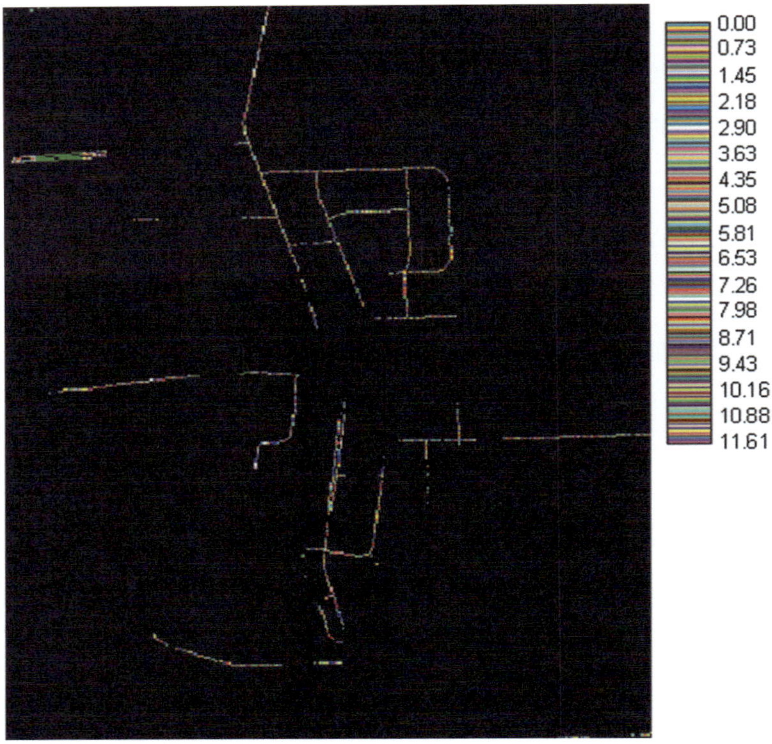

Fig. 3.25 Roads protected against flooding by applying the alternative 'Floodway 1' and weight set #1 using deterministic approach

environment respectively in a negative way. This aspect mostly addresses society and its citizens. For businesses, Business Impact Analysis (BIA) is recommended to be conducted that includes critical business functions and resources required for them and also accounting for interdependencies. Additionally, potential threats would also need to be identified.

After determining risks, potential hazards, and BIA, program's objectives must be established, and a planning process followed. This process would result in a well-developed plan that should be reviewed and revised regularly or as and when anything changes that calls for a review. Similar to any documented plan that needs to be implemented when required this plan must have clearly defined roles and responsibilities of personnel. Care must be taken to have clarity in statements with complete and updated information. Resource and logistics support, as well as interdependent functions must be clear for all personnel who may have to implement the plan on a short notice given the nature of the emergency. Communication channels, according to roles and responsibilities also must be clear with contact information up to date at all times.

3.7.1 Case Study – The 2004 Indian Ocean Tsunami

The first truly global tsunami in historical time is the event on 27 August 1883 following the eruption of the volcano Krakatoa in the Sunda Strait between Java and Sumatra in Indonesia. This event caused an estimated loss of life of 36,000. Due to the geographic extent this tsunami impacted and the oceans that it travelled through, this type of tsunami is referred to as a global tsunami. The second global tsunami in historical time based on definition, and the first one after modern instrumentation is put in place, is the tsunami of 26 December 2004 in the Indian Ocean (Murty et al. 2005).

In the aftermath of the highly destructive tsunami of 26 December 2004 in the Indian Ocean, dozens of teams worldwide surveyed the affected areas and carried out many nature, physical damage, fatality, economic, social, engineering and scientific analysis and interpretations. Most of the survey results were made available through the Internet. Preparedness, mitigation, and prevention strategies are usually developed by examining historical information. In light of this, it becomes very important and strategic to explore the inconsistencies in available data.

In this case study, a critical examination of three scientific parameters, namely travel times of the tsunami to various coastlines, amplitudes of the tsunami at different locations and the horizontal extent of coastal inundation by the tsunami waves was carried out. This examination highlighted fundamental inconsistencies in the data available on the web. Table 3.6 gives an account of arrival time (local) of the first three tsunami waves based on survey results posted on the web by Japanese teams headed by Dr. Y. Tsuji, Dr. T. Suzuki, Dr. S. Sato and Dr. K. Hirada. Table 3.6 lists the arrival times of the first three tsunami waves (in local time) at the southwest and southern coasts of Sri Lanka.

Early warning system would have mitigated the great loss of life. Socio-economic factors, such as high population density and lack of public awareness were also contributing factors. Since major tsunamis are rare, there was a lack of tsunami memory among the public and in the media. Early warning systems are available in India for more frequent natural marine hazards such as cyclones, monsoons, river floods etc. Besides lack of an early warning system and high population density, there are several physical oceanographic factors, which act together and made the tsunami of 2004 extremely violent on the coasts of Sri Lanka and South India.

The tsunami that originated offshore of Sumatra in Indonesia generally approached Sri Lanka from the east and hit the east coast of Sri Lanka roughly at the same time the tsunami struck the south coast of Sri Lanka and then travelled northward along its west coast. Hence, the arrival times should be earlier near the south eastern corner of Sri Lanka and the arrival times should increase monotonically as one proceeds westward along the south coast and then northward along the west coast.

The website from where this data was obtained is titled "Testified arrival time of tsunami". From this title inference can be made that the times listed on the website are likely to be obtained from eyewitness accounts. In principle, the travel times

Table 3.6 Arrival time (local) of the first three tsunami waves (adapted from the survey results of Shibayama et al. 2005)

Identification Number	Location	First wave	Second wave	Third wave
1	Galle Face Green	9:15–9:30	10:00–10:15	–
2	Dehiwela	9:45	1:0.45	–
3.1	Mount Lavinia 1	8:45	9:45	–
3.2	Mount Lavinia 2	9:45	10:15	–
3.3	Mount Lavinia 3	9:45	11:35	–
4	Moratuwa	10:30	11:00	11:05–11:10
5	Panadura	7.30	9:30	10:30–11:00
6	Beruwala	9:45	10:30–11:00	–
7	Bentota	9:55	10:30	11:30
8	Seenigama	9:45	10:30	–
9	Hikkaduwa	9:30–9:35	9:50–10:00	–
10	Galle Fort	9:00–9:30	9:30–9:45	9:50
11	Talpe	9:45	–	–
12	Matara	9:15	9:25	9:40
13	Tangalla	8:45–9:15	–	9:20
14	Hambantota	9:22	9:35	–

Table 3.7 Tsunami (first wave) arrival time (local) at certain locations on the Indian coast (based upon data from various websites in India)

S. N.	Location	Arrival Time of the 1st Wave
1	Visakhapatnam	09:05
2	Chennai	09:05
3	Tuticorin	09:57
4	Kochi	11:10
5	Azhikkal	12:30
6	Mormugoa	12:25

should increase monotonically as we proceed from location 14 to location 1. Table 3.6 shows that clearly this is not the case. The arrival times were inconsistent ranging from 7:30 to 11:30 AM. Tsunami, which is a long gravity wave, would not take 4 h to travel such short distances. Obviously one can put little faith in these reported travel times. Table 3.7 lists the tsunami arrival times at certain locations on the Indian coast, based upon data from various websites, including those of the SOI (Survey of India, Dehra Dun) and the NIO (National Institute of Oceanography, Goa, India).

This data also shows inconsistencies. It is unlikely that the tsunami arrived at Visakhapatnam and Chennai at exactly the same time and it took another 52 min extra to travel to Tuticorin. On the Kerala coast it is hard to imagine why there is so much time difference in the arrival times of the 1st wave at Kochi and Azhikkal.

Inconsistencies in tsunami amplitudes and inundation were also noticed. According to Shibayama et al. (2005), tsunami amplitudes rose up to 50 m in the

Table 3.8 Maximum tsunami amplitude (m) and horizontal distance of inundation (m) as reported by Shibayama et al. (2005)

Latitude (°N)	Longitude (°E)	Maximum tsunami amplitude (m)	Horizontal distances of inundation (m)
5°35′35.9″	95°19′45.3″	48.86	70
5°23′37.8″	95°15′10.3″	21.39	940

Banda Aceh area. Table 3.8 lists the maximum tsunami amplitudes and horizontal extent of inundation for Banda Aceh as reported by the same survey team.

To summarize, several inconsistencies and serious errors in the data posted on the web by various international survey teams were observed right after the occurrence of the 2004 Indian Ocean tsunami. In order to develop future mitigation strategies in the region, inconsistencies in scientific parameters such as, tsunami travel times, maximum amplitudes, and horizontal extent of inundation may lead to inaccurate and ineffective methods. After more than a decade of the tsunami, a network of instruments are in place to accurately measure and record physical variables associated geophysical and atmospheric processes.

3.8 Response, Recovery, Reconstruction, and Rehabilitation

This key element covers response capability and mutual agreement with neighboring regions (depending on the size and type of the event), assistance with recovery, and reconstruction. These are important aspects for the impacted communities to deal with their loss and remain optimistic about their future. The rehabilitation phase provides a rare opportunity to re-assess the situation, consider various options to relocate or build a better, stronger and more resilient community. Disaster aid – internal and/or international, bilateral (government to government or through NGOs) or multilateral (through the UN agencies) must be in place to reduce the impact of a disaster. The Government of India, in partnership with the United Nations Development Program (GOI-UNDP 2008, 2010), has developed a Disaster Risk Management Programme through disaster preparedness and vulnerability reduction. Their goal is to strengthen institutional capacity with specific emphasis on women and other marginalized groups. They have adopted a multi-hazard approach with an objective of achieving a sustainable disaster risk reduction in some of the most hazard-prone districts in selected states in India. Another example is from Fiji, where exposure to cyclones, floods, droughts, earthquakes, and tsunamis is widespread. Fiji has been able to develop a good disaster preparedness, response, and recovery plan in which NGOs are encouraged to actively participate in all the functions of disaster risk management (Freeman et al. 2002).

3.8.1 *Case of Vertical Evacuation Shelters in India*

It is important for mitigation and preparedness measures in coastal areas to consider the multiple hazards of the region. Additionally, the coastal states (Fig. 3.26) of India are not uniform in terms of intensities and occurrences of various hazards (GoI 2006). In most cases, women, children, marginalized communities, and the poor bear the brunt of disasters. Therefore, minority communities, civil society organisations, and NGOs should actively engage in the decision making process of disaster risk reduction measures (Orissa 2009; Kerala 2012; Rao 2012).

Fig. 3.26 The coastal states of India that could be affected by cyclones to various degrees (Source: https://upload.wikimedia.org/wikipedia/commons/9/9c/India_-_administrative_map.png)

Worldwide, construction of cyclone shelters has been a proven means of natural disaster preparedness as the vulnerable populations can be evacuated to these structures immediately after receiving cyclone warnings. For example, India has a 40-year history of construction and maintenance of cyclone shelters, notably from states such as Andhra Pradesh, Orissa, and Tamil Nadu (Rao 2012). Cyclone shelters constructed in these states have been proven effective and have become a source of local motivation for preparedness (GoI 2006). Every year during the April-May period and during the September-December period, cyclones are present in the Bay of Bengal. Every cyclone takes its toll on human and animal lives and imposes damage on property and infrastructure with associated loss of crops of some ecosystem services. The loss of livestock has serious consequences for villages because many of these affected communities are heavily dependent on the agricultural economy for sustaining their livelihoods (Orissa 2009). Figure 3.27 shows a typical village on the east coast of India, which is practically at the sea level and is extremely vulnerable to storm surge inundation. Figure 3.28 is an illustration of severe damage to coastal infrastructure.

Criteria for Cyclone Shelters the State of Orissa has a coastline of 480 km along the Bay of Bengal. The state is intercepted by a number of peninsular rivers systems. According to the Orissa State Disaster Mitigation Authority (Orissa 2009), a number of villages are sandwiched between rivers, lakes, and the Bay of Bengal. As a result, the vulnerability of the area to different disasters increases due to geographical and physical features. In planning the locations of cyclone shelters, information on water

Fig. 3.27 Waves crash onto the shore at a typical fishing harbour after cyclone Phailin hit (http://im.rediff.com/news/2013/oct/14odisha13.jpg)

Fig. 3.28 Storm surges not only cause loss of life but could also significantly damage the coastal infrastructure, as can be seen from this damaged coastal road

levels reached in past events is very crucial. It is necessary that the cyclone shelters are resistant to many hazards because coastal areas are prone to several natural disasters. This will significantly increase the life of the structure, its utility, and hence the economic efficiency of the investment made. Cyclone shelters are often used for a short period of time during natural hazard events, including cyclones, floods, and tsunamis. The sustainability of cyclone shelters depends on their use and maintenance during non-event periods. As huge investments are made to erect cyclone shelters, it is prudent that these structures are put to various other uses that will take care of maintenance of the shelter as well (NCRMP 2011).

Figure 3.30a shows a cyclone shelter in Orissa, which is located immediately adjacent to a school. This is a very useful feature because in the event of an emergency situation, the school children could be quickly moved into the cyclone shelter and protected. Figure 3.29 b–d shows the organization of the emergency shelter.

Cyclone shelters should not be looked at only as a means of mitigation but also as a means of development as these shelters provide facilities for a wide variety of sustainable uses, such as education and health care, and promote local development. Efforts are made to make sure that these guidelines are broad enough to promote local imagination and innovation without leaving any doubts/gaps that would create confusion on important issues. There is a difference of opinion among experts on what kind of normal use shelters should be put to. The consensus seems to be on uses that will not hinder the primary use (as a cyclone shelter) of the structure. Consideration of the local ethos and values in the design of cyclone shelter would likely

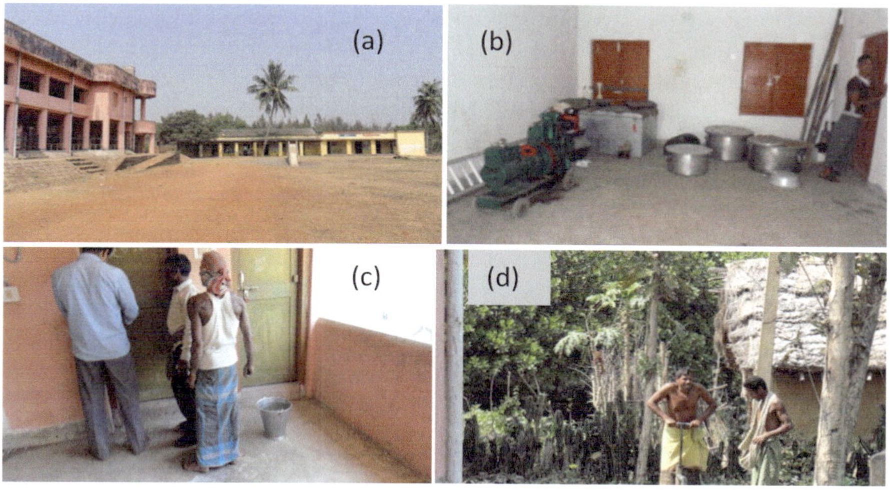

Fig. 3.29 A detailed look inside a cyclone shelter in coastal Orissa, India. Photos by author 2011

Fig. 3.30 A multi-purpose cyclone shelter in Orissa, India. Photo by author 2011

improve its use. For example, use of the building as community centre and for school will permit the use of the building as emergency shelter without any difficulty (Fig. 3.30). Providing separate toilets for both men and women would improve the 'usability' of the shelter by the communities.

Even strongly built structures can deteriorate in the absence of proper maintenance. Proper maintenance is possible through involvement of communities in the planning and design of the shelter as such a design would include the features necessary for the sustainable use of the shelter. The experience from villages in Orissa reveals that the communities have actively used these shelters when their opinions were considered when deciding its location and design parameters. Such a participatory approach not only brings a sense of ownership but also the pride of possession. For example, it is logical that a shelter can be used for running a school when features necessary for school are also made available in the shelter.

Design and Construction of Shelters the total height of the shelter above high tide line should meet the requirement of the design height of the storm surge. The local population and the fishermen also have a clear idea of the location of the high tide line. The height of the structure may be worked out as follows (GoI 2006):

- A minimum of 1 m should be ensured between the ground level and the high-tide level
- Raise the plinth about 1.2–1.5 m above the ground level
- Where needed, add 2.2–4.5 m high stilt depending on surge height
- Add one livable storey of 3.5 m height to the above level so arrived. This will be sufficient for most surge heights
- Design the roof to act as shelter space with parapet all around in case of larger storm surge in the area

Number and Location of Cyclone Shelters the number of cyclone shelters to be built largely depends on the number of vulnerable populations to be sheltered during emergencies. Emphasis should be given such that the cyclone shelters are located on the available elevated land. Guidance on the levels can be taken from large-scale maps or from a survey of India. In absence of an available elevated area, the structure may be elevated through the construction of a mound or shelter built on stilt. The Survey of India advises all jurisdictions to collect information within a 10 km radius of the coast, identify all villages therein, and provide the information on astronomical high-tide levels and ground levels within a level range of 0.5 m. They request this information of the coast to be sent to their office for future mitigation planning. Regenerating mangroves and raising shelter belt plantations will help reducing the fury of the storm surge. Suitable isolation distance must be provided between the shelter belt and the cyclone shelter to avoid damage to the foundation of the building by the roots of the shelter belt.

A Case of Shelters for Cattle it may be noted that thousands of cattle die during a major cyclone/storm surge event (Reddy 1991). In May of 1999, 115,492 livestock perished due to a cyclone in Andhra Pradesh. In the State of Andhra Pradesh (AP 2003; Rao 2012) it is being increasingly felt that there is a need to provide community cattle shelters in about 200 coastal areas. The State of Andhra Pradesh has proposed to construct cattle-shelters in association with fishponds. Of the ten acres acquired for each village, two acres would be used for constructing the cattle shelter and the remaining would be used for the formation of the fish tank. It is proposed that the earth excavated from the site of the proposed fish pond would be used for raising a platform up to a height of 1.5–1.8 m. The remaining excavated earth would be used for elevating the bounds of the cattle shelter to the desired level.

Existing and Future Plans the Ministry of Home Affairs of the Government of India, in consultation with the States and Union Territories (UTs) that are prone to cyclone risks have drawn up a National Cyclone Risk Mitigation Project to be implemented with assistance from the World Bank (GoI 2006). Under this Project, the following proposals were to be considered:

- Construction of cyclone shelters
- Shelter belt plantations
- Mangrove regeneration
- Construction of embankments to stop sea water inundation
- Construction of missing road links
- Commissioning of technical assistance/studies to sustain these initiatives in the States/UTs

In addition, four identified key components that are to be supported under the Project are:

- Upgrading the cyclone forecasting, tracking and warning system
- Cyclone risk mitigation investments
- Technical assistance for hazard risk mitigation capacity building
- Project management and monitoring

The vertical evacuation method and having shelters close to where the vulnerability exist have proven to be reasonably successful disaster mitigation measures. With modifications to suit local situations and circumstances, the vertical evacuation method should be seriously considered as an effective way to reduce risk in developed nations as well.

3.8.2 Case Discussion – California

A possibility for a mega earthquake hitting North America has been contemplated by scientists for a while. A BBC documentary (BBC 2012) explores the issue in great detail. Simulations have been conducted by various groups (Oregon State University, Stanford University) on the San Andrea plate and Cascadian subduction zone of California as well. The main interest is to assess the best and feasible evacuation strategy if a tsunami were to occur there. How long it would take for people in busy city cores to evacuate horizontally, primarily using road network? Would it cause crippling congestion on major streets and highways? Would it be better to climb up a tall building to save time? How much inland area would be inundated? Going vertical is catching support by the hour due to its advantages over horizontal evacuation (Weaver 2009). The Cascadia Subduction Zone can generate a magnitude 9 on the Richter scale causing earthquake and devastating tsunami (Roddey 2012).

Tucker (2013) proposes a tsunami evacuation park in Padang, Indonesia – right in the middle of the town. Based on his analysis of historical earthquakes and tsunamis in the Island of Sumatra where the 2004 tsunami changed Banda Ache for ever, it is learned that the Sunda trench is an active earthquake fault that is dragging the Eurasian plate down. The whole plate is moving in segments at a time when several hundred years of built up stress is released at once causing earthquakes. There is a particular concern about a segment that hasn't moved in the past 200 years and is believed to carry high risk of an earthquake occurrence. This observation has been done by studying coral reefs in the ocean for the past centuries. The last earthquake in the region was in 1797. The city of Pandang is in close proximity of this 'high risk gap'. The city has 900,000 people about half of those would be exposed to tsunami threat due to lack of any high grounds nearby and the low lying land. It is estimated that they will have a 25 min warning at the most just by feeling the shaking of the ground. If a building can be constructed in the heart of the city where people can conveniently go to get to a higher ground then many lives will be saved.

The 2004 Indian Ocean tsunami, which killed over 230,000 people and displaced 1.7 million across 14 countries, was a wake-up call for nations around the world, stimulating governments to address tsunami hazards. In the United States, Congress passed legislation to improve tsunami warning systems, preparation, and education. In the wake of the Japanese earthquake and tsunami of March 11, 2011, critical questions remain: How can we better understand the causes and characteristics of tsunamis? How can we use this information to provide better warnings? How can we prepare communities to more effectively respond and recover from potentially devastating events? (Bernard 2012). Evaluation of the successes of the responses to the past disasters has been a fascinating discussion on the Internet prompting experts to look for areas for improvements. Washington State's approach to tsunami vertical evacuation, Project Safe Haven (www.facebook.com/projectsafehaven) is an attempt to explore tsunami evacuation including vertical evacuation based on the M9.0 Tohoku earthquake and tsunami on March 11th.

3.8.3 Case of Nepal – Post-earthquake Reconstruction

Kathmandu's skyline evolved from 1930, 4 years before the devastating 1934 earthquake, to 1970 with tremendous speed due to the city's post-earthquake reconstruction (Fig. 3.31).

In 1934, Nepal experienced a major earthquake that claimed over 10,000 lives destroying one fourth of the capital city. Given the seismic history of this Himalayan nation, one can say that the government did not pay required attention towards planned urban development in the valley with implementation of seismic building codes. Initiatives such as GeoHazards International (2015) have raised awareness

Fig. 3.31 Kathmandu skyline in 1920s (left) and recent (right). Photos from http://thepicjournal.blogspot.ca/

regarding geophysical hazards, improved safety, and built the capacity of local governments and NGOs internationally. The country's population grew at a rate of 6.5% between 1997 and 2013, making the Kathmandu Valley one of the highest urban densities in the world. After the April 25 earthquake, Nepal, nonprofit organizations responded to the disaster to help the Nepalese people reconstruct their schools and homes. Although not all of them had sufficient expertise in building construction, they had the opportunity and capacity to influence Nepal's reconstruction to ensure that new infrastructure would be resilient to potential earthquakes in the future.

3.8.4 Case of Indian Ocean Tsunami 2004 – India

Following the 2004 Indian Ocean tsunami (damage shown in Figs. 3.32 and 3.33), there was a need to understand how the affected fishing community felt about the response to the disaster, how much help they received, and how they were coping with rehabilitation. This was done through a survey in the tsunami affected coastal communities in Southern India. The tsunami mainly affected the states of Tamil Nadu, Kerala, and Andhra Pradesh and the Union Territory of Pondicherry, all in south India, as well as the Andaman & Nicobar Islands of India in the Bay of Bengal. For various logistical reasons, no survey was conducted in the Andaman & Nicobar Islands. The survey was conducted during January 21 to February 19, 2005 and from March 1–8, 2005. A total of eight people, arranged into four teams simultaneously conducted the survey based upon a prepared questionnaire comprising a total of 16 questions. The total number of villages surveyed was 161, and the overall results of the survey are reported here. Among many observations, capacity building during the construction process, relocation and housing issues and tsunami education and awareness were prominent.

Ascertaining the psychological feelings of the victims of a disaster provide vital information for the authorities and disaster response agencies to better serve affected

Nagapattinam Harbour - 7

Fig. 3.32 Damage at Nagapattinam harbour from the Indian Ocean tsunami in 2004

Marina Beach-1-Chennai South

Fig. 3.33 Damaged Marina Beach, Chennai, India from the tsunami in 2004

population in future disasters. A major part of relief and recovery process is to make people believe that their voices are heard. Therefore, personal interviews helped to connect with villagers and also assess the level of public awareness of natural hazards and their impact on their lives. An examination of the affected regions, the

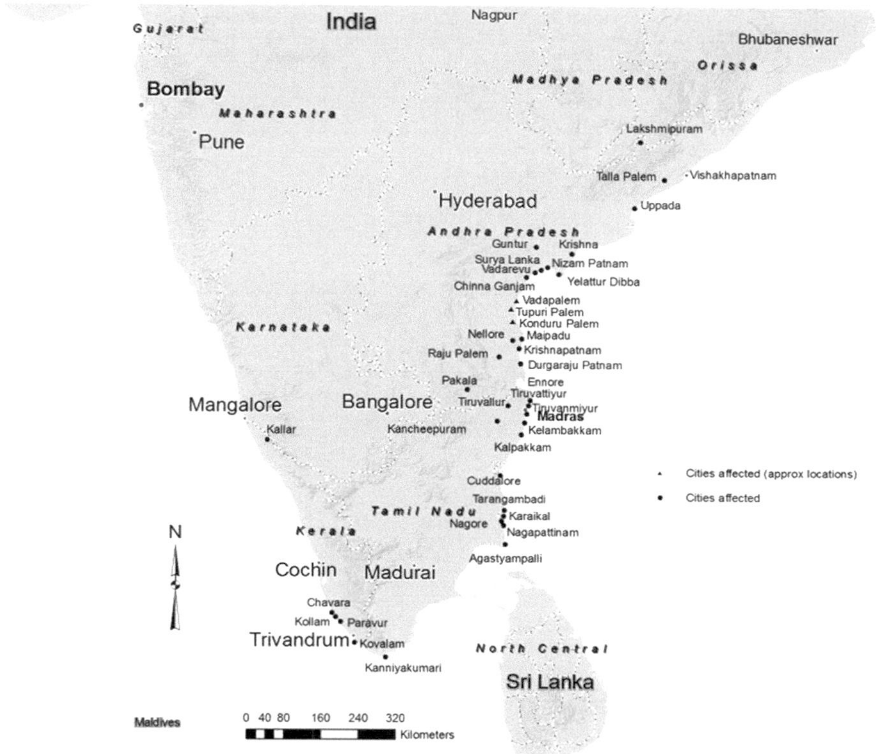

Fig. 3.34 Geographical locations in south India where the survey was conducted (Nirupama 2009)

population density and the livelihood patterns will enable the authorities to identify the vulnerable communities so that relief and rehabilitation can proceed in an efficient manner.

The social and human dimensions of the disaster that occurred are the main focus of this study. Major damage occurred in the States of Tamil Nadu, Kerala and Andhra Pradesh and the Union Territory of Pondicherry. The Andaman and Nicobar Islands lie closely north of the earthquake epicentre. The tsunami reached a height of about 7 m in the southern Nicobar Islands (Department of Ocean Development, Government of India). However, no personal interview was carried out in the Andaman & Nicobar Islands for logistical reasons. The personal interviews were carried out on the basis of a questionnaire from 21 January to 19 February 2005 and from 1 March to 8 March 2005. Four teams from Progressive Enterprises Ltd. of Hyderabad, India conducted the survey, simultaneously covering a total of 161 towns/villages in 45 days (see Fig. 3.34). The breakdown is shown in Table 3.9.

Two distinct groups were interviewed: one group comprising of general public, such as victim/affected families, eyewitnesses and the second group having more informed people such as experts identified by fishing associations and fishing

Table 3.9 Breakdown of personal interviews according to States, districts, and villages

States	District	Number of towns/villages surveyed
Tamil Nadu	Chennai north	8
	Chennai south	6
	Tiruvallur	4
	Kancheepuram	6
	Cuddalore	16
	Nagapattinam	16
	Kaniya Kumari	14
Total for Tamil Nadu		70
Pondicherry	Karaikal	5
Kerala	Alappuzha	2
	Kollam	5
	Thiruvananthapuram	1
Total for Kerala		8
Andhra Pradesh	Nellore	26
	Prakasam	23
	Krishna	21
	Guntur	8
Total for Andhra Pradesh		78
Total villages surveyed		161

communities. Where required, the interview team employed professionals who could translate the local language into English. The answers provided below are a literal translation (as provided by the translators) of what was said in the local languages (Fig. 3.35).

Based on the responses, following noteworthy observations made:

Most people knew about cyclones but not about storm surges. According to them even though these events are more frequent, they were much less severe than the tsunami event. They use local terms for cyclones or typhoons. The fishermen in the area always took cyclones as routine events, but they could feel that the 2004 tsunami was something they had never seen before.

People recognized the tsunami event as a natural hazard and did not give any religious explanation, such as, ‘act of God’ etc. Very few people mentioned that the sea Goddess was displeased and angry and punished them. Irrespective of the above two thoughts, the survivors all along the coast performed rituals to Goddess *Gangamma* (Goddess of sea/water) within 1 week of the disaster in the midst of their misery and distress.

The fishermen/villagers living near the seashore/ports got regular cyclone warnings in advance through the local Government and media (newspaper, radio, and television) in the local language of the area. The District Collectors’ office communicates to the local bodies and they in turn alert the villagers. The warnings are provided 24–72 h in advance to the vulnerable areas depending on the intensity and the expected place of landfall of the cyclones. However, in this particular case no

Devanampattinam-1-Cuddalore

Fig. 3.35 Local fishermen being interviewed based on a questionnaire, with the help of a translator after the 2004 tsunami in Cuddalore village, Tamil Nadu, India

warning/advise was provided by any source. Had there been a warning in advance, even a few minutes, most of them felt, they could have saved many of their women, children, sick, and old people. Their most valuable item was fishing gears which many were sorry to have lost.

Despite the fact that they had never heard the name 'tsunami' before and they were convinced that this type of high sea wave may come in future, they didn't worry about when and how frequent would that may be. They were confident that if they had to face a similar event again, they would be much better prepared. Most of the villages have public and privately owned cyclone shelters, tall structures, and safe buildings within a distance of 1/2 kilometer–2 kilometers of the beach. However, there were many villages where the safe shelters were either not in place or located at more than 1½ kilometer or both. In some places, the distance to the shelters was more than what fishermen could travel before the tsunami wave swept them. In couple of villages, there were obstructions such as a canal in between the beach and the shelters. Some cyclone shelters were too small to accommodate large number of people. At Nagapatnam harbor, which has an adjacent town with high density of population, a large number of people could not escape the tsunami waves as they were near the harbor and the shoreline with fishing activity and harbor work.

The fishermen/villagers had good understanding that mangrove areas along the coast, sand dunes, both natural and man-made, and suitable protection walls etc. would safeguard their villages to a great extent. However, they wonder why the Government was unable to maintain them. They were aware of the illegal exploitation of seacoast for commercial purposes and sand mining by the moneyed people and feel helpless about the situation. They didn't think that increasing population was responsible for the degradation of coastlines. After the year 1960, State Governments had implemented plans for safeguarding the existing mangroves, shelterbelts, sand dunes and sea walls as well as plans to increase mangrove coverage but the Government has been unable to stop the illegal activities and strictly implement Coastal Zone Regulations.

The fishermen/villagers are physically and mentally attuned to living near the sea, earn their livelihood and face natural hazards, such as cyclones and flooding. The individual families and the village communities continually improve their protection and mitigation methods. The Government and NGOs make regular visits and provide them training to enhance their coping capacities. The village *Panchayats* (community level governance body) arrange for some young men and women's training in rescue and recovery work by the Government/NGOs. Above all they believe that the community always rises to the occasion and help those who get hit by the disaster.

The very first response and help came from the nearby communities that were not affected by the disaster. This included rescuing the injured and people stuck up in odd/dangerous places, the sick, children and women. They moved the injured and sick to the nearest doctor/hospitals and arranged for safe drinking water, food and clothing to the victims. Interestingly, the very rigid caste system in the society disappeared during the emergency, and so did other differences among different communities. By the afternoon of 26 December 2004 official machinery began attending emergencies and NGOs started reaching the disaster struck places bringing the rescue and relief operations to a full swing by the evening. The next 2 days saw overwhelming activity in erecting temporary shelters, providing food, safe water and emergency medical assistance. The Government and private teams, with the help of locals, started clearing up the debris as well. On the second day the priority was to identify the dead and handover them to the families for burial. Photographs were taken for the unidentified bodies and mass cremation/burial was carried out. The next priority was establishing communications, providing electricity and mass sanitation. After a week the Government identified the families that suffered deaths, injuries and loss of belongings. NGOs, industrialists, business community and philanthropic organizations adopted most effected villages for providing temporary accommodation and relief for 1–2 months till the Government takes decided on long term measures for the community to move back to their villages and restart their lives.

The Relief Package was provided by the Tamil Nadu Government for temporary relief comprised of Indian Rupee (INR) 4000 (=US$60 @ 2005 rate) in cash for the dead, INR 2000 (=US$30 @ 2005 rate) for injured, 65 kg rice, 5 l of kerosene oil for cooking stove, clothing, bed sheets, and cooking pots. Many villages in the southern states of India have active Fishermen Cooperative Societies so an additional aid of INR 2000 (=US$30 @ 2005 rate) in cash was provided by them for each family. Government of Andhra Pradesh provided 100 kg of rice and INR 100 (US$1.5 @ 2005 rate) in cash for each affected family. The Government of India, in consultation with the State Governments, announced a relief package for implementing immediate and long-term measures for restoring normalcy in the region.

Total relief package =	INR 27.3 billion (US$410 million; all numbers based on 2005 value)
Immediate relief package =	INR 8.62 billion (US$131 million) Housing = INR 7.5 billion (US$113 million) Fishing community = INR10.93 billion (US $165 million) Repair of catamarans = INR 32,000 each (US$483) For fiber boats = 35% subsidy and 65% soft loan at 7% interest per annum For mechanized boats = 35% or a maximum of INR 500,000 (US$7540) as subsidy and 65% as soft loan
Long term relief package =	Tamil Nadu = INR 5.6647 billion (US$855 million), Andhra Pradesh = INR 341.6 million (US$5.16 million), Pondicherry = INR 609.8 million (US$9.2 million), Kerala = INR 755.6 million (US$11.4 million). In addition, Government of India, the World Bank, the Asian Development Bank (ADB) and United Nations have estimated and pegged the long-term tsunami rehabilitation requirements at US$1.2 billion. Out of this, Tamil Nadu will get US$ 868.3 million, Andhra Pradesh US$72.6 million, Kerala US$157.7 million, and Pondicherry US $114.4 million. A portion of it will be borne by Government of India and the rest will be provided by the above organizations as soft loan. In addition, the World Bank has extended an aid under IDA credit scheme – US $465 million to Tamil Nadu and Pondicherry. ADB has extended an aid of US$ 200 million to Tamil Nadu and Kerala.

Most of the fishermen lost their boats and fishing nets and some of the boats suffered damages that were repairable. Few fishermen got their boats repaired (mere minimum) from the nearby workshops and started fishing activity providing work and relief to some. Majority of the fishermen were poor and could not afford to get their boats/nets replaced or repaired immediately. They were looking for help from all levels of Government, NGOs and philanthropists. Few big industrialists came forward and made arrangements for repairing the boats and replacing the nets for selected and poor fishermen. Since most of the mechanized boats are equipped with engines and motors manufactured by Kirloskar Brothers & Co, the company undertook the repair work without charge for the poor fishermen. The State Governments directed their engineering workshops to provide necessary services for the repair of the damaged boats and engines for free. All these efforts and measure put together might have helped around 20% of the fishermen community to restart their livelihood.

The fishermen would have liked to have their equipment to earn their livelihood either purchased or repaired for them. They would have preferred receiving the fishing gear itself rather than money to buy it. They fear that if the Government decides to release money, only 10% of it may reach them, which would not be sufficient for the purpose.

Note that the States of Andhra Pradesh, Pondicherry, and Kerala received milder tsunami waves, therefore, the relief activities were not as intense as in Tamil Nadu. For this reason, they received less in terms of immediate relief by which the fishermen were displeased.

3.9 Knowledge Management and Sustainable Development

Institutional knowledge must be preserved for better learning and understanding. An approach of sustainable development would allow for the use of local resources (human, social, environmental) and thus contributes to local economy. Interestingly, in developing nations, NGOs play an active role in risk reduction activities in the region. The so-called 'knowledge network' involving civil society, the scientific community, and to some extent, the market is gaining popularity among people in India.

An approach suggested by Cardona (2006) for the Americas, and which can also be applied to other regions, is to use a system of indicators to measure a country's risk management performance. As shown in Eq. (3.5), the Risk Management Index (RMI) is based on a set of indicators that represent organization, development, capacity, and institutional actions taken to reduce vulnerability and losses, to prepare for crisis, and to recover efficiently from disasters.

$$RMI = (RMI_{RI} + RMI_{RR} + RMI_{DM} + RMI_{FP})/4 \tag{3.5}$$

Where,

RMI_{RI} = risk identification, includes objective and perceived risks;
RMI_{RR} = risk reduction measures including prevention and mitigation;
RMI_{DM} = measures of response and recovery; and
RMI_{FP} = governance and financial protection measures.

Resilience Building and Community Participation the final element in the cycle of disaster risk management is to work toward building resilient communities with community participation and community owned programs. For an effective and helpful risk management program, it is critical that communities make risk-based choices to address vulnerabilities and mitigate disaster impact. Resilience building must become the foundation of future risk management programs. A well designed communication strategy can be instrumental in the successful implementation of policy and other measures. In Asian countries, a community based holistic approach is gaining popularity as people feel responsible for their safe future (Padmanabhan 2008).

On account of knowledge management, we must explore available research and literature to be aware, not ignorant in order to participate in a society that is trying to live a sustainable development model. In this regard, a few example of research are given in following sections.

3.9.1 Case of Canadian Coasts – Arctic, Western, and Eastern

Canada has coastlines (Fig. 3.36) on three of the four oceans on the globe – the Pacific, Atlantic and Arctic oceans. The Pacific and Atlantic oceans are connected to the Arctic Ocean in the north, but still they are three distinct oceans, and need three individual tsunami warning systems. Tsunamis in the Arctic Ocean are not as well documented as in the Pacific and Atlantic oceans. From what is known, tsunamis in the Arctic Ocean are rare and probably are small in amplitude. Due to very low population density, around the Canadian Arctic, it is not a current priority for a tsunami warning system for Arctic Canada. For the Pacific Ocean, a tsunami warning system is in existence since 1948. In at least one sense, the warning aspects of the tsunami warning system for the Pacific coast of Canada, is relatively simple and straight forward, because it involves only the federal government and the provincial government of British Columbia. For the Atlantic Ocean, A tsunami warning system is now being established. The warning aspects will be somewhat more complex for eastern Canada, since it not only involves the federal government, but also five provinces, namely, Newfoundland and Labrador, Nova Scotia, New Brunswick,

Fig. 3.36 Canada's coastlines with Pacific, Atlantic, and Arctic oceans http://www.iiasa.ac.at/~marek/fbook/01/geos/ca.html

Prince Edward Island and Quebec. The Alaska tsunami warning center (ATWC) in Palmer, Alaska, provides tsunami warnings for both Pacific and Atlantic Canada.

Arctic Canada significant earthquakes in Canada are discussed in Chap. 1. Tsunamis in the Arctic Ocean are poorly documented and there are no reports of major ocean-wide tsunamis in historical time in the Arctic. What little evidence is there, seem to suggest that tsunami amplitudes in the Arctic Ocean probably will be small (Murty 1977).

Pacific Canada Fig 3.37 illustrates the very active region in western USA and Canada. Following the disastrous Aleutian earthquake tsunami of April 1, 1946, the USA established the Pacific tsunami warning center (PTWC) in Ewa Beach on Oahu

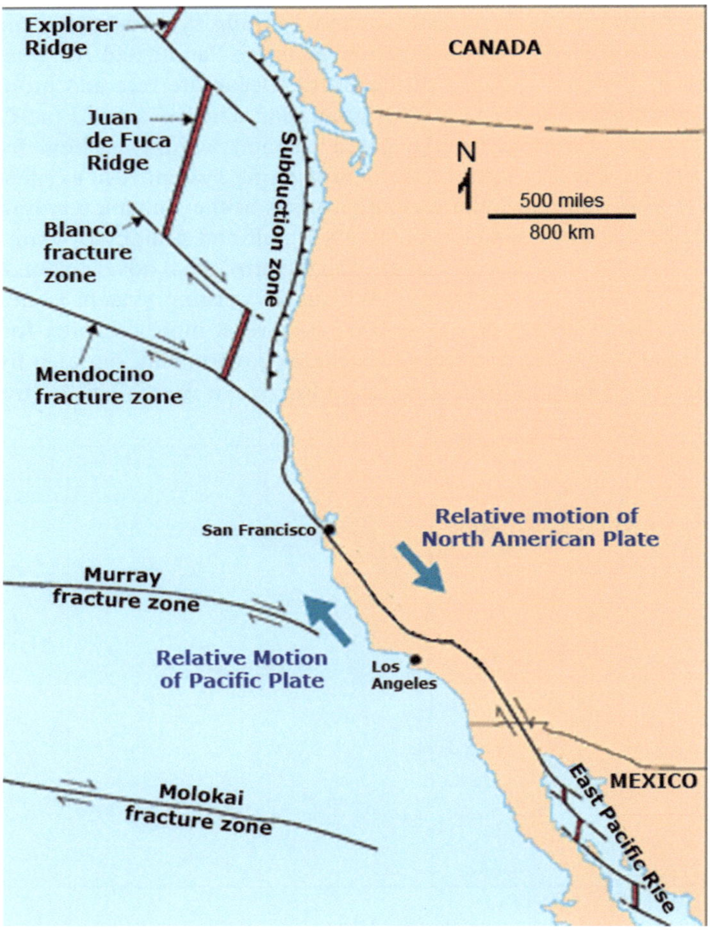

Fig. 3.37 Tectonically active region in the Pacific Ocean (USGS 2015)

Island of Hawaii. In 1965 the IOC started coordinating the activities of the Pacific tsunami warning system for some 26 nations around the rim of the Pacific Ocean, including Canada. After the disastrous Alaska earthquake and tsunami of 28 March 1964, the USA established the Alaska tsunami warning center (ATWC) in Palmer, Alaska, in 1967. The Pacific coast of Canada receives tsunami warning from the Palmer center. In terms of logistics, the warning aspects of this system for Canada are relatively straightforward, as it involves only the federal government and the provincial government of British Columbia.

In terms of scientific knowledge, it may be noted that, during the Alaska earthquake tsunami of March 1964, outside of Alaska, the greatest tsunami amplitude occurred, not at the open coast, but at Port Alberni, located at the head of the Alberni inlet, on Vancouver Island. The tsunami from the Pacific Ocean, with amplitude of about 0.5 m was amplified to 5.2 m at Port Alberni through quarter wave resonance (Murty 1977). While it is somewhat unlikely that the Alberni inlet can amplify tsunamis coming from the south Pacific (for example, the Chilean earthquake tsunami of 22 May 1960), its geographical orientation and geometry is such that, it can magnify tsunamis from Alaska and the Aleutians.

Atlantic Canada the Atlantic Ocean and the mid-Atlantic ridge (Fig. 3.38), which is a divergent plate boundary, do not give rise to tsunamigenic earthquakes. Usually there are no ocean-wide tsunamis in the Atlantic, and most tsunamis are local, for example, the Caribbean Sea. The Lisbon earthquake tsunami of 1755 was supposed to have had several meters of amplitude in the Caribbean, but was not significant in the western Atlantic.

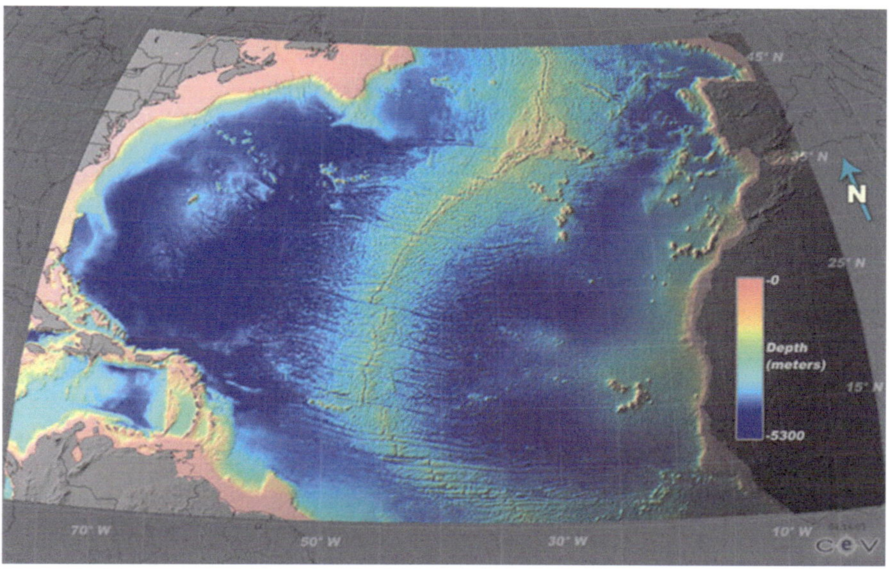

Fig. 3.38 Mid-Atlantic ridge in the Atlantic Ocean (Lost City 2015). http://www.lostcity.washington.edu/science/geology/midatlanticridge.html

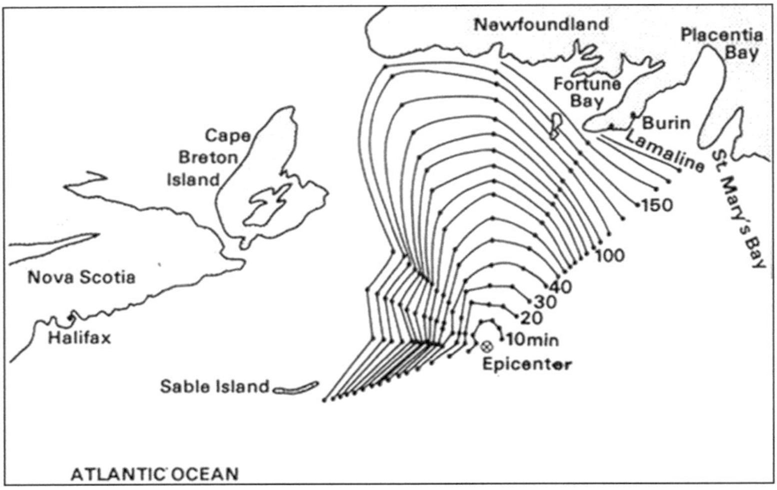

Fig. 3.39 Travel time contours of Grand Banks earthquake tsunami (in minutes) (Murty et al. 2005)

There have been three tsunamis that were generated in Canada and impacted mostly Canada only. Figure 3.40 shows the travel time contours of the Grand Banks earthquake tsunami of 18 November 1929 that was reported to have killed 28 people. Quarter wave resonance amplification played a major role in amplifying the tsunami in some of the bays and gulfs on the south coast of Newfoundland. It can be seen from Fig. 3.39 that the tsunami energy could not propagate towards Nova Scotia, mainly because of extensive sand banks in between.

Greenberg et al. (1993, 1994) and Ruffman et al. (1995) numerically modelled the tsunami in the Halifax harbour, due to a large chemical explosion on 6 December 1917. It is not clear how many people died from the tsunami, as opposed to the explosion itself. Figure 3.40 shows the tsunami travel times. In the Halifax harbour narrows, the tsunami achieved amplitudes of up to 14 m, but it quickly dissipated as soon as it entered the Atlantic Ocean.

Tsunamis can also be generated in the St. Laurence estuary from earthquakes and landslides. Some 8400 years B.P., at the end of the last glaciation, there was a large discharge of glacial melt water (Teller et al. 2005) from the huge glacial Lake Agassiz into the Labrador Sea through the Hudson Strait. Figure 3.41 shows the possible location of sand deposits today from this tsunami, which could have achieved amplitudes between 2 and 5 m.

Basically, for the Atlantic Ocean, a tsunami warning system is now being established. The scientific as well as the logistical issues for this system must be different from the Pacific system because tsunamis that impact eastern Canada are of local origin and other countries cannot provide a tsunami warning to Canada. We have to rely more on our own efforts and systems. Furthermore, for eastern Canada,

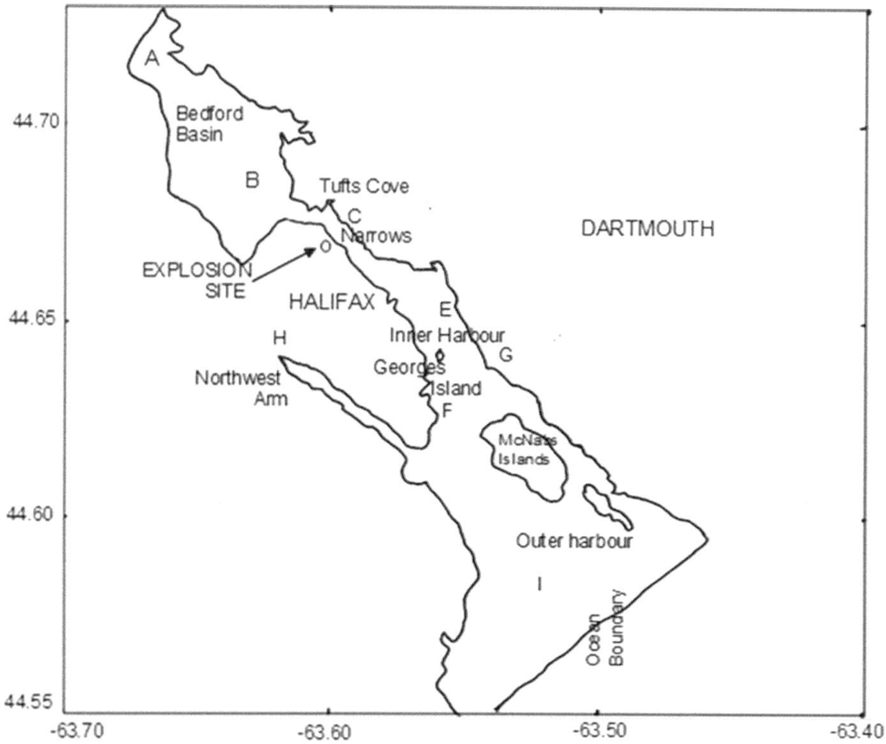

Fig. 3.40 Halifax Harbour location map. The letters refer to place names (Greenberg et al. 1993, 1994)

even the logistics have to be more complicated because, not only the federal government, but five provinces (Newfoundland and Labrador, Nova Scotia, New Brunswick, and Quebec) need to be involved. Nirupama et al. (2006) have developed an atlas for tsunami travel time charts based on known earthquake epicentre locations in the region.

3.9.2 Case of Climate Change and Sunspots in Canada

During 2009 there has been an average of 1.7 sunspots per month in comparison to the monthly average of 70 since 1930. This is the lowest sunspot activity since 1913 when there was an average of 1.4 sunspots per month. Whatever influence the sun has on the earth's climate, it will surely be felt in 2009 (Garnett et al. 2006; Garnett 2009). In May 2009, the five major cities on the Canadian prairies averaged 2.0 °C below normal. During June 2009, 30 prairie stations averaged 1.1 °C below average,

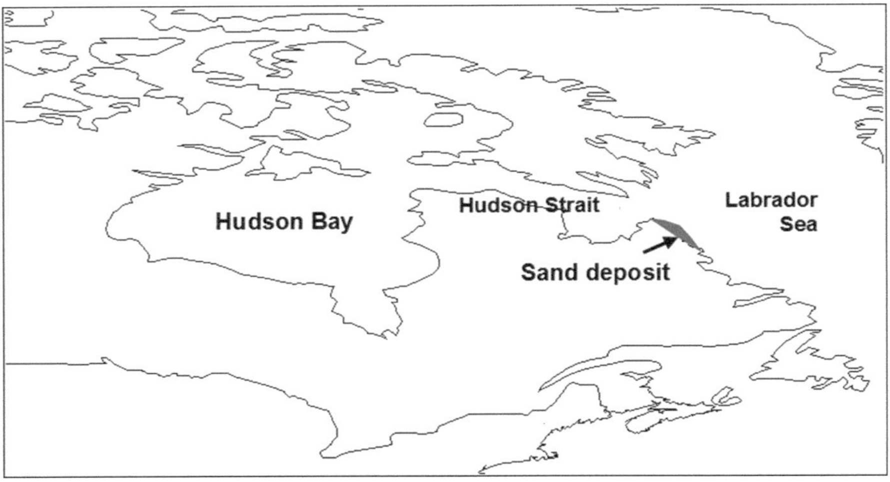

Fig. 3.41 Location of possible sand deposits in the Labrador Sea from a tsunami, some 8400 years B.P (Teller et al. 2005)

with Winnipeg July 2009 temperatures being 1.3 °C below the mean. This follows the coldest prairie winter since 1978/1979 based on data maintained by Garnett (2009).

Balfour Currie of the University of Saskatchewan's Institute of Space and Atmospheric Studies discovered that Canadian prairie annual precipitation tends to be above (below) normal centred on sunspot minimal (maximum). Currie was awarded the Companion of the Order of Canada for his work on the aurora and other solar influences, and his work is consistent with the forecasting approach of Agro-Climate Consulting. More recent research has shown that globally, low sunspot activity is associated with increased cosmic rays, ions, condensation nuclei and low-level clouds (below 3.2 km). Also, the Canadian prairies tend to be wetter than normal May through July. Further, European research has shown that low sunspot activity is associated with reduced solar irradiance and colder than normal lower surface temperatures. In our discussion of Canada's weather in 2009, we must not ignore the sun, one of the drivers of our climate and source of life on earth.

Large-scale atmospheric circulations and anomalies have significant influence upon seasonal weather over many parts of the world. Garnett et al. (2006) examined the impact of sunspot activity and large-scale atmospheric features on regional seasonal weather, as well as implications for crop yield and agronomy for the Canadian prairie. The atmospheric variables analyzed included the stratospheric quasi-biennial wind oscillation (QBO), El Niño/Southern Oscillation (ENSO), and North American snow cover (NAS) on Canadian summer rainfall, regarded as a key variable for establishing grain yield. The study was based on correlation coefficients between selected variables on 55 years of atmospheric, crop yield and climatic data for over 50 weather stations over the Canadian Prairie region. The study revealed

Fig. 3.42 Propane explosion site. Photo by Fernando Morales/The Canadian Press

that high (low) sunspot activity, an easterly (westerly) phase of the QBO, persistent La Niña (El Niño) conditions and heavier (lighter) than normal NAS in seasons leading up to the summer months are associated with low (high) summer rainfall.

3.9.3 Case Study – Propane Explosion in Toronto, Canada

The province of Ontario in Canada requires municipalities and provincial ministries to ensure that a consistent, accountable, and robust system of emergency management is established and maintained throughout the province. Risk management is a key process when addressing government or /continuity of operations/ business continuity. Vulnerability represents the fragility or weakness of the unit under consideration, be it a building, a place, a person, a group or settlements (Martínez and López 2010). The probability of a disaster impacting a community increases with increased vulnerability from previous events, as vulnerability is dynamic in nature where various factors continuously interact in time and space (Wilches-Chaux 1993). Currently, the Province of Ontario has identified 37 types of hazards in its *Provincial Hazard Identification and Risk Assessment Report* (HIRA 2004), in which hazards are classified according to their general source: natural, technological, or human-caused. According to Emergency Management Ontario doctrine (EMO 2010), risk identification, analysis, and evaluation are keys to development of mitigation measures.

A propane explosion occurred in a Toronto neighborhood in August 2008. The case was studied by Armenakis and Nirupama (2013) to illustrate how to estimate spatial risk in communities around the explosion site (Fig. 3.42) using GIS.

According to a report released by Ontario Propane Safety Review (Birk and Katz 2008), people and businesses across Ontario consume roughly 650 million kilograms of propane every year. There are over 330 Propane Storage and handling sites in GTA, within high density population areas. Propane is an odorless, colorless gas at room temperature. It is normally transported and stored under pressure for convenience, as well as to save storage space. To be used safely, it must be taken out of its container as either a gas or liquid, and its pressure must be reduced by means of a pressure regulator. This may also involve piping, tubing, valves and other fittings. The Canadian Standards Association (CSA 2015) recommends that location and safety of critical infrastructures, such as emergency services, hospitals, schools, electricity, police, water treatment plants, transportation, major highways, and government buildings are vital to disaster management. The Propane Safety Review report (Birk and Katz 2008) lists a total of 40 recommendations. More than 20% of them are related to location and zoning requirements for propane storage sites; and two are specifically on the use of GIS technology for risk and safety management. The report also emphasizes on need for the Technical Standards and Safety Authority (TSSA) to make available to municipalities and planning boards the locations of facilities and the defined hazard distance around each, either as maps or, if the community prefers, GIS data (Armenakis and Nirupama 2013).

Liquid propane expands by about 270 times in volume. A propane explosion results in fire and projectiles causing blast wind damages, fireballs, and flying objects. All these can cause injuries, death and property damages. The hazard distance is derived based on the size of the tanks in the facility and is called the '1-psi overpressure distance'. An explosion causes outward air pressure from the location of the larger tank and its effects are measured in terms of excess in the normal atmospheric pressure. This 'overpressure' is measured in pounds per square inch (psi). For facilities with up to 9.5 ton (5000 US water gallons) tanks, the '1-psi overpressure distance' is about 320 m. As explosions are unpredictable for safety reasons the recommended evacuation distance is normally 2.5 times the '1-psi overpressure distance' which would be 800 m or more (Birk and Katz 2008).

In the Toronto propane explosion event, the evacuation distance ordered by the authorities was 1.6 km due to the densely populated neighborhood. Figure 3.43 illustrates the three zones of 320 m, 800 m, and 1600 m. Based on the proximity of the center of the propane site and therefore the likelihood of projectiles landing in each distance zone the relative spatial severity impact factor for each hazard zone was assumed to be 0.6, 0.3 and 0.1, respectively (Armenakis and Nirupama 2013).

Data Acquisition the data used for the estimation of the spatial disaster risk for each hazard zone were the population, the landuse and critical infrastructure. The data were extracted based on the spatial coverage of the dissemination areas (DA). The demographic, economic, physical and critical infrastructure data that fall within each zone were extracted using the spatial analysis operations of overlay and clipping between the zone and the DA spatial layers. Through this process, out of the 7012 DAs in the GTA, parts of 7 DAs were extracted for Zone 1, parts of 16 DAs for Zone 2 and parts of 43 DAs for Zone 3.

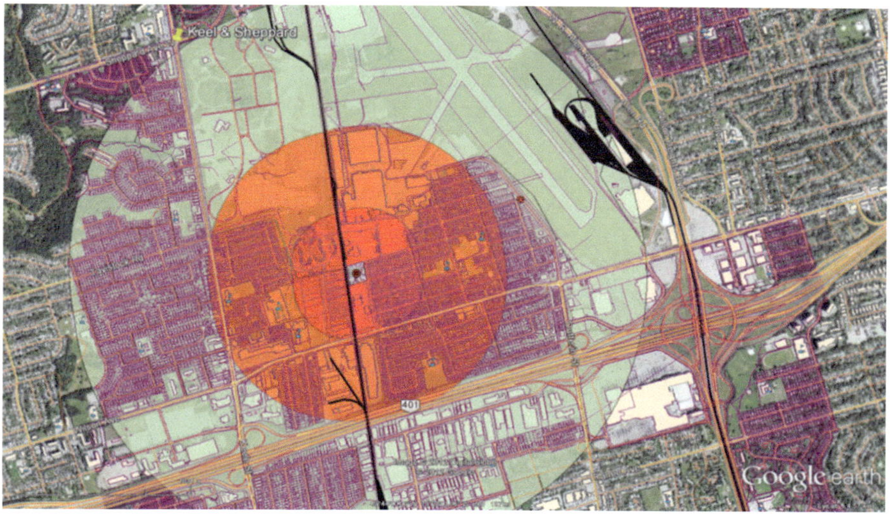

Fig. 3.43 The three zones (1, 2, and 3) – hazard zone 1 in red, zone 2 in orange, zone 3 in green (Armenakis and Nirupama 2013)

The percentage of the clipped DA's polygonal area (%DA) in each zone with respect to its total area (DA Area) was calculated as:

$$\%DA = \frac{Zone\ DA\ Area}{DA\ Area} \tag{3.6}$$

where, *Zone DA Area* is the clipped DA polygonal area in each zone; and *DA Area* is the total polygonal area of the DA.

Based on this and the assumption of uniform density, the corresponding demographic, economic and physical parameters in each clipped DA were determined. The critical infrastructure information, namely fire stations, health centres, education centres, other buildings, railway, and roads were extracted through the spatial overlay query of 'selection by location'. It performs the selection of the features from a spatial infrastructure layer that are contained within each of the three zones. Afterwards, the critical infrastructure elements from each zone were reduced to critical infrastructure per DA polygonal area in each zone assuming uniform density. The vulnerability indicator was determined using derived demographic, economic, physical and critical infrastructure data for each DA portion falling within each zone.

Vulnerability Estimation the total vulnerability was estimated based on social, physical, economic, and critical infrastructural vulnerabilities (Chakraborty et al. 2005). Demographic data were used according to DA falling within each hazard zone.

3.9.3.1 Social Vulnerability

Social vulnerability is associated with a lack of resources to mitigate, cope with, or recover from disaster. Social vulnerability can become clearly visible as the result of the impact of a disaster. Ferrier and Haque (2003) and HRVA (2004a, b) clearly define groups of people that are especially at risk. Guided by the 2006 Census in Canada, multiple social indicators were identified to be used as: *age* (A (very young; $\leq$6 years) and elderly; $\geq$65 years)); *single* (DW; divorced/widow); single parent families (*SP*); migrants for at least past 5 years (*MG*); allophones (*L*); immigrants between 2001 and 2006 and those who came when they were 45 or more in age (*IM*); visible minority (*VM*); and persons without high school education (*ED*). The population POP (Zone %DA) falling within each DA portion in each Zone is calculated as percentage of the total DA population (POP_{DA}) as:

$$POP\ (Zone\%DA) = \ \mathrm{POP}_{DA} \times \%DA \tag{3.7}$$

The social vulnerability ($SV_{\%DA}$) was determined using the social indicators described above as:

$$SV_{\%DA_{ij}} = \frac{(A + DW + SP + MG + L + IM + VM + ED)_{\%DA_i}}{\sum_{i=1}^{n} POP\left(Zone_j\%DA_i\right)} \tag{3.8}$$

where, i = *number of DA per zone*; j = *number of zone*

3.9.3.2 Economic Vulnerability

The economic vulnerability (*EV*) for each zone was determined using the indicators of unemployed (*UE*) persons and the families with annual income of less than \$50,000 (*F_50*) as:

$$EV_{\%DA_{ij}} = \frac{(UE + F\_50)_{\%DA_i}}{\sum_{i=1}^{n} POP\left(Zone_j\%DA_i\right)} \tag{3.9}$$

where, i = *number of DA per zone*; j = *number of zone*

In Zone 1, about 57 families earn a total income of less than \$50,000 a year and in Zone 2 the number goes up to 990 families. The average unemployment rate over the small area of seven DAs that falls within hazard distance is 1.4%. On the other hand, the unemployment rate in evacuation zone, spread over 17 DAs is 3.7%.

Figure 3.44 shows just how densely populated the impact area was.

Fig. 3.44 The very densely populated impact region (Armenakis and Nirupama 2013)

3.9.3.3 Physical Vulnerability

In terms of exposure to unsafe conditions, the population of the study region was found to be dangerously close to potential source of threat. Gas and oil transmission, distribution, and storage facilities pose considerable risk to their surroundings. Hazards associated with propane include explosions, fire and projectiles, which are rare but they can have severe impacts if occurred (Birk and Katz 2008). The 2006 census indicates that most of the area was developed prior to 2001; however, about 12 new dwellings were developed between 2001 and 2006 – even though the propane business at 54 Murray Road started operations in 2004. In the same period, about 39 new properties were developed within Zone 2. The physical vulnerability ($PV_{\%DA}$), based on the condition of dwellings and their construction history, was determined as:

$$PV_{\%DA_{ij}} = \frac{(MJ + C)_{\%DA_i}}{\sum_{i=1}^{n} D(Zone_j\%DA_i)} \tag{3.10}$$

where, MJ is dwellings requiring major repairs; C represents construction of the dwellings and other buildings prior to 1960; D is number of buildings in $\%DA$ polygonal area;

i = number of DA per zone; j = number of zone.

3.9.3.4 Critical Infrastructure

The critical infrastructure elements included fire stations, hospitals, community centres, schools, places of worship, road and rail network, and a large employer, Bombardier facility. These were extracted for each zone and each DA polygonal area in each zone assuming uniform density. The critical infrastructure ($CI_{\%DA_{ij}}$) per clipped DA polygonal area was determined as:

$$CI_{\%DA_{ij}} = CI_j \times \frac{\%DA_{ij}}{A_j} \tag{3.11}$$

where, $CI_j$is the number of critical infrastructure element in Zone j; $\%DA_i$ is the polygonal area of DA_i in Zone j; and A_j is the polygonal area of Zone j.

Figure 3.45 shows some of the critical infrastructure elements in the area.

3.9.3.5 Spatial Disaster Risk Estimation

Different methodologies exist for assessing disaster risk of natural hazard events, ranging from qualitative to quantitative (FEMA 1993; Collins et al. 2009). In this study, the probability term of the conventional risk equation (e.g., Long and John 1993; HRVA 2004a, b; EMO 2004; Martínez and López 2010), has been replaced by a spatial hazard index as:

$$R = H \times V \tag{3.12}$$

where, R = spatial risk; H= relative impact of hazard; and V = vulnerability.

The spatial risk index R_{ij} for each DA within each hazard zone is estimated as a layer overlay operation of the two weighted hazard and the weighted vulnerability spatial layers (Fig. 3.46).

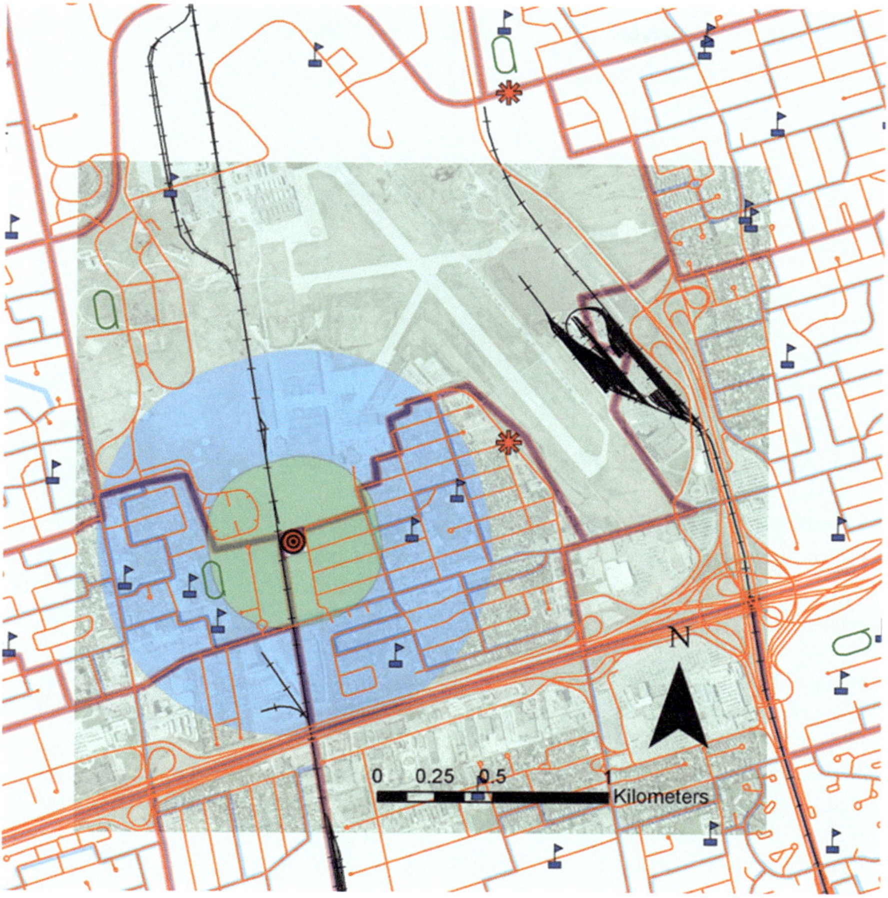

Fig. 3.45 Critical infrastructure – schools as blue flags, fire stations as re asterisk, major roads, highway, and rail networks

$$R_{ij} = H_j \times 0.25 \sum_{kij} V_{ij}^{k} \tag{3.13}$$

where,
i refers to DA
j refers to hazard zone
k refers to the type of vulnerability
R_{ij} refers to the spatial risk index in DA i located in zone j
H_j refers to the relative hazard zone index
V_{ij}^{k} refers to the k type vulnerability in DA i located in zone j
0.25 refers to the average of the total four individual vulnerability types

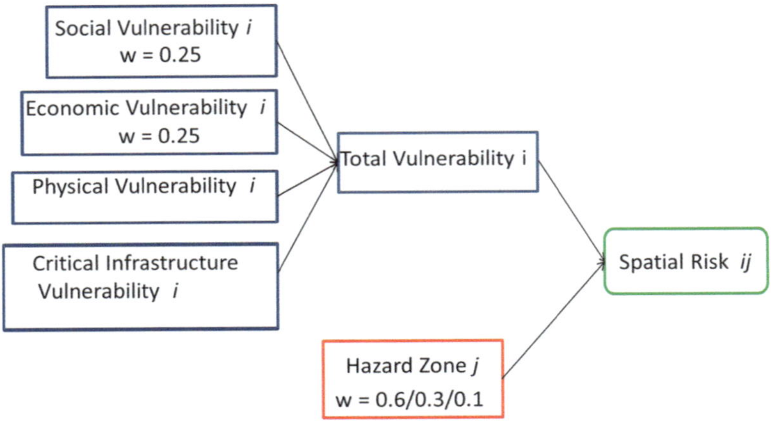

Fig. 3.46 Spatial risk estimation model (Armenakis and Nirupama 2013)

ArcGIS was used to implement the methodology and visualize the results. In addition, the results were exported to Google Earth map browser. Figure 3.47 shows the study area and estimated spatial risk indices in each hazard zone in Google Earth.

The case of propane explosion in Toronto highlights the usefulness of the method that accounts for vulnerable population living in the proximity of potentially hazardous industrial complexes as well as impact on critical infrastructure in the area. As the method uses analytical tools to prioritize spatial risks, it can support spatial decision making regarding risk based landuse planning and disaster management. The accuracy of the method depends upon the actual distribution of demographics and infrastructure within each DA. Furthermore, in reality, the boundaries between different impact zones are fuzzy.

To summarize, disaster risk management involves overall understanding and realization of potential hazards, identification of vulnerable people and property, risk evaluation, institutional support, and the adoption of a culture invested in preserving institutional knowledge. Various qualitative and quantitative methods can be used for risk assessment for the purpose of the development of a disaster risk management framework. The use of indicators to capture a sense of the central components in a holistic risk management process is worth examining. It is, however, safe to say that in recent years, most nations have shown an increasing trend toward developing comprehensive disaster management programs. They have broadened their national disaster management programs to encompass risk assessment, risk control, mitigation, preparedness, political will, economic feasibility, response, recovery, resilience building, and strategic and sustainable development activities. The success of such a framework or program may depend on the commitment of stakeholders such as communities, professionals, academics, and policy and decision makers.

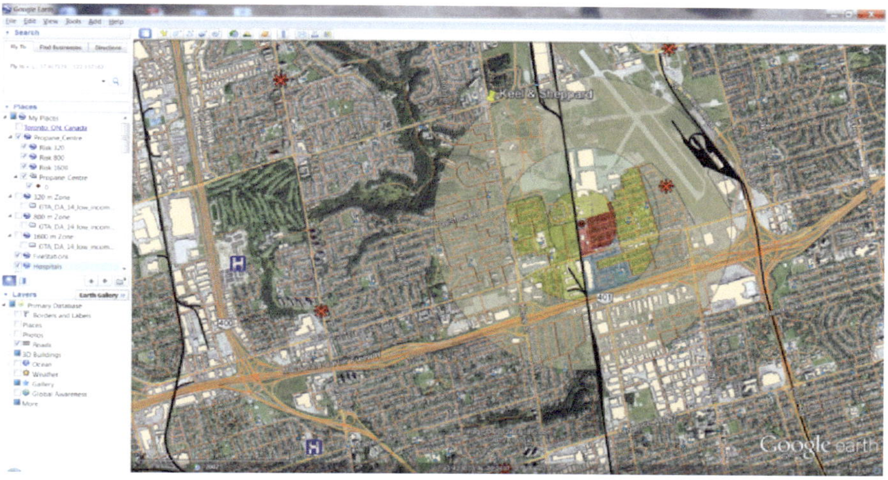

Fig. 3.47 Spatial risk in each hazard zone. (red = very high risk; yellow = high risk; blue = medium risk; greyish green = low risk) (Color figure online)

3.10 Exercise

Select any recent disaster and apply spatial risk assessment methodology discussed in this chapter using the example of propane explosion in Toronto.

References

Alexander D (1999) Natural disasters. Kluwer, Londra. 632p

AP (2003) Andhra Pradesh cyclone hazard mitigation project. www.deltares.nl. Accessed 1 May 2012

Armenakis C, Nirupama N (2013a) Estimating spatial disaster risk in urban environments, geomatics, natural hazards and risk. Taylor & Francis 4(4):289–298

Armenakis C, Nirupama N (2013b) Prioritization of disaster risk in a community using GIS, special issue on sociological aspects of natural disasters. Nat Hazards 66:15–29

Aurora BS (2013) The Tandava: Nature's dance of fury in Lord's own valley. India Today. http://indiatoday.intoday.in/uttarakhand/deluge.jsp. Accessed 21 Mar 2016

Bagla P (2013) The making of Himalayan tsunami, NDTV media, July 18, 2013. http://wwwndtvcom/article/india/uttarakhand-the-making-of-the-himalayan-tsunami-393901. Accessed 30 June 2016

Banai R (1993) Fuzziness in geographical information systems: contributions from the analytic hierarchy process. Int J Geogr Inf Syst 7:315–329

BBC (2012) MegaQuake could hit North America. https://www.youtube.com/watch?v=vEgLjgnv_3c. Accessed 21 Mar 2016

Bernard E (2012) Tsunamis: are we underestimating the risk? Oceanography 25(2):208–218
Birk M, Katz S (2008) Report of the propane safety review. Ministry of Small Business and Consumer Services, Ontario. pp52
Birkmann J (2006) Indicators and criteria for measuring vulnerability: theoretical bases and requirements. In: Birkmann (ed) Measuring vulnerability to natural hazards: towards disaster resilient societies. United Nations University Press, Tokyo. 524p
Canadian Standards Association (CSA) (2015) http://www.csagroup.org/. Accessed 21 Mar 2016
Canton LG (2007) Emergency management: concepts & strategies. Wiley, Hoboken
Cardona OD (2006) A system of indicators for disaster risk management in the Americas. In: Birkmann J (ed) Measuring vulnerability to natural hazards: towards disaster resilient societies. United Nations University Press, Tokyo, pp 189–209
Carver SJ (1991) Integrating multi-criteria evaluation with geographical information systems. Int J Geogr Inf Syst 5:321–339
CDEMG (2005a) Chatham Islands Civil Defense Emergence Management Plan 2011-2016. http://wwwcicgovtnz/. Retrieve 21 March 2016
CDEMG (2005b) Chatham Islands Civil Defense Emergence Management Plan. http://www.cic.govt.nz/CDEMPlansReports.html. Accessed 30 Mar 2012
Chakraborty J, Tobin GA, Montz BE (2005) Population evacuation: assessing spatial vulnerability in geophysical risk and social vulnerability to natural hazards. Nat Hazards Rev 6(1):23–33
Climate Himalaya (2013) Disaster in Uttarakhand, India: huge death toll. Climate Himalaya. http://chimalayaorg/2013/06/19/disaster-in-uttarakhand-india-huge-death-toll/. Retrieve 21 March 2016
Collins TW, Grineski SE, Lourdes M, Aguilar R (2009) Vulnerability to environmental hazards in the Ciudad Jua´ rez (Mexico)–El Paso (USA) metropolis: a model for spatial risk assessment in transnational context. Appl Geogr 29:448–461
Cutter S (2001) The changing nature of risks and hazards. In: American hazardscapes: the regionalization of hazards and disasters. Joseph Henry Press, Washington, DC. 211p
Cutter SL (2003) The vulnerability of science and the science of vulnerability. Ann Assoc Am Geogr 93(1):1–12
CWC (2013) Flood forecasting site, Central Water Commission. http://www.india-water.com/ffs/index.htm. Accessed 12 June 2014
DMMC (2013) Uttarakhand Disaster Mitigation and Management Centre, Government of Uttarakhand. http://dmmc.uk.gov.in/. Accessed 12 June 2014
Dobhal DP, Gupta AK, Mehta M, Khandelwal DD (2013) Kedarnath disaster: facts and plausible causes. Wadia Institute of Himalayan Geology, Dehradun
EMO (2004) Provincial hazard identification and risk assessment. Emergency Management Ontario, Toronto
EMO (2010) Emergency management doctrine of Ontario, Emergency Management Ontario, Canada, pp 40
FEMA (1993) Risk assessment approaches. Federal Emergency Management Agency, Washington, DC
Ferrier N, Haque CE (2003) Hazards risk assessment methodology for emergency managers: a standardized framework for application. Nat Hazards 28:271–290
Freeman PK, Martin LA, Linnerooth-Bayer J, Mechler R, Pflug G, Warner K (2002) Disaster risk management: national systems for the comprehensive management of disaster financial strategies for natural disaster reconstruction. SDD/IRPD, regional policy dialogue. Inter-American Development Bank, Washington, DC
Garnett RE (2009) Ray Garnett climate & crop letter, Agro-Climatic Consulting, Winnipeg, Canada. 4(5)
Garnett RE, Nirupama N, Murty TS, Haque CE (2006) Correlates of Canadian prairie rainfall-implications for crop yield. Clim Res 32:25–33
GeoHazards International (2015) Post-earthquake reconstruction in Nepal. http://wwwgeohazorg/#!post-earthquake-reconstruction-in-nepal/clzhv. Accessed 30 June 2016

GoI (2006) Guidelines for design and construction of cyclone/tsunami shelters. Government of India and UNDP Disaster Risk Management Programme, Ministry of Home Affairs, Government of India, India

GOI-UNDP (2008) Through participation of communities and local self governments. National Disaster Management Division, Ministry of Home Affairs, Government of India. http://www.ndmindia.nic.in/EQProjects/goiundp2.0.pdf. Accessed Mar 2012

GOI-UNDP (2010) Disaster risk management programme: evaluation and review of lessons learnt. Ministry of Home Affairs, Government of India

Government of Michigan (2001) The Michigan hazard analysis. Michigan Emergency Management. https://www.michigan.gov/documents/msp/Doc1_394216_7.pdf

Green C (2004) The evaluation of vulnerability to flooding. Disaster Prev Manag 13(4):323–329

Greenberg DA, Murty TS, Ruffman A (1993) A numerical model for the Halifax harbour tsunami due to the 1917 explosion. Mar Geod 16(2):153–168

Greenberg DA, Murty TS, Ruffman A (1994) Modelling the tsunami from the 1917 Halifax harbour explosion. Sci Tsunami Haz 11(2):67–80

Helm P (1996) Integrated risk management for natural and technological disasters. Tephra 15 (1):4–13

HIRA (2004) Provincial hazard identification and risk assessment, emergency management Ontario. Govt Ontario, Canada

HIRA (2011) Emergency management Ontario. Ministry of Community Safety and Correctional Services, Toronto, Canada

Hoggan DH (1996) Computer-assisted floodplain hydrology and hydraulics. McGraw-Hill, New York

HRVA (2004a) http://hrva.embc.gov.bc.ca/toolkit.html

HRVA (2004b) Hazard, risk, and vulnerability analysis toolkit. Provincial emergency preparedness, British Columbia, 66p

Hydrologic Engineering Center 2001 HEC-RAS River Analysis System. Hydraulic Reference Manual,Version 3.0. United States Army Corps of Engineers

IMD (2013) India Meteorological Department. http://imd.gov.in/. Accessed 10 June 2014

ISDR (2004) Living with risk: a global review of disaster reduction initiatives – volume 1. UN International Strategy for Disaster Reduction publication, Geneva. 429p

Kerala (2012) Kerala to get three cyclone shelters. The Hindu http://www.thehindu.com/news/states/kerala/article2978305.ece. Accessed 1 May 2012

KGS Group (2000) Red River Basin – Stage-damage curves update and preparation of flood damage maps. Report prepared for International Joint Commission

Khaladkar RM, Mahajan PN, Kulkarni JR (2009) Alarming rise in the number and intensity of extreme point rainfall events over the India region under climate change scenario. Indian Institute of Tropical Meteorology Pune, India, RR-123

Krenz G, Leitch J (1993) A river runs north – managing an international river. Red River Water Resources Council, Fargo

Kundzewicz ZW (2002) Non-structural flood protection and sustainability. Water Int 27(1):3–13

Long MH, John JI (1993) Risk-based emergency response. Paper presented at the ER93 conference on the practical approach to hazardous substances accidents. St. John, New Brunswick, Canada

Lost City (2015) The Mid-Atlantic ridge and spreading. http://www.lostcity.washington.edu/science/geology/midatlanticridge.html. Accessed 21 Mar 2016

Martínez EV, López FC (2010) A cybercartographic tool for supporting disaster prevention planning processes and emergency management in Mexico city. In: Showalter PS, Lu Y (eds) Geospatial techniques in urban hazard and disaster analysis. Springer, Heidelberg, pp 255–271

McKinney DC, Maidment DR (1993) Expert geographic information systems for Texas water planning. ASCE J Water Resour plan Manag 119:170–183

Mukherji B (2013) Monsoon speediest in over 50 years. The Wall Street Journal, India, June 18, 2013

Murty TS (1977) Seismic sea waves-tsunamis. Canadian Bulletin of Fisheries and Aquatic Sciences, Ottawa, Canada, 198:1–337
Murty TS, Nirupama N, Nistor I, Hamdi S (2005) Far field dispersion characteristics of the tsunami of 26th December 2004. Indian. J Earthq Eng 42(4):213–217
National Disaster Management (NDM) (2012) Ministry of Home Affairs. India. http://www.ndmindia.nic.in/. Accessed 14 Mar 2012
NCRMP (2011) National cyclone risk mitigation project. http://www.thehindu.com/2011/01/29/stories/2011012966692000.htm. Accessed 21 Mar 2016
NDMA (2013) National Disaster Management Authority, Government of India. http://ndma.gov.in/ndma/index.htm. Retrived 10 June 2014
NIDM (2013) National Institute of Disaster Management. http://nidm.gov.in/default.asp. Accessed 12 June 2014
Nirupama N (2009) Socio-economic implications based on interviews with fishermen following the Indian Ocean tsunami. Nat Hazards 48(1):1–9
Nirupama N (2012) Risk and vulnerability assessment – a comprehensive approach. Int J Disaster Resilience Built Environ, Emerald 3:2
Nirupama N (2013) Disaster risk management. In: Bobrowsky P (ed) Invited chapter in encyclopaedia of natural hazards. Springer. Dordrecht/New York, pp 164–169
Nirupama N, Etkin D (2009) Emergency managers in Ontario: an exploratory study of their perspectives. J Homel Secur Emerg Manag 6(1), Article 38. https//:doi.org/10.2202/1547-7355.1464
Nirupama N, Simonovic SP (2007) Increase of flood risk due to urbanisation: a Canadian example. Nat Hazards, Springer 40:25–41
Nirupama N, Murty TS, Nistor I, Rao AD (2006) Persistent high water levels around Andaman & Nicobar Islands following the 26 December 2004 tsunami. Sci Tsunami Haz 24(3):183–193
Nirupama N, Sharma R, Verma K, Panigrahi S (2014) A multi-tier hazard-part I: description of the event. Nat Hazards 76(1):259–269
Noson L (2009) Hazard mapping and risk assessment. Asian Disaster Preparedness Center. http://www.adpc.net
Orissa (2009) Multipurpose cyclone shelter under ICZMP. Orissa State Disaster Management Authority, Bhubaneswar
Padmanabhan G (2008) Conference on community-based disaster risk reduction, UNICEF, Kolkata, India, 26–28 November 2008
Pereira JMC, Duckstein L (1993) A multiple criteria decision-making approach to GIS-based land suitability evaluation. Int J Geogr Inf Syst 7:407–424
Ramachandran R (2013) Why Kedarnath happened. Frontline, July 26, 2013. http://www.frontline.in/the-nation/why-kedarnath-happened/article4894867.ece?homepage=true
Rao M (2012) State-of-the-art cyclone shelters for AP. http://www.thehindu.com/todays-paper/tp-national/tp-andhrapradesh/article3347335.ece. Accessed 21 Mar 2016
Reddy AVS (1991) The concept of cyclone shelters for livestock. Natl Inst Rural Dev 3(3). http://www.fao.org/livestock/agap/frg/FEEDback/lrrd/lrrd3/3/reddy.htm
Roddey J (2012) Cascadia subduction zone. https://nisquallyquakewordpresscom/tag/cascadia-subduction-zone/. Accessed 21 Mar 2016
Ruffman A, Greenberg DA, Murty TS (1995) The tsunami from the explosion in Halifax harbour. In: Ruffman A, Howell CD (eds) Ground zero. Nimbus Publishing Ltd., Halifax, pp 327–344
Schneider T, Basler E, Partner AG (2006) A delicate issue in risk assessment. In: Ammann, Dannenmann, Vulliet (eds) RISK21 – coping with risks due to natural hazards in the 21st century. CRC Press, Boca Raton
Sharma R, Verma K, Panigrahi S, Nirupama N (2014) A multi-tier hazard-part II: meteorological analysis. Nat Hazards 76(1):271–281
Shibayama T, Okayasu A, Wijayaratna N, Sasaki J, Suzuki T, Jayaratne, R (2005) The December 26, 2004 Sumatra Earthquake Tsunami, Tsunami Field Survey around Southern Part of Sri

Lanka. http://www.drs.dpri.kyoto-u.ac.jp/sumatra/SriLanka_survey_result/srilanka_survey_ynu_e.html
Shrubsole D (2001) The cultures of flood management in Canada: insights from the 1997 Red River experience. Can Water Resour J 26(4):461–479
Simonovic SP (2002) Two new non-structural measures for sustainable management of floods. Water Int 27(1):38–46
Simonovic S, Nirupama N (2005) A spatial multi-objective decision making under uncertainty for water resources management. J Hydroinf 7(2):117–133
Slovic P (2000) The perception of risk. Earthscan Publications, London. 473p
Smith K (2004) Environmental hazards: assessing risk and reducing disaster. Routledge, London. 306p
Smith K, Petley DN (2009) Environmental hazards: assessing risk and reducing disaster. Routledge, London. 306p
Sphere India (2013) Coordinating disaster management by local NGOs. http://www.sphereindia.org.in/. Accessed 21 Mar 2016
Sundaramoorthy A (2013) Lessons learnt from landslides of Uttarakhand disaster. Geological Survey of India, Dehradun
Teller JT, Murty TS, Nirupama N, Chittibabu P (2005) A possible tsunami in the western part of the Labrador Sea about 8,400 years B.P. Sci Tsunami Haz 23(3):1–14
Theophilus E (2013) A river pulse: a discussion paper on the flood events in June 2013, Mahakali basin, Uttarakhand. Himal Prakriti September 2013
Tim US (1997) Integrated spatial technologies for non-point source pollution control and environment management. Proceedings of the Eleventh Annual Symposium on Geographic Information Systems, Vancouver, BC, Canada, pp 143–149
Tkach RJ, Simonovic S (1997) A new approach to multi-criteria decision making in water resources. J Geogr Inf Decis Anal 1(1):25–43
Tucker B (2013) Tsunami evacuation park in Padang, Indonesia, Global Platform for disaster risk reduction, Fourth session, Geneva, Switzerland, 19–23 May 2013
Twigg J (2007) Characteristics of a disaster-resilient community. DFID disaster risk reduction interagency coordination Group, 39p
USGS (2015) Geologic history of the San Andreas fault system. http://geomaps.wr.usgs.gov/archive/socal/geology/geologic_history/san_andreas_history.html. Accessed 21 Mar 2016
UWI (2013) Uttarakhand flash floods: situation update, United Way of India (in association with NGO, Pragya), July 2, 2013
Weaver J (2009) Going vertical, Stanford News Report, December 14, 2009. http://news.stanford.edu/news/2009/december14/vertical-evacuation-study-121409.html. 20 December 2015
Whyte AV, Burton I (1982) Perception of risk in Canada. In: Burton I, Fowle CD, McCullough RS (eds) Living with risk. University of Toronto, Toronto, pp 39–69
Wilches-Chaux G (1993) La Vulnerabilidad Global. In: Maskrey A (ed) Los Desastres no San Naturales. La Red, Colombia
Wisner B, Blaikie P, Cannon T, Davis I (2004) At risk: natural hazards, people's vulnerability and disasters. Routledge, London. 469p
Wolfe DS (1997) Integrated resources management using spatial allocation: a case study. Proceedings of the Eleventh Annual Symposium on Geographic Information Systems, Vancouver, BC, Canada, 424–427

Chapter 4
Disaster Resilience

Disaster resilience is a term that lends a more positive note to the term used in previous decades, 'disaster vulnerability.' In similar vein, capacity building is now becoming a more popular term and concept in the context of disaster risk management. Resilience can be defined at micro or macro scales, depending on the objective of the research; for example, at the household, community, county, sub-national, national, regional, or global level. Resilience to disasters involves addressing the root causes of what makes people and communities susceptible to potential threats. Experts in the field have used a variety of terms to define resilience, namely robustness; the ability to bounce back after adverse events; the critical infrastructural capacity to withstand external shocks; and recovery and rehabilitation components embedded in the concept of resilience. Resilience also considers different forms of capital, namely human, social, physical, financial, natural and political. This chapter includes various definitions of and perspectives on resilience, vulnerability and sustainability in communities. Case studies from around the world are an essential component of the chapter.

4.1 Resilience

Every disaster has a significant social impact on the population. People's vulnerability is directly related to their sociological aspects, which are generally overlooked in disaster management (Nirupama and Armenakis 2013). In our ever dynamic system of human-nature interaction, a number of great definitions, interpretations, and discussion on resilience as well as adaptation have been developed. In this context, perspectives such as social, economic, physical, and environmental need to be explored and examined with subtle nuances for developing policy and capacity building strategies. While the social construction of risk focuses on vulnerability as the tendency of exposed elements to harm people and cause damage to properties and the environment (Birkmann et al. 2013; Vogel et al. 2007; Adger 2006), efforts

N. Agrawal, *Natural Disasters and Risk Management in Canada*, Advances in Natural and Technological Hazards Research 49,
https://doi.org/10.1007/978-94-024-1283-3_4

are being made to integrate the conceptual frameworks of vulnerability, resilience, and adaptation into sustainability science (Lei et al. 2014; Turner et al. 2003; Turner 2010; Miller et al. 2010; Endfield 2012). A discussion on resilience is incomplete without discussing vulnerability and adaptation. Vulnerability identification and assessment had been the focus of risk reduction studies and methodologies for decades, but realization of the importance of resilience is gaining traction in recent years. Resilience which originates from the Latin word *resilio,* meaning 'to jump back' (Klein et al. 2003; Timmerman 1981) is a multi-faceted concept that is widely used in the context of disasters, ecosystems, and health. A chronological arrangement of the three terms would be vulnerability, resilience, and adaptation wherein vulnerability can be considered as a negative arm and adaptation can be viewed as a positive arm of resilience. Paton (2006) conceptualizes resilience on three different levels: physical, social, and behavioral. The physical level attributes attention to the physical integrity of the built environment, including: building codes, land use planning, and retrofitting buildings to withstand disasters.

Resilience building at individual, community, and institutional level must be perceived with a lens of cost benefit analysis and therefore, adaptation must always be a part of the equation. Tables 4.1, 4.2, and 4.3 summarize various definitions of vulnerability, resilience, and adaptation respectively.

Resilience at community level can be enhanced effectively if communities are encouraged to engage and better understand its meaning and benefits. People's perceptions are often subjective, not objective, and therefore, it is vital that a two-way dialogue between communities and professionals is established. Simple steps such as providing education to all citizens, vocational training, equal wages for women, diverse income opportunities for all, health care and affordable healthy food choices for all, etc. will help in enhancing resilience in societies. A resilient community is an essential component for sustainable development and disaster risk management. Empowerment is a process and not the endpoint. In particular, communities can attain significant resiliency if women are empowered – educationally, technologically, health-wise, socially, and of course economically. Human security is imperative in that it can be enhanced by addressing capital assets as listed in Table 4.4 (Collins 2009).

A rising need for better understanding resilience of social, economic, and the environment in the current environment of climate change, severe and frequent disastrous events has been realized in recent decades. Complexities in modern society, globalization, rapid urbanization, terrorism, technological advancements and most of all climbing cost of loss due to disasters are some of the reasons for growing interest in better understanding of resilience. The lack of our ability to foresee threats that we may be faced with limits our ability to prepare or prevent for potential risks. In this case, the concept of resilience and a good understanding of it become important (Boin et al. 2010). For a sustainable future and reducing losses from disasters, enhanced resilience at all levels – local, regional, national, and international – is required. Livelihood is a vital component of sustainability, and in this context, Sustainable Livelihood Framework proposed by Chambers and

Table 4.1 Various definitions of vulnerability

Author(s)	Definition
Downing et al. (1997)	Vulnerability means an environmental sensitivity. There are a number of factors related to vulnerability such as demographic, economic, social and technical factors, and the economic dependences
Kasperson and Kasperson (2001, 2005)	Vulnerability is the flip side of resilience: when a social or ecological system loses resilience, it becomes vulnerable to change that previously could be absorbed
IPCC (2001, 2007)	Vulnerability is the degree to which a system is susceptible to, or unable to cope with, adverse effects of climate change, including climate variability and extremes
Turner et al. (2003)	Vulnerability is the degree to which a system, subsystem, or system component is likely to experience harm due to exposure to a hazard, either a perturbation or stress/stressor
Cutter et al. (2003)	Social vulnerability is a measure of both the sensitivity of a population to natural hazards and its ability to respond to and recover from the impacts of hazards
Wisner et al. (2004)	Vulnerability means the characteristics of a group or individual in terms of their capacity to anticipate, cope with, resist, and recover from the impact of a hazard
Adger (2006)	The key parameters of vulnerability are the stress to which a system is exposed, its sensitivity, and its adaptive capacity
Birkmann (2006)	Social vulnerability refers to the inability of people, organizations, and societies to withstand adverse impacts from multiple stressors to which they are exposed
UNISDR (2009)	Vulnerability, the characteristics and circumstances of a community, system or asset that make it susceptible to the damaging effects of a hazard
Zhou et al. (2010)	Vulnerability places stress on system's response to hazard or hazard potential, which determines the likelihood of loss from hazards. Exposure and sensitivity are two aspects of vulnerability
Han (2011)	Vulnerability cannot be explained solely either by exposure or response capacity (including both short-term coping and long-term adaptive capacities), but are the result of interactive change of both, or the covariance between them

Lei et al. (2014)

Conway (1992) and further discussed in Carney (1998, 2002) and Carney et al. (1999) is an important and relevant concept (Fig. 4.1). Within the livelihood framework, the term sustainability is often linked to the ability to cope with and recover from stresses and shocks as well as to maintain the natural resource base (Birkmann 2006).

Disaster resilience is a shared responsibility among individuals and all sectors. It is directly related to capacity building that starts with individuals, translates to communities and further spreads to local, provincial, and federal levels – kind of a

Table 4.2 Various definitions of resilience

Author(s)	Definitions
Holling (1973)	Resilience is defined as the amount of disturbance that can be sustained by a system before a change in system control or structure occurs. It could be measured by the magnitude of disturbance the system can tolerate and still persist
Timmerman (1981)	Resilience is the ability of human communities to withstand external shocks or perturbations to their infrastructure and to recover from such perturbations
Holling (1996)	Resilience is the buffer capacity or the ability of a system to absorb perturbation, or the magnitude of disturbance that can be absorbed before a system changes its structure by changing the variables
Kimhi and Shamai (2004)	Social resilience is understood as having three properties: resistance, recovery, and creativity, in which (1) resistance relates to a social entity's efforts to withstand a disturbance and its consequences; (2) Recovery relates to an entity's ability to pull through the disturbance; (3) Creativity is represented by a gain in resilience achieved as part of the recovery process, and it can be attained by learning from the disturbance experience
Carpenter et al. (2001) and Resilience Alliance (2009)	The Resilience Alliance consistently refers to social–ecological systems (SES) and defines their resilience by considering three distinct dimensions: (1) the amount of disturbance a system can absorb and still remain within the same state or domain of attraction; (2) the degree to which the system is capable of self- organization; and (3) the degree to which the system can build and increase the capacity for learning and adaptation
Folke et al. (2002)	The capacity to buffer perturbations, self-organize, to learn and adapt. Resilient systems contain the experience and the diversity of options needed for renewal and redevelopment. Sustainable systems need to be resilient
Adger (2006)	Resilience refers to the magnitude of disturbance that can be absorbed before a system changes to a radically different state as well as the capacity to self-organize and the capacity for adaptation to emerging circumstances
UNISDR (2009)	The ability of a system, community, or society exposed to hazards to resist, absorb, accommodate to and recover from the effects of a hazard in a timely and efficient manner, including through the restoration of its essential basic structures and functions
Walker et al. (2009) and Folke et al. (2010)	Resilience is the capacity of socio–ecological systems (SES) to continually change and adapt yet remain within critical thresholds. Adaptability is part of resilience
Zhou et al. (2010)	From a geographic perspective, disaster resilience can be defined as the capacity of hazard-affected bodies (HABs) to resist loss during disaster and to regenerate and reorganize after disaster in a specific area in a given period. Resilience

(continued)

Table 4.2 (continued)

Author(s)	Definitions
	can be classified as inherent resilience (IR) and adaptive resilience (AR)
Han (2011)	A resilience thinking requires not only changing the focus from modifying hazard events to reducing vulnerability, but also essential to embrace and internalize variability and uncertainty in decision making
IPCC (2012)	Resilience is the ability of a system to anticipate, absorb, accommodate, or recover from the effects of a hazardous event in a timely and efficient manner

Lei et al. (2014)

quilt of resilience (Cutter 2011). The term 'communities' carries a broad meaning encompassing local neighborhoods, family units, cities, regions, provinces, and countries. Furthermore, monitoring and evaluating community resilience would be a site-specific endeavor.

Birkmann et al. (2013) outline a framework called MOVE (Methods for the Improvement of Vulnerability Assessment in Europe; www.move-fp7.eu) to enhance the disaster risk management (DRR) perspective by integrating new understanding of coupling, adaptation and resilience. In so doing, this builds a potential ground for closer linking between the concepts and assessment methodologies being developed in DRR and climate change adaptation (CCA). Climate change adaptation increasingly places emphasis on improving the capacity of governments and communities to address existing vulnerabilities to current climate variability and climatic extremes (Thomalla et al. 2006). Lack of resilience or societal response capacity is determined by limitations in terms of access to and mobilization of the resources of a community or a social-ecological system in responding to an identified hazard. This includes pre-event risk reduction, in-time coping and post-event response measures. Compared to adaptation processes and adaptive capacities, these capacities focus mainly on the ability to maintain the system in the light of a hazard event impacting the system or element exposed. In this sense, the capacity to anticipate, the capacity to cope and the capacity to recover can include significant changes to existing practices around a referent hazard event/scenario but does not include learning based on the potential for future change in hazard and vulnerability contexts. However, the concept of resilience also includes learning and reorganization processes and therefore is positioned as a sub-component of the adaptation box. Compared to the key factor 'lack of resilience', which refers to existing capacities, the adaptation box also deals with the ability of a community or a system to learn from the past disasters and to change existing practices for potential future changes in hazards as well as vulnerability contexts. An example of community resilience focusing on women is discussed in the following section.

Table 4.3 Various definitions of adaptation

Author(s)	Definitions
Burton et al. (1978)	Adaptation refers to the process, measures, or structural change in order to reduce or offset the potential disasters associated with climate change, or the use of the opportunities brought about by climate change, which include reducing the vulnerability of social, regional, or activities on climate change and its variability
Stakhiv (1993)	The term adaptation means any adjustment, whether passive, reactive or anticipatory, that is proposed as a means for ameliorating the anticipated adverse consequences associated with climate change
Smith (1996)	Adaptation to climate change includes all adjustments in behavior or economic structure that reduce the vulnerability of society to changes in the climate system
Smit et al. (2000)	Adaptation refers to the adjustments of ecological–social–economic system for the actual or foreseeable climate stimulate their effects or impacts
Adger (2006)	Adaptations include changes in the rules and governance of disaster risk, change in organizations, and promotion of self-mobilization in civil society and private corporations
Brooks (2003); Young et al. (2006)	Adaptation means adjustments in a system's behavior and characteristics that enhance its ability to cope with external stresses. Adaptation will allow a system to reduce the risk associated with these hazards by reducing its social vulnerability
UNISDR (2009)	The adjustment in natural or human systems in response to actual or expected climatic stimuli or their effects, which moderates harm or exploits beneficial opportunities
Walker et al. (2004, 2009)	Adaptability, a manifestation of adaptation, has been defined as "the capacity of actors in a system to influence resilience"
Folke et al. (2010)	Adaptability is part of resilience. It represents the capacity to adjust responses to changing external drivers and internal processes and thereby allow for development along the current trajectory
McLaughlin (2011)	Adaptation to climate is the process through which people reduce the adverse effects of climate on their health and well-being, and take advantage of the opportunities that their climatic environment provides. The term adaptation means any adjustment whether passive, reactive, or anticipatory
IPCC (2012)	In the context of climate change, adaptation is the adjustment in natural or human systems in response to actual or expected climatic stimuli or their effects, which moderates harm or exploits beneficial opportunities

Lei et al. (2014)

4.2 Community Resilience

A famous quote by Mahatma Gandhi "You must be the change you wish to see in the world" is fitting in the context of resilient communities in that we should participate in building resilience and not depend solely on government assistance at all times. The new trend, worldwide, is shifting towards designing

Table 4.4 Capital assets that can be enhanced to improve human resilience

1	Human Capital	Skills, knowledge, health, ability to work: all required to make use of the other assets and to make security
2	Social Capital	Family, clan, neighborhood, religious/other association networks and connectedness; help strengthening social groups and relationships required to build a safety net, gain positive livelihood
3	Physical Capital	Infrastructure, goods and services e.g. roads, water and electricity supply, shelter, information technology, emergency services, engineering equipment: all these enhance livelihood and security
4	Natural Capital	Land, water (sea, fresh water, rainfall, etc.), soil, forests, air quality required to achieve livelihood and environmental security
5	Financial Capital	Financial resources available to a person: savings, credits, remittances and pensions, and which are more secured from varied sources
6	Political Capital	Personal rights, including gender equality, political representation and regular access to vote on who has power and how the powers influence one's livelihood and security

Collins (2009)

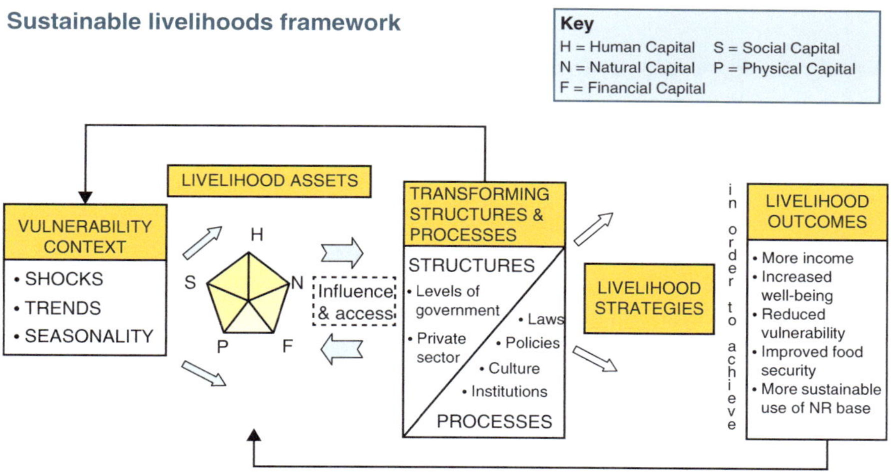

Fig. 4.1 Sustainable livelihood framework (Carney 1998)

disaster resilient societies and developing the digital archive system to pass the lessons from current disasters to future generations.

Managing and alleviating social and psychological harm among people, in the face of recurring disasters in the world, is very important. Not every community holds the same level of resilience; and, with the inevitable increase in disasters on a global scale (EMDAT 2009), it is important that emergency managers and government authorities alike understand why some communities may have outstanding adaptive capacity and others very little. In social context, resilience is conceptualized as a social resource whereby an equitable distribution of costs and benefits associated with hazard reduction and readiness activities is withheld within a given society.

The behavioral dimension is about encouragement and a sustained adoption of preparatory actions that determine the ability to respond and adapt to adverse hazard effects within communities. In addition, resilience requires encompassing the cultural and environmental dimensions of the given population. As its complexity unfolds, resilience can be considered as a vast web of social interactions, characteristics, and capacities that enable a community to live with the hazards they face (Crowley and Elliott 2012). Vibrant leadership, shared goals and values, established institutions and organizations, positive socio-economic trends, constructive external partnerships and linkages, and the availability of resources and skills, all together and singly, are characteristics that increase resilience on a community level (Gardner and Dekens 2007). In the following section, community resilience has been further explained and demonstrated using three different case studies and research studies.

4.2.1 Case Study Based on Events in Canada, Japan and New Zealand

A study (Nirupama et al. 2015) examining three earthquake events in different parts of the developed world with a focus on community resilience is discussed in following sections. The 2012 earthquake in Haida Gwaii (formerly Queen Charlotte Islands), Canada, the 2011 Tohoku earthquake in Japan, and the 2010 earthquake in Christchurch, New Zealand can be compared through the lens of community resilience in order to better understand the concept.

The November 2012, earthquake in Canada's West Coast region of Haida Gwaii, was a Richter magnitude 7.7 earthquake, said to be the most severe Canadian earthquake in more than 60 years. The small community of Haida Nation was reported to be shocked and having difficulty coping. Although no significant damage occurred, the event shook the community. In comparison, during the March 2011 Tohoku earthquake in Japan, Japanese people demonstrated tremendous resilience in dealing with their despair; for example, affected citizens waited patiently in lines to receive emergency relief supplies. Their previous experiences with various disasters may have shaped their perception and equipped them to fortify a component of resilience in their daily lives. One year later, examples of elderly fishermen continuing to live in temporary shelters and making payments on their lost homes are clear evidence of a need to develop a more supportive system that would enhance people's resilience; by doing so, individuals will be more adept at coping with adverse situations.

In New Zealand, while floods are commonly known, volcanic eruptions and earthquakes are the most underrated hazards (Britton and Clark 2000). It is generally understood that New Zealand is an earthquake prone area (Stirling et al. 2012). However, New Zealand is also recognized for being one of the most prepared earthquake communities in the world (Crowley and Elliott 2012). Although Canada has a long documented history of earthquakes predominantly on its East and West Coasts dating back as early as the 1700s, it is often criticized for its subpar disaster prevention and mitigation strategies.

New Zealand records more than 14,000 earthquakes a year but only about 150 are usually felt. Following the September 2010 earthquake near Christchurch, and the February 2011 earthquakes in Canterbury, New Zealand, the Ministry of Civil Defense and Emergency Management partnered with local universities, and a Christchurch based research group, to conduct an inquiry into community resilience. A sample of 100 randomly selected Christchurch residents participated; the interviews were designed to help develop an understanding of the competencies that supported community resilience in response to the earthquake.

By conducting a content analysis of the interviews, it became clear that an organic spontaneous process of community connecting was often how individuals coped with the event. Individuals within communities willingly came together and aided one another. This process was described as naturally occurring and was a key component to maintaining physical and psychological resilience. Further, an individual's attitude, outlook, physical mobility, sociability, and connectedness with others were found to be the most significant factors in determining individual resilience. Those who held outdoor lifestyles or who volunteered for community organizations such as St. John Ambulance, or civil defense, were those who held significant individual resilience.

In general, self-activation, sufficiency, responsibility, and management were among the top characteristics to prove one's resilience. For a better understanding of how individuals within communities in New Zealand make sense of hazard information, Becker et al. (2009) conducted a series of interviews. Several participants mentioned that having children in school influenced their level of preparedness. The children would often come home and discuss what they had learned in school and engaged their parents in preparedness activities for their household. In addition, people who felt they held responsibility for others, whether it is a spouse, child, or their community in general, were more likely to prepare for potential disasters. Interestingly, a number of individuals thought that the responsibility of being able to cope following a disaster was shared between the individual, the community, and organizations, with each having particular roles to play in the aftermath of disasters. The study pinpointed three types of information that individuals shared on how to prepare for disasters: (i) passive (e.g. brochures, television), (ii) interactive (community, school, and workplace activities), and (iii) experiential information (experiencing a hazard). Participants tended to recall experiential information better than interactive information, and interactive information more often than information conveyed in a passive format. Passive communication resulted in a generally poor recollection of hazard information.

The earliest Canadian earthquake on the West Coast of British Columbia, recalled in Huron folklore, and reported in literature, occurred in the 1700s (Cassidy et al. 2010; Lamontagne 2008) in the subduction zone of Cascadia fault. While reflecting on this massive earthquake and examining vulnerability and resilience, at the time, many long-held cultural practices made Native communities sufficiently resilient to these natural hazards.

Vancouver Island is situated in a seismically active region off the South-West Coast of British Columbia, where the oceanic Juan de Fuca plate subducts beneath the continental North American plate in the Cascadia Subduction Zone. In this zone, the probability of a widely felt earthquake exceeds 25% over 10 years and increases to over 90% in 100 years (Seemann et al. 2011). However, despite the risk associated with living in a city prone to earthquakes, the number of people living in Vancouver has more than doubled in the past three decades (Kovacs and Kunreuther 2001). Furthermore, millions of tourists travel to the region each year.

With more and more people flocking to the shores of Vancouver Island, and the staggering probabilities of a widely felt earthquake, the need for building community resilience is undeniable. The 2012 M 7.7 earthquake in the Haida Gwaii region was the second largest event ever recorded in the region. The largest recorded event (M 8.1) occurred in 1949, and among other significant earthquakes was the 1929 event of M 7.1 (Lamontagne 2008). Persons residing in Haida Gwaii were reportedly surprised and shaken up by the event, in view of the fact that they were ill prepared, both, physically and mentally for such an experience, even though the area is well-documented as being seismically active. The media reported cases of widespread panic in the impacted area.

In the afternoon of March 11, 2011, an M 9.0 reverse fault megathrust earthquake occurred along the subduction zone under the north-western Pacific Ocean where the Pacific and North American tectonic plates meet. It was the most powerful earthquake ever recorded to have hit the country, and the fourth most powerful to have occurred anywhere in the world since 1900 (JMA 2011; USGS 2011). The earthquake generated a tsunami of epic proportions with a maximum height of 17 m in some places. The powerful tsunami waves engulfed the low-lying coastal areas of the north-eastern coastal area of Honshu, causing severe flooding and bringing millions of tons of debris to the coast. About twenty prefectures suffered catastrophic loss of lives with 15,365 deaths, 8206 missing or injured, and infrastructure (Matanle 2011). The protective walls at the Fukushima Daiichi nuclear power plant could not withstand the force of the tsunami causing the meltdown of nuclear reactors, which wreaked havoc for both the communities in the area and the government.

Japan has seen earthquakes in the past (The Great Kanto Earthquake of 1923, and the 1995, the Great Hanshin Earthquake) and recovered from them. The Japanese are a seemingly resourceful and knowledgeable people and possess deep wells of self-discipline and determination; these attributes certainly assist them to cope with disasters that allow them to recover. In order to address the issue of an aging society, especially in rural communities (Shaw and Takeuchi 2011), the Japanese government has made a declaration that people with disabilities, seniors over 65 years of age, and very young children of up to 6 years of age, are easily affected by disasters. Furthermore, a relatively new concept of 'voluntary self-help' with an intention of building local capacity and resilience is gaining popularity among Japanese people (Fig. 4.2). In addition, after the disaster, the Japanese Prime Minister, Kan established (Han 2011) three principals for sustainable recovery, including: a hazard resistant society, establishing a social system that allows people and the environment to live in harmony, and promoting compassion and care for those vulnerable (Matanle 2011; Mitsuyoshi 2011). This approach validates the very psyche of Japanese people, who

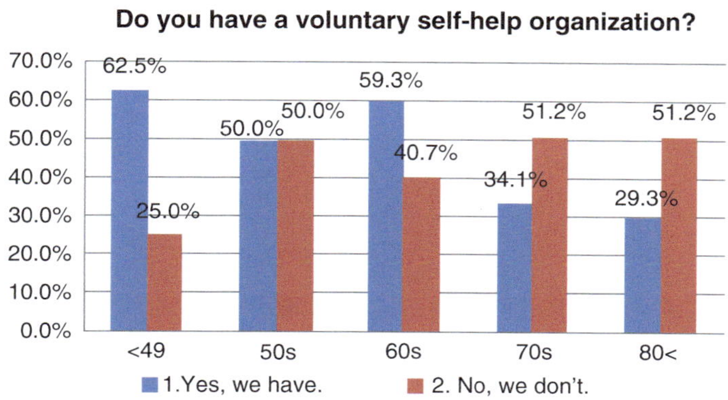

Fig. 4.2 Recognition of voluntary self-help organization by age bracket (Shaw 2012)

showed tremendous patience, generosity, and consideration for each other, despite having lost all their material possessions in the March 2011 disaster.

4.2.2 Case Study Based on a Women's Group in Toronto, Canada

A study exploring community resilience through engaging a group of immigrants in Toronto, Canada in a focus group setting (Nirupama and Maula 2013) focuses on the importance of people's perception of hazards, risk and resilience. A focus group session (Appendix 2) was conducted involving a group of 41 women at the South Asian Women's Centre (SAWC) in Toronto, Canada. Although the sample size was small, the population was representative of a significant type of immigrant group of Toronto. The participating women were engaged in discussion on emergencies and crises, and their (participants') perception of the threat associated with them. After explaining the objective of the research, which was to explore ways to improve community resilience, the group understood the importance of the exercise and expressed their full support during the session. The focus group session was transcribed and analyzed using Microsoft Excel and NVivo qualitative software.

The SAWC participants agreed that they felt vulnerable due to variety of reasons, including age, poor health, inadequate income, and mobility challenges in some cases. Figure 4.3 clearly shows that only 5% of the participants had a personal vehicle while the majority depended on public transit. Figure 4.4 illustrates that participants offered varied responses regarding their sense of belonging to the community they lived in, and whether or not they feel that they are prepared to deal with emergencies. While few responded they felt comfortable with their level of safety, many acknowledged they felt exposed to risk. In terms of realizing the

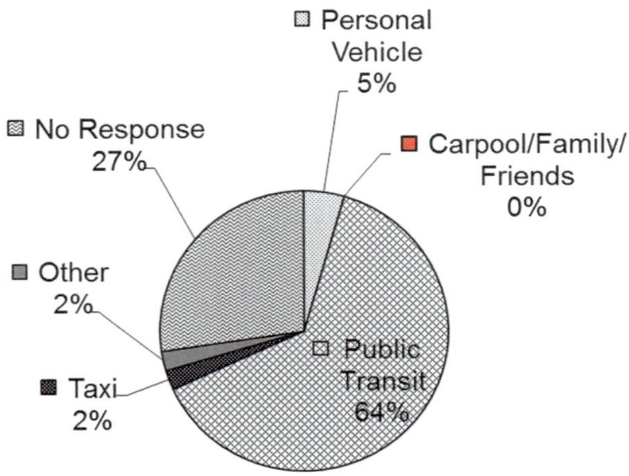

Fig. 4.3 Main source of transportation used by the participants

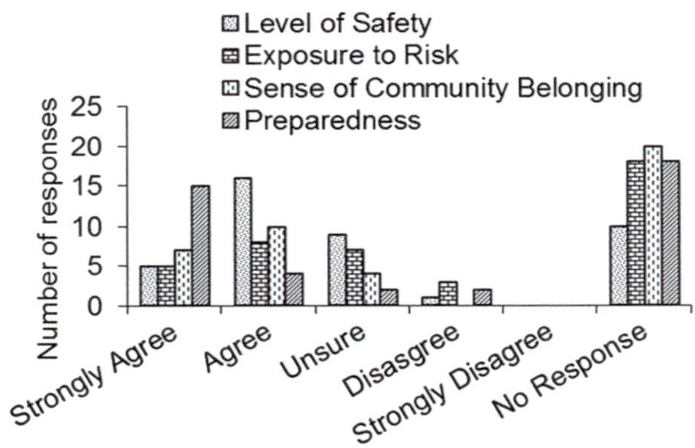

Fig. 4.4 People's perception of their safety, exposure to risk or threat, sense of belonging with their community, and preparedness to deal with emergencies

importance of preparedness, an overwhelming affirmation is clearly indicated in Fig. 4.4 that the focus group participants felt exposed to risk and poorly prepared to deal with crisis. Interestingly, a large number of participants were simply not sure about these issues associated with emergency preparedness and risk, also highlighted a serious concern regarding emergency management knowledge and coping capabilities from this sample group.

Through further probing into the awareness of community participation programs available in participants' local community centre, most of the respondents said either "*yes*" or "*no response.*" Similar responses were seen when asked whether or not they had any interest in municipal elections. In another (related) question, a few participants said that they took interest in local government affairs, but had no actual involvement. This aligns with the theory of a ladder of citizen participation (Arnstein 1969) in which various levels of participation is described as – Nonparticipation,

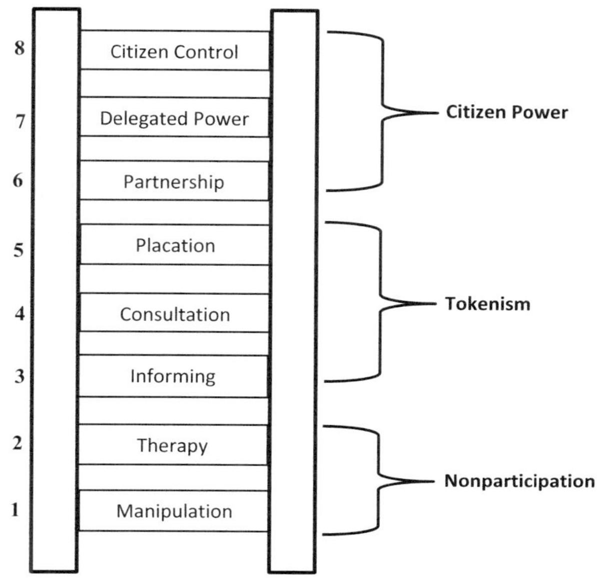

Fig. 4.5 A ladder of citizen participation (Arnstein 1969)

Tokenism, and Citizen Power (Fig. 4.5). The bottom rungs of the ladder are manipulation and therapy, used to substituting for genuine participation. Middle rungs progress to levels of tokenism that allow a voice to vulnerable people but their views will, most likely, not be heeded by the powerholders. The top most rung insures full citizen power.

The participants showed confusion when asked about potential threat near their homes until they were explained what a potential threat meant and then they were able to identify some (Fig. 4.6). In regard to their choices and preferences for seeking help in case of an emergency, a clear consensus in favour of family and friends was evident (Fig. 4.7).

Issues such as education, low income, poor health, limited access to resources came to light during the focus group session, convincing us of the merits of having a direct conversation with the very people who are the main stakeholders in their communities. The outcome of this study serves to identify people's vulnerabilities and the potential hazards they may be exposed to, which are essential components needed to develop strategies for increasing people's resilience. Community participation is key to successful implementation of any program that is developed and funded for future disaster risk reduction. It is noteworthy that throughout the focus group session, participants were confused and distrustful of the government for possible misuse of the information that was being collected there. Similar studies involving diverse population covering wider geographical locations in the Toronto area are being conducted for a comprehensive grasp on the topic of enhancing community resilience.

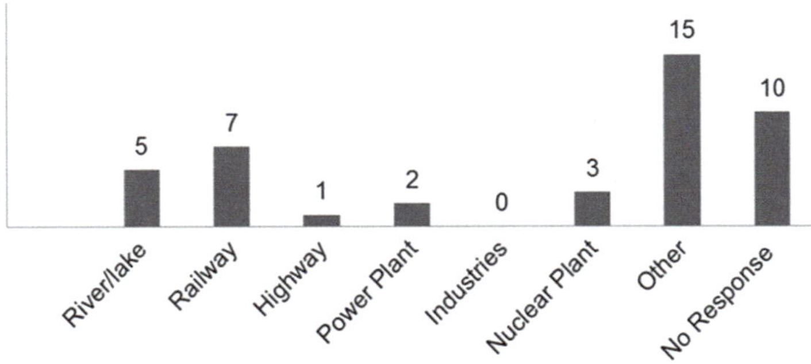

Fig. 4.6 Proximity to potential risks. Numbers on each bar represent number of respondents

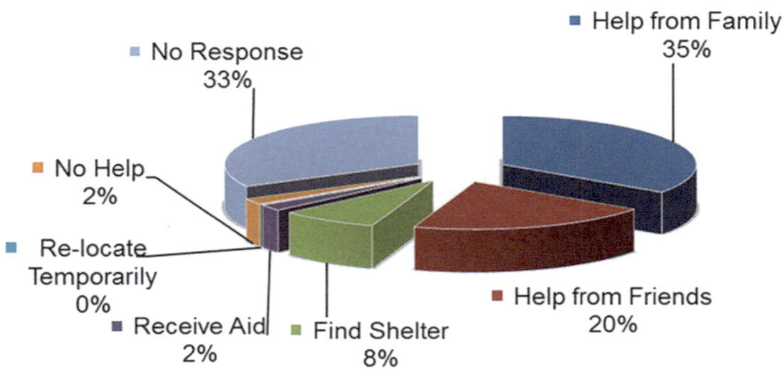

Fig. 4.7 Options and preferences for seeking help when faced with an emergency

4.2.3 Case Study Based on Women's Training Group in Pakistan

This study explores the impact of vocational training for women in rural Pakistan on the enhancement of their community resilience (Khalafzai and Nirupama 2011). It is well understood that sustainability arises from empowering women as households are the smallest unit of communities. Disaster risk reduction and sustainable development are tightly linked with knowledge dissemination. Information and Communication Technologies (ICT) have a great potential for effective learning, knowledge dissemination, and development worldwide. They provide an overarching enabling platform for sustainable development processes. The study is about the Community Technology Learning Centers (CTLC), an innovative project commissioned by the Government of Pakistan to empower marginalized rural women in sixteen districts of the country. The impact of the project and its comparison with a similar intervention in Uganda are some of the issues discussed here in order to illustrate and validate the hypothesis that ICT greatly contribute to the process of women empowerment. All of the above is deeply connected with sustainable development, communities, their resiliency, and their ability to cope better in the event of disasters.

Former Secretary General of the United Nations, Kofi Annan (Annan 2003) made a remark during his tenure that said:

> New ICT are among the driving forces of globalization. They are bringing people together, and bringing decision makers unprecedented new tools for development. At the same time, however, there is a real danger that the world's poor will be excluded from the emerging knowledge-based global economy.

As women continue to be assigned the jobs with the lowest skilled level of work and lowest remuneration (Morgan et al. 2004) in many places in the world, ICT is helping in empowering women and hence make them resilient. The process of empowerment is dynamic and multidimensional in nature, and various approaches and frameworks have been developed by academics and practitioners to assess it. Increased women's access to and control over productive resource and greater participation in decision making processes will bring about empowerment for them (Johnson 2005; Murthy 2002). Women are not simply a group amongst several disempowered groups such as, poor, ethnic minorities, disable individuals, etc. Social relations such as household power relations are central to women's disempowerment. Training in ICT has contributed to expanding income generating opportunities by providing new forms of employment opportunities in Nakaseke (Mijumbi 2002). For example, the Credit Society of Medchal in Andhra Pradesh, India has been training its women members on digital literacy and encouraging them to start computer training centers as business propositions (Subrahmanian 2003). These efforts have eventually led to more resilient and sustainable households.

The Pakistan government was one of the first in the world to set up the National Commission for Human Development (NCHD) devoted to achieving the Millennium Development Goals (UN Millennium Summit 2000). In 2004, sixteen CTLC in sixteen remote districts were established across the country (Fig. 4.8), particularly focusing on improving the living conditions of rural women. The local governments provided their buildings/spaces, financial grants, and political and moral support. Microsoft supplied the hardware, software and a curriculum designed to promote digital literacy. For the purpose of sustainability and accounting for the needs of rural women, the curriculum was taught by local women trained as master trainers. While imparting basic digital skills, the master trainers were also engaged in various activities to instill necessary life and employment skills among the participants. The concept of 'access to resources' (Wisner et al. 2004) can be seen here in applied form, validating the theory of disaster risk management.

The women also received training in data transcription and were engaged (paid work) in a data entry project, which not only provided them with an opportunity to earn a decent livelihood, but also offered an extraordinary chance to put their skills to practice. The CTLC project also imparted training to the female school teachers of the local governments. In 3 years, over 2500 women were successfully trained. A follow up study was conducted by the first author to assess the impact of the intervention. The data was collected during 2006 and analyzed in 2009.

Fig. 4.8 The 16 remote districts shown in white spots where the Community Technology Learning Centres (CTLC) established across Pakistan in 2004 (Khalafzai and Nirupama 2011)

In order to assess the impact of the project, two structured questionnaires were administered (Appendix 3). The first questionnaire was directed towards CTLC graduates, while the second was developed to obtain responses from the graduates' parents/relatives/friends. To measure empowerment, the questionnaires incorporated a number of variables that were developed in accordance with the Empowerment Framework (Chen 1997). The framework measures empowerment by categorizing different variables into four conceptual pathways: (i) the material pathway covering the changes in access to or control over material resources such as level of income; (ii) the cognitive pathway would be concerned with how far participants' skills, knowledge and awareness of the wider environment have changed; (iii) the perceptual pathway included changes in self-confidence, self-esteem and vision of the future as well as recognition by other; and (iv) the relational pathway pertained the impact that an intervention may have had in changing decision-making roles, dependence on others and mobility.

National language, Urdu was used to reduce language barriers, as different dialects and languages are spoken in the study area. For accuracy and consistency, the data was collected through personal interviews. Out of the total number of project participants (N = 2500), a random sample of sixty eight (n = 68) respondents was selected from seven districts in four provinces for statistical analysis. The random sample of smaller number was chosen for the purpose of feasibility, and age and time constraints. Generally, the participants were between the ages of 16 and 40 years, with diverse socio- economic and cultural backgrounds and education levels ranging from middle school to postsecondary level. The authors, however, would like to expand this research and analysis to at least two more random samples.

The survey results revealed that about 87% of respondents agreed that they got their work opportunities as a result of the ICT training they received. Data transcription work was found to be in demand and paid a decent salary. The analysis provides ample evidence to suggest that the ICT skills assisted the participating women in expanding their economic independence, security, and resiliency. For instance, a trained woman received an ICT related job with the local government with a decent starting salary of approximately $80 USD per month. She also found additional part time work that paid her about $40 USD per month.

An overwhelming 73% agreed to high-very high increase in knowledge and information, only 27% observed a ‘medium’ level change, and interestingly, no one chose the ‘low’ option. The increase in amount of knowledge and information helped changing self-perception of the women. To some extent, it also brought awareness of the implications of coercion amongst them. 83% CTLC graduates agreed that they significantly improved their lives, 11% felt that their lives improved just a little, and 6% thought that their lives were not at all improved.

The findings of the study suggest that the life skills learnt by the women noticeably increased their sense of rights and duties, both as citizens and as members within their families, thus improving their social relations. The theoretical concept of coping capacity in disaster management literature (Wisner et al. 2004) is validated here. About 40% responses from the parents/relatives/friends of the participating women suggested ‘very high’ level of positive change, 47% said ‘high’, and only 13% were a ‘medium’. Two main causes for the enhancement of confidence level were the quality of the training and exposure to ICT.

In regards with *capacity or ability* improvement, majority (88%) of parents/relatives/friends of the participating women said that they had observed ‘a lot of change’, only 12% reported ‘little change’. Enhancement of capacity/ability contributed to participating women’s economic independence, expansion of mental spaces, emergency preparedness, and empowerment. About 81% said that she was more conscious and sensitive to her rights, and 19% indicated ‘little change’. The finding suggested that easy access to information through the Internet and worldwide web played crucial role in reducing women's physical barriers. Participating women were asked whether they would be allowed by their parents/husband/guardian to move to another town to avail a job opportunity. Only 17% said that they would accept the job offer in another city, while 28% indicated that they would accept the job but could not move to another city, and 55% said that they would accept the job offer but they would not be allowed to move to another city alone. Contrary to other variables, the finding suggests that women were not able to significantly break the physical impediments, primarily due to the socio-cultural predicaments. However, there is some evidence suggesting that the ICT did help some graduates in acquiring jobs in nearby towns. The women who commuted from nearby villages to the CTLC while travelling 2–3 h daily to attend the training reflects their keenness to learn new technologies and the travelling in itself was an expansion in their physical space.

Similar to the Pakistan case discussed above, in 2002 the government of Uganda began implementing an ICT policy, and offered rural women direct access to information they needed to improve their productivity and socio-economic status. The project consisted of one introductory and three content sections. The introductory section comprised of a guide to using the computer and the CD-ROM; Section one, entitled 'Starting with what we have', emphasized the need to identify what they have and what they can build on; Section two was about making money from a product or service; and Section three was about expanding business opportunities.

The CD ROM project was conceived with a focus on broader issues having national priority (Mijumbi 2002). Table 4.5 describes a brief comparison of both Pakistan and Uganda initiatives. It is noteworthy that both the projects adopted the approach of 'women in development', indicating that empowerment of women is imperative to sustainable development, resilience building, and disaster preparedness. Neither CTLC project nor the CD-ROM project questioned the sources or nature of women's subordination and oppression, while accepting existing social structure in a society. Encouraging an understanding and implementation of the concepts of social relations and structure of domination seem to have taken precedence over strategic integration of women into development. Women always have been important economic actors in their societies and that the work they do both inside and outside the household is central to the maintenance of those societies (Rathgeber 2007).

It is clear from the above example that women who participated in the CTLC project obtained jobs and started micro-enterprises. They are now more confident, better informed, conscious about their rights and surroundings, and interested in accessing virtual information sources. They have improved capacity and keenness to acquire new occupations and expanded their socioeconomic, physical and political spaces, realized their potential, and hence become more resilient.

Table 4.5 Comparison of the CD ROM Project in Uganda and the CTLC Project in Pakistan

	Differences	
Similarities	CD ROM project of Uganda	CTLC project of Pakistan
Similar objectives: to impart ICT training to empower and underserved rural women with localized sustainable content	Broader scope and funding base	Pilot project by NCHD
	Differed strategies were employed to achieve similar objectives	Focus on imparting quality ICT skills and life skills
Rural women participants	Encouraging small business and increasing food productivity, thus promoting sustainable development	Focus on imparting quality ICT skills and life skills
Few ICT training facilities with low coverage		Aimed at data transcription
Lack of basic English language skills required to learn digital technologies	Did not rely on the internet and web applications	
Women are keen to learn digital technologies	Attempted to engage women in telecentres	
Trained women were found to be training fellow women		
By employing ICT skills, the women of both projects have improved their living conditions and earnings through ICT jobs and enterprises		

4.3 Resilience of the Built Environment

Resilience of the built environment is a vital component in our overall resilience. A two part case study by Nirupama et al. (2014) and Sharma et al. (2014) of the 2003 hydrometeorological hazard in the Indian Himalayan region illustrates the point. A detailed version of this example is given in Chap. 1 in Landslides section. During June 14–17, 2013, a rare weather system converged over critical infrastructure. A synoptic situation of cloud formation and a cloud burst with precipitation higher than during a normal period occurred, followed by heavy riverine flooding triggered landslides in the region. The media (BBC, CNN) reported more than 5700 deaths and damage estimates reaching $500 USD million. An area covering 95,830 km^2 (400 villages) was destroyed, 1000 bridges and 695 water distribution plants damaged, 600 villages completely cut off from roads, and 70 hydroelectric stations and 505 dams damaged. The ancient and pious Hindu shrine in the town of Kedarnath in the region experienced extensive growth in the infrastructure due to its popularity among Hindu pilgrims. Four holy sites in the impact area of Uttarakhand (Kedarnath, Badrinath, Gangotri, and Yamnotri) are visited by millions of Hindu pilgrims every year. Figures 4.9 and 4.10 illustrate a comparison of the town of Kedarnath in India 50 years ago and present. In Kedarnath alone, the pilgrim population has been growing at the rate of approximately 10 fold in 50 years (Sacred Yatra 2013).

Fig. 4.9 The region of Kedarnath about 50 years ago (By Not known [Public domain], via Wikimedia Commons)

Fig. 4.10 Kedarnath before the landslide of June 2013 (http://www.sacredyatra.com/kedarnath-pictures.html)

The extent of damage and the widespread and long term disruption of essential services occurred due to a complex interaction between weather variations and human activities pertaining to deforestation and unplanned growth in infrastructure. About 10 million people living in that area are extremely vulnerable in the face of similar disasters in a difficult terrain such as the Himalayan Mountains where rescue operations and rehabilitation efforts are extremely difficult to carry out (Dhobal et al. 2013). Tourists count as vulnerable groups that risk their lives by staying in shantytowns and cheat hotels where construction codes are practically ignored. If another major event similar to this one were to occur in coming decades, the potential loss of life may be 10 fold as compared to the June 2013 event. The damage to infrastructure could be about 6 fold for a similar event.

4.4 Decision Support Tool For Estimating Resilience

It is apparent that the need for the integration of disaster resilience management into planning, design and operational policies is strong. Recently, researchers (Bruneau et al. 2003; Chang and Shinozuka, 2004; Cutter et al. 2008) are focusing on merit in defining resilience quantitatively. All work done on the quantification of resilience to date, however, has used a static resilience measure; that is a single measure calculated over the duration of the disaster (Bruneau et al. 2003). Simonovic and Peck (2013) are the first to quantify resilience dynamically in time and space. They calculate the metric using simulation linked to a geographic information system (GIS) for temporal and spatial analysis. A dynamic resilience metric allows for prioritization of regions and systems that require adaptation upgrades. It also allows for the comparison of adaptation options that improve community resilience and the functioning of critical facilities in the event of a disaster.

The concept of ResilSIM, a web-based decision support tool (with mobile access), has been proposed by Irwin et al. (2016) for estimating the dynamic resilience of an urban center to hydro-meteorological events. The basic concept of ResilSIM is based on the metric developed by Simonovic and Peck (2013). The tool uses fundamental hydraulic principles to simulate hydro-meteorological events under climate change scenarios in conjunction with publicly accessible spatial datasets to estimate the resilience metric. The users will be able to virtually employ different adaptation measures and assess how they upgrade or downgrade urban resilience, thereby assisting decision makers in selecting and prioritizing community upgrades and protection measures. Data requirements are determined based on publicly accessible data of the municipality that the tool is designed for. ResilSIM must be programmed separately for individual municipalities.

ResilSIM is designed to use publicly available datasets required for the computation of resilience and the potential for the implementation of various adaptation options. A demonstration of a generic methodology for the ResilSIM has been shown for two major urban centers: London and Toronto, Ontario, Canada. Both cities are located in the Great Lakes-St Lawrence lowlands climate region of Canada. The regional climate is characterized by prevailing winds from the West, humid air from the Gulf of Mexico and cold, dry air from the North in addition to the presence of the Great Lakes and their interactions with the lower atmosphere (USEPA 2012). Lake effect precipitation is common during the fall and winter seasons (Lapen and Hayhoe 2003; Sousounis 2001), and convective rainfall and thunderstorms are typical of the summer season (Ashmore and Church 2001). Although both cities experience similar climates, they are subject to different types of flooding. London is most susceptible to riverine flooding, while Toronto is prone to a combination of riverine and urban flooding. The latter is caused by high intensity precipitation events that overwhelm the capacity of the municipal drainage system, resulting in the pooling of floodwater on the impervious surface. An explanation of the flood generation processes is provided for each city.

4.4.1 Data Requirement

The resilience metric combines several performance measures that represent the physical, social and economic impacts to an urban system. The data used to compute the physical performance measure are obtained from a few sources including the City of London, the Municipal Property Assessment Corporation (MPAC 2016), the City of Toronto (2016), and CanVec+ (2016). The data required to compute the social and economic performance measures are Census profiles acquired from Statistics Canada. These datasets were chosen because they are frequently available from municipalities in the Province of Ontario; CanVec + and Canadian Census profiles are available across the country. Table 4.6 provides a summary of the required and publically available datasets.

The Municipal Property Assessment Corporation (MPAC) is a non-profit organization funded by Ontario municipalities. Its objective is to assess and classify all properties in compliance with the Ontario government's Assessment Act. The City of London has derived building "envelopes" (outlines) from topographic information and assigned land use classifications and descriptions supplied by MPAC to all of the properties within their jurisdiction. This type of dataset is very useful to the ResilSIM application; however, it is not made available by all Ontario municipalities such as the City of Toronto. In the physical system of the ResilSIM tool, the buildings that are assigned to all commercial, industrial, and residential land use categories are retained for analysis. The critical facilities with the following descriptions are also retained and used in model development: ambulance stations, fire halls, hospitals, police stations, and schools. For the City of Toronto, building envelopes for critical facilities are obtained from the Open Data source that is available online. For detailed information on ResilSIM, see the Blue Book series publication by the Institute for Catastrophic Loss Reduction, London, Ontario (Irwin et al. 2016).

The engineering infrastructure data that are employed in the physical component of the model are obtained from the CanVec + catalog that is produced and maintained by Natural Resources Canada (NRCan). CanVec + is a digital cartographic reference product that is comprised of a variety of topographic entities in a vector format. With CanVec+, NRCan aims to provide uniform topographic data across Canada that is updated frequently to offer the best available resources at the highest resolutions. The entities are available for download from: ftp://ftp2.cits.rncan.gc.ca/pub/canvec+ (last accessed 2016 January).

The data used for the calculation of the social and economic performance measures are Census profiles acquired from Statistics Canada. The Canadian Census program provides a statistical representation of the country's socio-economic

Table 4.6 Summary of data used in the development of ResilSIM

System	Dataset	Format	Source
Physical	Buildings (landuse)		
	Commercial	shape-file	MPAC, City of London
	Industrial	shape-file	MPAC, City of London
	Residential	shape-file	MPAC, City of London
	CRITICAL FACILITIES (description)	shape-file	
	Ambulance Station	shape-file	MPAC, City of London, City Toronto
	Fire Hall	shape-file	MPAC, City of London, City of Toronto
	Hospital, private or public	shape-file	MPAC, City of London
	Police Station	shape-file	MPAC, City of London, City of Toronto
	School (elementary or secondary, including private)	shape-file	MPAC, City of London, City of Toronto
	Engineering infrastructure		
	Domestic waste facilities	shape-file	CanVec+
	Gas and oil facilities	shape-file	CanVec+
	Industrial solid waste facilities	shape-file	CanVec+
	Pipeline	shape-file	CanVec+
	Pipeline (sewage/liquid waste)	shape-file	CanVec+
	Power transmission line	shape-file	CanVec+
	Railway	shape-file	CanVec+
	Road segments	shape-file	CanVec+
	Transmission stations/lines	shape-file	CanVec+
Economic	Unemployed persons	shape, csv	StatsCan
	Families w/ annual income < $50,000	shape, csv	StatsCan
Social	Age (<6; > 65)	shape, csv	StatsCan
	Single (divorced/widowed)	shape, csv	StatsCan

(continued)

Table 4.6 (continued)

System	Dataset	Format	Source
	Single Parent	shape, csv	StatsCan
	Migrants	shape, csv	StatsCan
	Allophones	shape, csv	StatsCan
	Immigrants	shape, csv	StatsCan
	Visible Minorities	shape, csv	StatsCan
	Persons w/o high school education	shape, csv	StatsCan

Irwin et al. (2016)

environment every 5 years; the last year being 2011. Canadian Census boundaries are available as shape-files for a variety of geographic levels; the smallest of which are the dissemination areas. In the presented research it is recommended to compute resilience for dissemination areas in order to provide the highest level of information detail. The Census profile datasets are stored in comma separated value files (csv-files) that are accessible from: https://www12.statcan.gc.ca/ (last accessed 2016 January). The Census profile data are assigned to their respective dissemination areas by matching identification codes. This function is performed in an ArcGIS environment (https://www.arcgis.com/features/, last accessed 2016 January). Although it is recommended to compute resilience for each dissemination area, other larger Census boundaries (such as Census Tracts) may be used to define the study area in which several resilience metrics are estimated.

In brief and without mathematical details, if adaptive capacities are introduced into the urban system, ResilSIM demonstrates improved community resilience. ResilSIM provides a list of adaptation options that can be applied to improve the system performance with respect to the physical, social and economic units of resilience. The options are listed in Table 4.7 and they are divided into two groups: (i) real-time adaptation measures that are implemented during the flooding event; and (ii) proactive adaptation measures that are implemented in advance of the flooding event.

Real-time adaptation options can be implemented in response to a flood warning that has been issued by the regional conservation authority. In the province of Ontario, regional conservation authorities are responsible for operating flood forecasting models and providing flood alerts to municipalities located within their watershed. Flood forecasting tools use near real-time estimates of precipitation (from rain gauge or radar instruments) as input to hydrologic models to estimate surface water flows and subsequently, accumulated water elevations in drainage basins. Once municipalities are warned of an imminent flooding event, government

Table 4.7 Adaptation options available on ResilSIM

	Adaptation option
Real-time	Implementation of temporary dyking measures (ex. sand bags) to maintain roads and access routes to buildings and critical facilities;
	Pumping out of flooded area – divert floodwater to adjacent open areas such as parks;
	Allocation of resources (monetary, technological, informational, and human resources) to clean up after the flooding event;
	Evacuation and relocation of people belonging to vulnerable social and economic groups;
Proactive	Implementation of lot-level flood protection measures to prevent floodwater from entering buildings, thereby maintaining structural function;
	Maintenance of drainage infrastructure (through the removal of debris) in order to optimize drainage capacity and reduce the effective flood depth;
	"Twinning" critical infrastructure (water and power supplies) such that if one infrastructure piece fails in the event of a hazard, there is a secondary source.

Irwin et al. (2016)

officials from several groups including communications, fire, paramedic, police, public health, and transportation services must be available to act in accordance with their local emergency response plan. Other municipal departments, namely those responsible for critical engineering infrastructure such as power, water supply, water treatment, and solid waste management are often assigned responsibilities during the recovery phase of the flooding event that typically begins 2 days after the disaster has ceased. The engineering departments are also most likely involved with the proactive adaptation options that are implemented in advance of the flood. The ResilSIM tool may be employed once a flood alert has been issued in order to select the real-time adaptation options that result in the highest value of resilience. The tool may also be used to create detailed emergency response plans that outline the best real-time adaptation options to be implemented for different regions of the city. Alternatively, ResilSIM can be used to select from the best proactive adaptation options.

An explanation of how each adaptation option affects the resilience calculation is provided below:

(i) Temporary diking measures (ex. earth berms and sand bags) are used to prevent water from flooding roads and access routes to buildings and critical facilities. This, in turn, improves building function that is accounted for using several different impacts of the physical performance measure (length of road inundated by the flood and the economic damages incurred by critical facilities as well as communications, industrial, and residential buildings). This measure is more easily employed in municipalities that are subject to riverine flooding. Sand bags may also be employed as flood proofing measures that protect structures and when used in this context, are accounted for by the physical performance impacts that measure the magnitude of flood inundation of the buildings, critical facilities and engineering infrastructure.

(ii) Pumping out floodwater from vulnerable regions and diverting it to open areas such as parks and stormwater management ponds is an adaptation option that reduces the magnitude of flood extent and inundation over a region. Since all impacts of the physical, social, and economic performance measures are driven by the magnitude of flood inundation and flood extent, this adaptation option may have a significant effect on the overall value of resilience.

(iii) The allocation of resources (monetary, technological, informational, and human resources) to clean up after the flooding event increases the rate of recovery of an urban system to a normal level of functioning. When more resources are assigned to a certain activity (ex. deployment of personnel, equipment, and financial support for dykes), the rate of improvement to the relevant impacts of system performance would be higher and consequently the community would be more resilient. The ResilSIM tool applies different rates of recovery to the impacts of system performance depending on the proportion of resources that are made available to the region.

(iv) Evacuation and relocation of vulnerable social and economic groups requires the establishment of reception centres such as schools and community halls that act as a safe and protected shelter for evacuees to be transported to and reside in during the flood. Police services are typically responsible for evacuation and for protecting the properties of those who are evacuated until the flood recedes; looting is common during this time. The "evacuation and relocation" adaptation measure directly affects all impacts of the social and economic performance measures.

(v) Lot-level protection measures may be employed proactively to prevent floodwater from surrounding and entering buildings, thereby maintaining structural function and integrity (ex. the physical performance impacts that measure the magnitude of flood inundation of the buildings, critical facilities, and engineering infrastructure). Lot-level protection measures include the installation of backwater valves and downspout disconnections (that may be done in conjunction with the installation of a rain barrel) that mitigate basement flooding due to sewer surcharge events; in addition to lot re-grading and the sealing of windows and foundation cracks to prevent basement flooding attributed to infiltration and overland flows.

(vi) Maintenance of drainage infrastructure (through the removal of debris) may be conducted to optimize drainage capacity and reduce the effective flood depth. Municipalities may wish to implement annual programs where drainage infrastructures, particularly structures located in regions that are vulnerable to flooding, are maintained so they can operate at their full potential. This may be accomplished through the ResilSIM tool using fuzzy set theory; a fuzzy membership function is used to represent the level of infrastructure maintenance or alternatively, the proportion of designed infrastructure capacity that is available for conveying stormwater. By following the methodology proposed herein, this adaptation option can only be employed for cases of urban flooding.

(vii) "Twinning" of critical infrastructure (water and power supplies) means that there is a backup or secondary source in the event that one infrastructure piece fails as a result of the hazard. It is an example of building redundancy into the urban system. Using the ResilSIM tool, if one critical infrastructure entity is inundated by the flood and there is a secondary source that can provide the same services within the region, then there is no loss in system performance with respect to the critical infrastructure that is inundated.

In summary, the term resilience literally means "to jump back" in Latin, implying a capability to return to a previous state. In the context of disasters, returning back to exactly pre-disaster state is practically impossible due to physical, social, psychological, and environmental changes that would have taken place during the unfolding of the event. In post-disaster reality, everyone including citizens, local, provincial, and federal governments (depending on the severity of the event), essential service providers such as electricity, telecommunication, transportation, health care, etc. need their ability and capacity to bounce back to new normal as soon as they can. Therefore, it is imperative that people are part of the solution which is to enhance resilience and build capacity to combat adverse events. Resilience, being a measure of how well people and societies can adapt to a changed reality and capitalize on the new possibilities offered, the definition of resilience must embody the notion of adaptive capacity (Klein et al. 2003; Paton 2006).

The concept of ResilSIM: a web-based decision support tool used to estimate urban resilience in the event of a flood can be put to practice to assist decision makers (engineers, planners, and government officials) in selecting the best options for integrating adaptive capacity into a community in order to protect against the hazardous impacts of a flooding event. The proposed first generation of the model is employed in two Canadian cities: (i) London, Ontario to estimate the resilience corresponding to riverine flooding events; and (ii) Toronto, Ontario to estimate the resilience corresponding to urban flooding events. The current structure of the tool is quite basic; however, it provides a foundation for other researchers to improve upon.

4.5 A New and Comprehensive Approach to Evaluate Resilience

In light of rapidly increasing number of natural disasters around the world, some of the commonly raised and compelling questions include: how communities can be educated with relevant and useful knowledge on disaster resilience so they would want to participate in the discussion; how can people make a difference in minimizing damage from disasters for themselves; is people's perception important in building coping and adaptive capacity (Nirupama and Maula 2013); and finally, how to incorporate these issues in planning tools (Simonovic and Nirupama 2005; Welle and Birkmann 2015; Irwin et al. 2016). A resilient community is one where people take proactive steps to engage with local leaders and community networks. To help communities enhance their adaptability, coping capacity, and overall resilience for natural disasters, it is important that people understand the risks that may affect them so they can be prepared to protect their assets and livelihoods (Armenakis and Nirupama 2013).

The method discussed in this section allows for estimation of disaster resilience by integrating quantitative and qualitative approaches. The methodology is first of its kind, and it is hoped that this approach (Agrawal et al. 2017) will be adopted for the development of a unified strategy for disaster resilience in Canada.

4.5.1 A Canadian Case Study to Demonstrate the Approach

Relationships between the makeup of the communities, their priorities and general vulnerabilities, and the geomorphology of the region are the basis of this methodology. To incorporate people's perception, we have used survey data collected from four different locations in the City of Brampton in the Greater Toronto Area (GTA) of Canada (Fig. 4.11). The survey data reflects how people perceive risks from natural hazards, how they would cope in emergencies, and how they generally

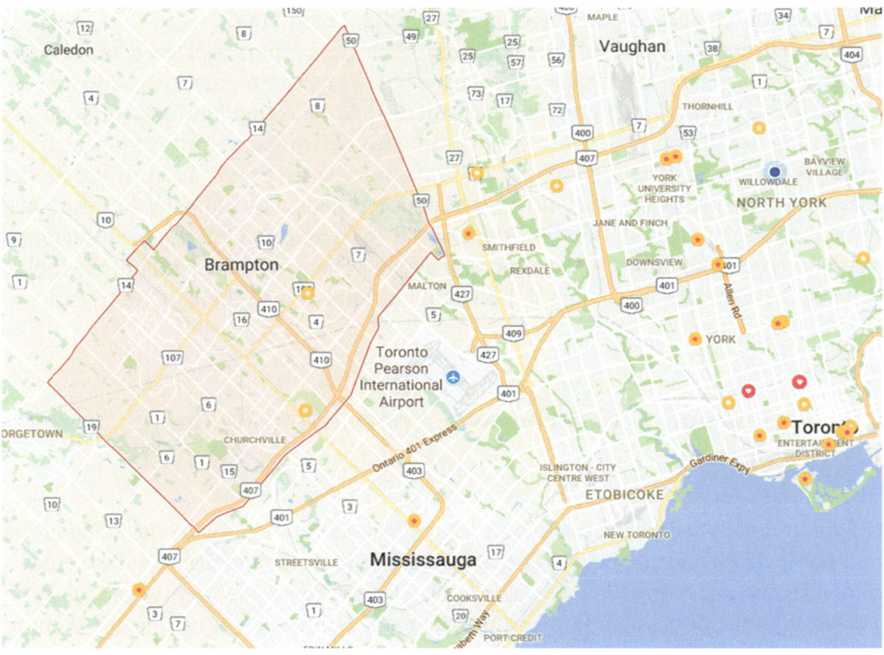

Fig. 4.11 (top) part of the Greater Toronto Area (GTA) for perspective purposes showing the City of Mississauga and the City of Toronto in the south and southeast of Brampton; (bottom) the four survey locations in the City of Brampton in the GTA. Survey location 1: South Fletcher's Sportsplex community centre, Survey location 2: Brampton multicultural community centre; Survey location 3: New Birth Tabernacle, a non-denominational faith gathering center; and Survey location 4: MJ'S BBQ & Suya, a local restaurant

participate in their local environment. People's perceptions are treated as a representation of the entire City for demonstrating the methodology leading to the qualitative or perceived assessment of community resilience (Table 4.8). We have used the 2011 census of Canada for demographic information; Municipal Property Assessment Corporation (MPAC) average property values; slopes and terrain of the region; and landuse, especially for critical infrastructure and facilities to represent the quantitative or objective assessment of community resilience (Table 4.9). The GIS is used to carry out data processing and analysis according to Dissemination Area (DA) map of the study area. Figures 4.12, 4.13, 4.14, 4.15, 4.16, 4.17, 4.18, 4.19 and 4.20 show step-by-step processing of the dataset.

Equations (4.1, 4.2) are developed to estimate qualitative aspect of resilience.

$$\begin{aligned} \textit{Lack of Resilience}_{Perceived} = \sum 20 \times \textit{Exposure}_{Perceived} + 40 \\ \times \textit{Susceptibility}_{Perceived} + 40 \\ \times \textit{Lack of Coping Capacity}_{Perceived} \end{aligned} \tag{4.1}$$

$$\textit{Resilience}_{Perceived} = 1 - (\textit{Lack of Resilience}_{Perceived}) \tag{4.2}$$

Here, various parameters, namely *Exposure*, *Susceptibility*, and *Lack of Coping Capacity*, are appropriately extracted from the responses to the survey questions by assigning binary values to them. These parameters are assigned weights according to their influence. Each parameter is comprised of several variables. Table 1 explains

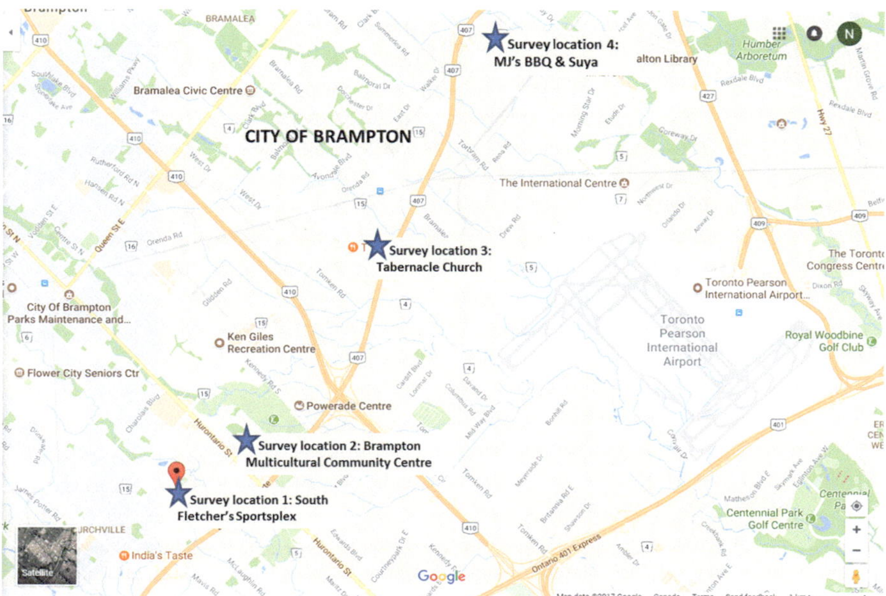

Fig. 4.11 (continued)

Table 4.8 Information extracted from the four surveys toward perceived resilience along with assigned weights and binary values

Parameter	Guidance to assign binary values 0 or 1 based on responses to survey questions			
Exposure	**Question**	**Binary value = 0**	**Binary value = 1**	**Remarks**
Weight = 20%	I live near hazardous situation such as river, chemical plant etc.	If false	If true	Add all binary numbers to get total *Exposure*
Susceptibility Weight = 40%	**Question/variable**	**Binary value = 0 if response is as below**	**Binary value = 1 if response is as below**	**Remarks**
	My home is	Owned	Rented or other	Add all binary numbers to get total *Susceptibility*
	# members in household (crowdedness)	< 4	>4	
	Language (English)	very well/ good	moderate/ poor/ blank	
	Employment	Full time/self employed	Part time/ retired	
	Job satisfaction	Very satisfied/satisfied	Somewhat satisfied/not satisfied/ blank	
	Age	>20 and <65	<20 and >65	
	Disability	No	Yes	
Lack of Coping Capacity Weight = 40%	**Question/variable**	**Binary value = 0 if response is as below**	**Binary value = 1 if response is as below**	**Remarks**
	Education	College and higher	Less than college	Add all binary values to get total *Lack of Coping Capacity*
	Income	>50 K	<50 K	
	Transportation	Personal vehicle	Public/ ride share	
	Social network	Very important	Important/ somewhat/ other	
	Disaster experience	Yes	No/ blank	
	Disaster preparedness option	Family/ friend	Public shelter/ blank	
	Voted in past election	Yes	No/ blank/ n/a	

Note: Survey locations 3 and 4 are being combined for convenience. Thiessen polygons are developed for all survey locations to delineate representative geographic regions. This step facilitates data processing, analysis, and visualization based on Dissemination Areas falling within the Thiessen polygons

Table 4.9 Objective resilience related parameters and variables

Parameter	Guidance to assign binary values 0 or 1			
Exposure Weight = 20%	**Potential hazardous situations**	**Binary value = 0**	**Binary value = 1**	**Remarks**
	Highways	If no exposure to potential hazard	If within a 800 m buffer zone of any of the potential hazardous situations	Add all binary assigned values to get total Exposure
	Railway tracks			
	River and creeks			
	Industries			
	Oil and gas storage facilities and pipelines			
	Dump sites – stockpiles			
	Low lying areas (terrain/ slope)			
Susceptibility Weight = 40%	**Variables**	**Binary value = 0**	**Binary value = 1**	**Remarks**
	Residence type	Detached/ semi	Rented apartment	Add all binary assigned values to get total Susceptibility
	Age of property construction	Post 1980	Pre 1980	
	Language	very well/ good	moderate/ poor/ blank	
	Employment	Full time/ self employed	Part time/ retired	
	Age	>20 and <65	<20 and >65	
	Disability	No	Yes	
	Property value	> 400 K	≤400K	
Lack of Coping Capacity Weight = 40%	**Variables**	**Assign value = 0**	**Assign value = 1**	**Remarks**
	Education	College and higher	Less than college	Add all binary assigned values to get total Lack of Coping Capacity
	Income	>50 K	≤50K	
	Disaster preparedness: hospital	≤1 km	>800 m away	
	Disaster preparedness:	≤ 1 km	>800 m away	

(continued)

Table 4.9 (continued)

Parameter	Guidance to assign binary values 0 or 1			
	ambulance service			
	Disaster preparedness: health emergency services	≤ 1 km	>800 m away	
	Disaster preparedness: police station	≤ 1 km	>800 m away	

Information extracted from the 2011 census and other sources as given in Table 3

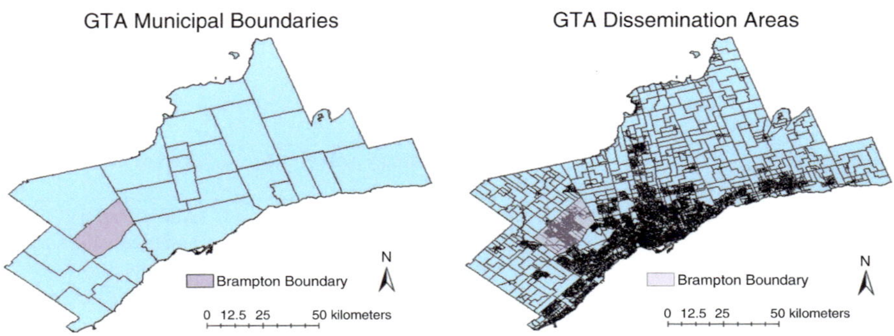

Fig. 4.12 GTA municipal boundaries (L) and dissemination areas (R)

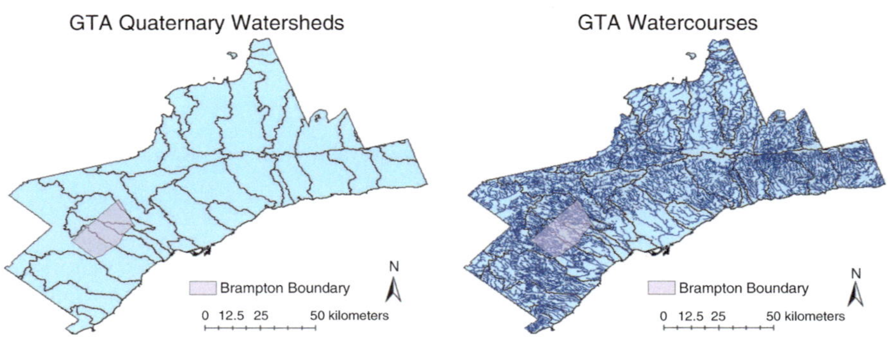

Fig. 4.13 GTA watersheds (L) and watercourses (R)

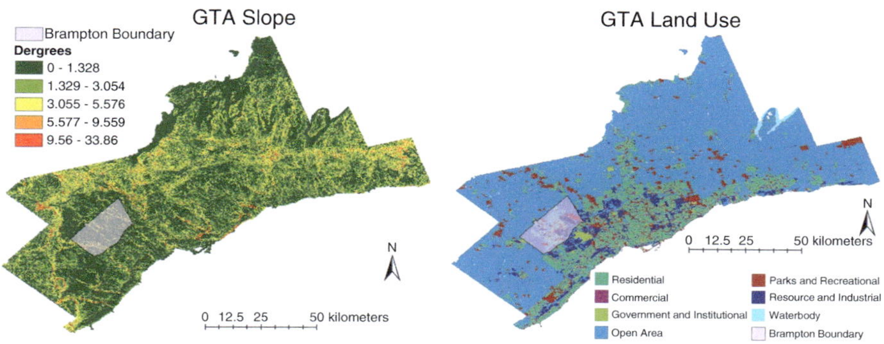

Fig. 4.14 GTA terrain (L) and landuse (R)

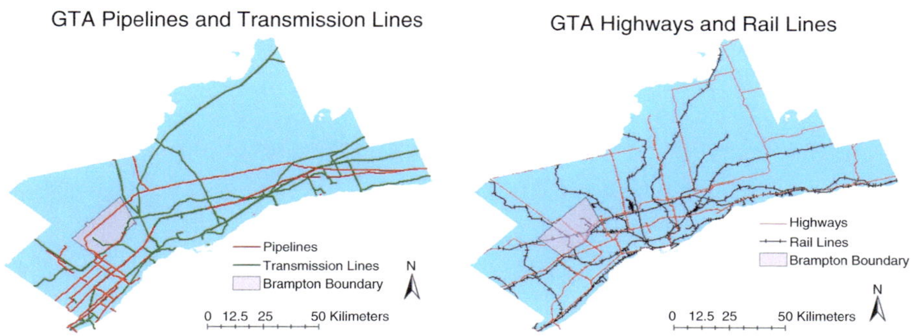

Fig. 4.15 GTA pipelines and transmission lines (L); highway and railways (R)

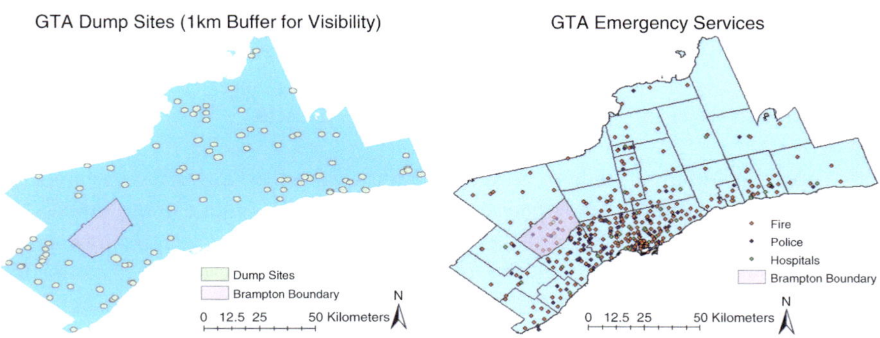

Fig. 4.16 GTA dump sites (L); emergency services (R)

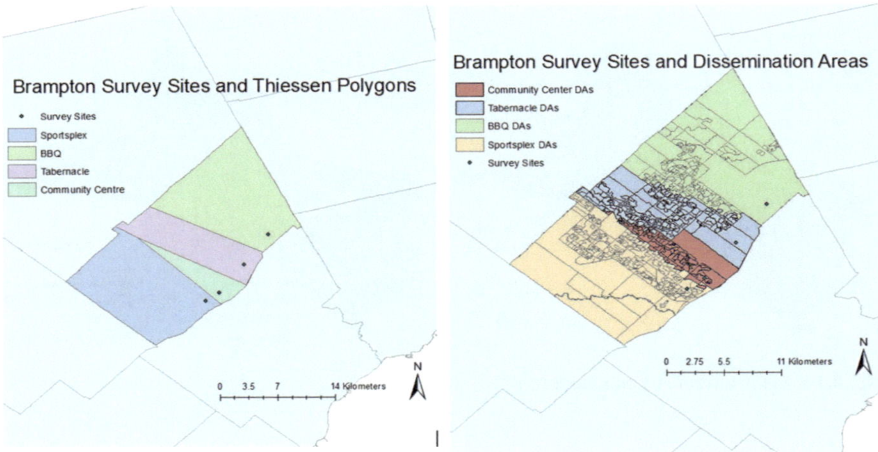

Fig. 4.17 The city of Brampton survey sites and Thiessen Polygons (L); survey sites superimposed on dissemination areas (R)

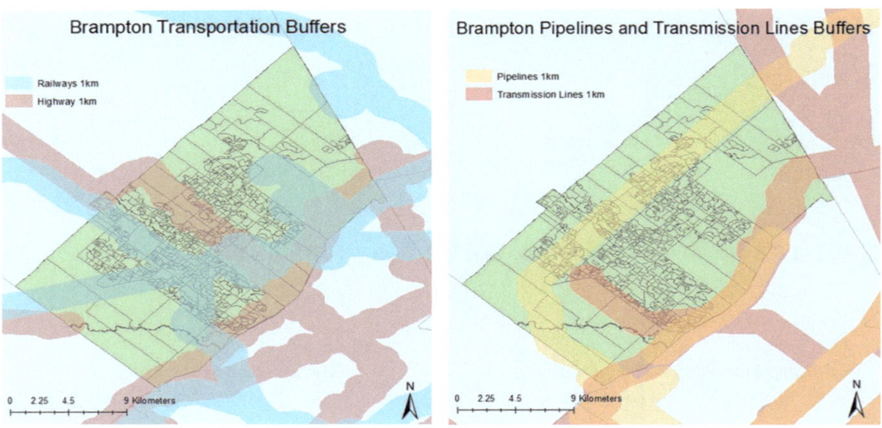

Fig. 4.18 Transportation line buffers (L); Transmission line buffers (R)

the process including a guide to assigning binary values to individual variables as part of the methodology developed to account for people's input in the process of resilience assessment. For example, parameter *Susceptibility* is comprised of seven variables – residence status in terms of owned or otherwise, crowdedness factor based on the number of members in the household, ability to understand English or French language, status of employment, job satisfaction factor, age factor assuming the very young and the very old would be more susceptible than the rest, and

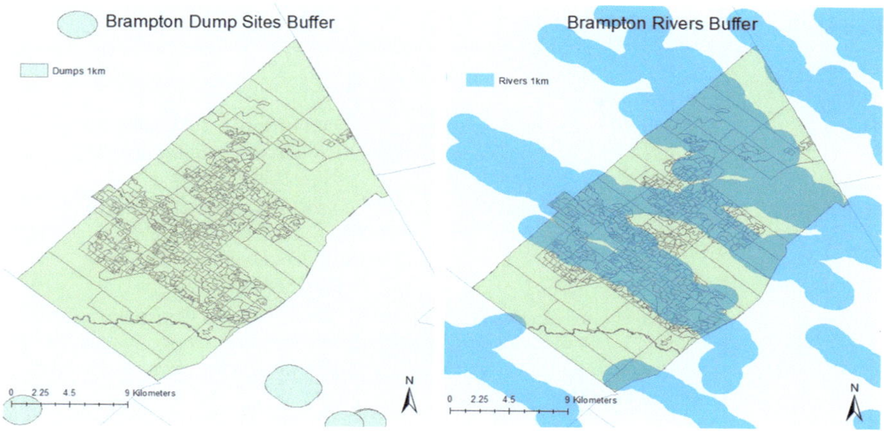

Fig. 4.19 Dump sites buffers (L); Buffers around the rivers (R)

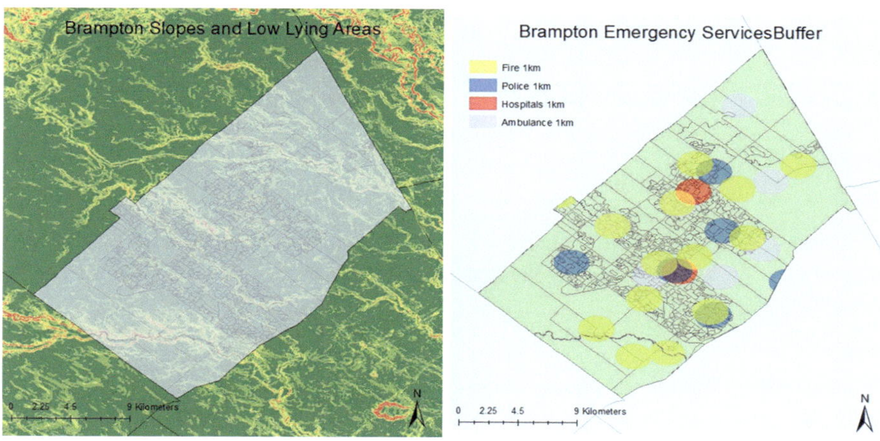

Fig. 4.20 Slopes and low lying areas in Brampton (L); Buffers around emergency services (R)

disability situation. Similarly, *Lack of Coping Capacity* contains following variables – level of education, annual family income, transportation (personal or public), presence or absence of social network support, disaster experience as it impacts perception on disaster preparedness, importance of disaster preparedness, and engagement in local politics as a proxy to willingness to participate in local issues.

Overall resilience is calculated using Eqs. (4.3, 4.4, 4.5 and 4.6), as given below (Table 4.10):

Table 4.10 List of data type and sources

Exposure		
Data	Description	Source
Highways	Major highways in the Greater Toronto Area	DMTI Spatial, 2015 http://geo.scholarsportal.info/#r/details/_uri@=2347499980
Rail Lines	Rail lines in the Greater Toronto Area	Ontario Ministry of Natural Resources, 2012 http://geo.scholarsportal.info/#r/details/_uri@=1862671914
Industrial Sites	Includes: chimneys, cranes, gas and oil facilities, liquid depot/dumps, mines, storage tanks, wells, and wind-powered devices	Canada, Federal Government Open Data Program, created by request, 2016 http://open.canada.ca/en/open-data
GTA Pipes and Transmission Lines	Natural gas pipelines and electrical transmission lines in GTA	DMTI Spatial, 2014 http://geo.scholarsportal.info/#r/details/_uri@=3261862604$DMTI_2014_CanMapRL_Topo_PTL_ALL_PROV
Slope and DEM		DMTI Spatial, 2015 (retired) http://geo.scholarsportal.info/#r/details/_uri@=658779033
Major Rivers	Only major rivers	Provided by Toronto and Region Conservation Authority
Susceptiblity		
Data	Description	Source
Home Ownership	Whether or not primary occupants owns or rents the home	Statscan, 2011 census data via http://dc1.chass.utoronto.ca/census/
Age of Construction	Year residence was constructed	Statscan, 2011 census data via http://dc1.chass.utoronto.ca/census/
Language Skills	Allophone or not	Statscan, 2011 census data via http://dc1.chass.utoronto.ca/census/
Employment	Part time, full time, or retired	Statscan, 2011 census data via http://dc1.chass.utoronto.ca/census/
Age	Older than 65 or younger than 6	Statscan, 2011 census data via http://dc1.chass.utoronto.ca/census/
Disability		Statistics Canada, by request, 2017
Property Value	Greater or less than $500,000	online real estate listings (Remax), geocoded by address in ArcGIS Online
Coping capacity		
Data	Description	Source
GTA Fires Stations		http://geo.scholarsportal.info/#r/details/_uri@=3739967620

(continued)

Table 4.10 (continued)

Exposure		
Data	Description	Source
GTA Police Stations		http://geo.scholarsportal.info/#r/details/_uri@=3739967620
GTA Hospitals		http://geo.scholarsportal.info/#r/details/_uri@=3570906326
Ambulance Stations		Addresses gathered from publicly available information at municipal websites, wikipedia, and Google; geocoded using ArcGIS Online
Income	Above or below $50,000/year	Statscan, 2011 census data via http://dc1.chass.utoronto.ca/census/
Education	University education or not	Statscan, 2011 census data via http://dc1.chass.utoronto.ca/census/
Miscellaneous		
Data	Description	Source
Watercourses	Contains all linear water features down to the scale of small streams	Ontario Ministry of Natural Resources, 2011 http://geo.scholarsportal.info/#r/details/_uri@=73897213
Watersheds	Quaternary watersheds in the GTA	Ontario Ministry of Natural Resources, 2011 http://geo.scholarsportal.info/#r/details/_uri@=45500785
Dissemination Areas	Shapefile of dissemination areas	University of Toronto Census Analyser http://dc1.chass.utoronto.ca/census/
Land Use	Includes: commercial, government and institutional, open area, parks and recreational, residential, resource and industrials, and waterbodies	DMTI Spatial, 2014 http://geo.scholarsportal.info/#r/details/_uri@=2785150059$DMTI_2014_CanMapRL_Topo_LUR_ALL_PROV
Census Tracts	2011 Canadian Census Tracts	University of Toronto Census Analyzer http://dc1.chass.utoronto.ca/cgi-bin/census/2011nhs/displayCensus.cgi?year=2011&geo=ct
GTA Municipalities	Municipal boundaries within the Greater Toronto Area	DMTI Spatial, 2014 http://geo.scholarsportal.info/#r/details/_uri@=4044335176$DMTI_2014_CanMapRL_Streets_MUN_ALL_PROV
Education	locations of schools, from elementary to university	DMTI Spatial, 2015 http://geo.scholarsportal.info/#r/details/_uri@=4062179246

$$
\begin{aligned}
Lack\ of\ Resilience_{Objective} = \sum 20 \times Exposure_{Objective} + 40 \\
\times Susceptibility_{Objective} + 40 \\
\times Lack\ of\ Coping\ Capacity_{Objective}
\end{aligned} \tag{4.3}
$$

$$
Resilience_{Objective} = 1 - (Lack\ of\ Resilience)_{Objective} \tag{4.4}
$$

$$
Resilience_{combined} = Resilience_{Perceived} + Resilience_{Objective} \tag{4.5}
$$

$$
Resilience = \frac{Resilience_{combined} - MIN\ (Resilience_{combined})}{(MAX\ (Resilience_{combined}) - MIN\ (Resilience_{combined}))} \tag{4.6}
$$

4.5.2 Findings – Resilience Maps

Findings are illustrated through a series of maps shown in Figs. 4.21, 4.22, 4.23, 4.25 and 4.26 representing the step-by-step process of calculations of perceived and objective parameters (exposure, susceptibility, and lack of coping capacity) discussed in Eqs. (4.1, 4.2, 4.3, 4.4, 4.5 and 4.6). Figure 4.27 shows the final map of resilience by integrating the perceived and objective resilience. It is important to point out that the method needs improvement from many points of views, but it is a reasonable preliminary attempt to integrate qualitative and quantitative approaches to assess resilience of a community. The methodology can also be applied to disaster risk assessment.

This study was made possible by following: funding contribution from National Science & Engineering Research Council (NSERC) of Canada under the CREATE program; support from Alan Normand, Manager Emergency Management of the City of Brampton for providing logistic support to conduct the survey of the

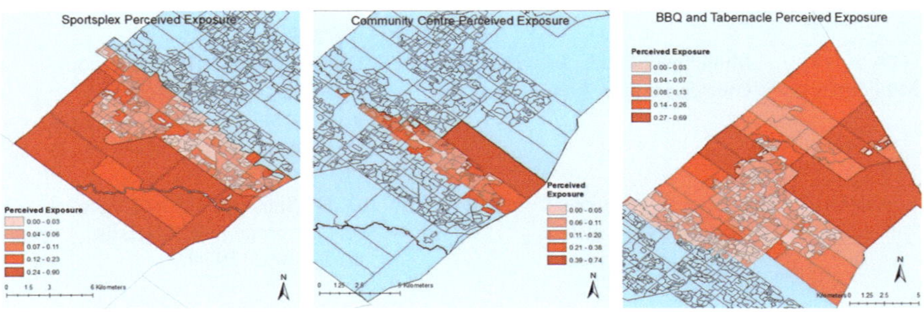

Fig. 4.21 Assessment of Perceived Exposure for the three survey locations. Zero being the lowest exposure

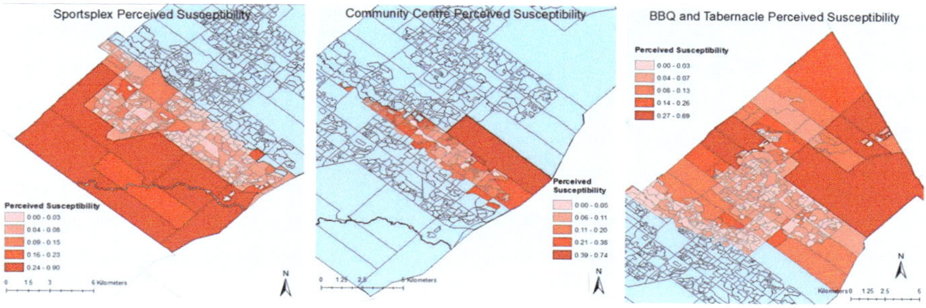

Fig. 4.22 Assessment of Perceived Susceptibility for the three survey locations. Zero being the lowest susceptibility

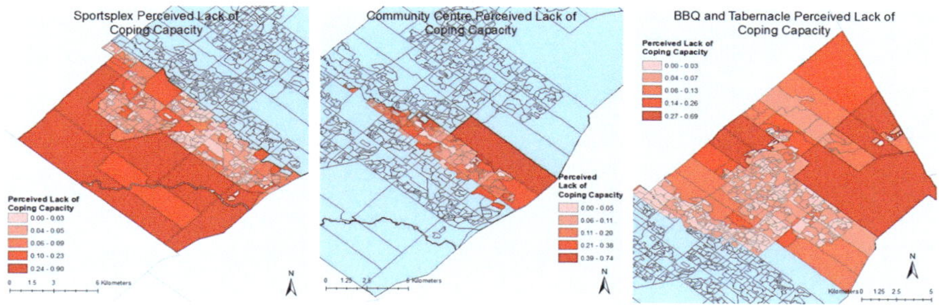

Fig. 4.23 Assessment of Perceived Coping Capacity for the three survey locations. Zero being the lowest value of Coping Capacity

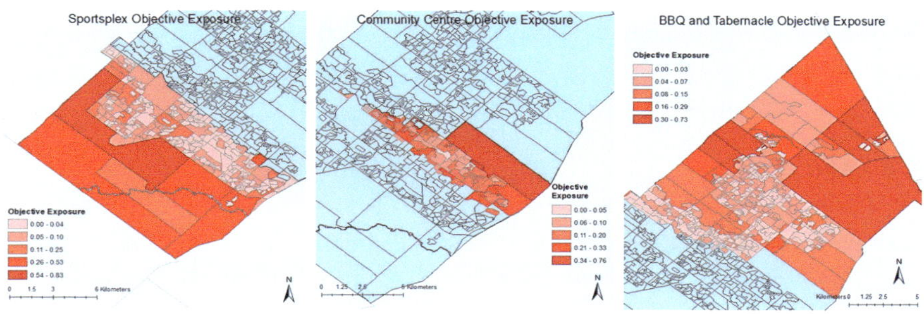

Fig. 4.24 Assessment of Objective Exposure for the three study locations. Zero being the lowest value of Exposure

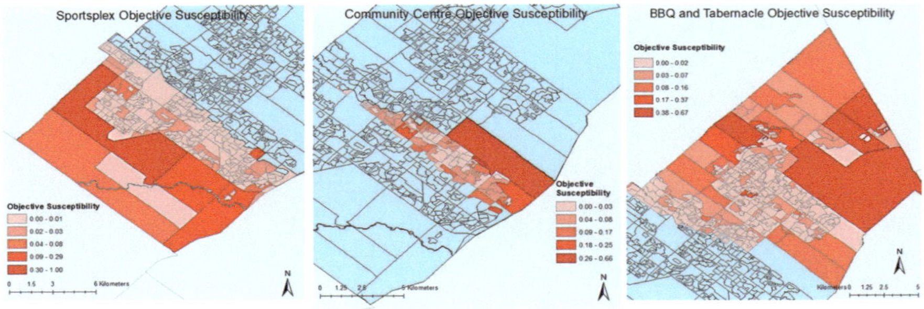

Fig. 4.25 Assessment of Objective Exposure for the three study locations. Zero being the lowest value of Susceptibility

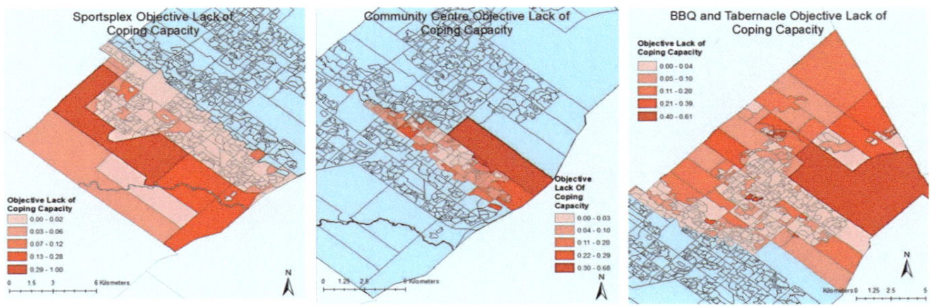

Fig. 4.26 Assessment of Objective Lack of Coping Capacity for the three study locations. Zero being the lowest value of Lack of Coping Capacity

population of Brampton; Judith Jubril and Sarah Maude, Graduate students, MDEM program, York University for their assistance with carrying out the survey; and Mark Elliot, Graduate student, MES program, York University for the processing of the data using GIS and creating final maps.

4.6 Exercise

Identify similarities and differences in various definitions of resilience discussed in this chapter. Analyze your observations in light of, the role of community, economic aspect, and political ideology.

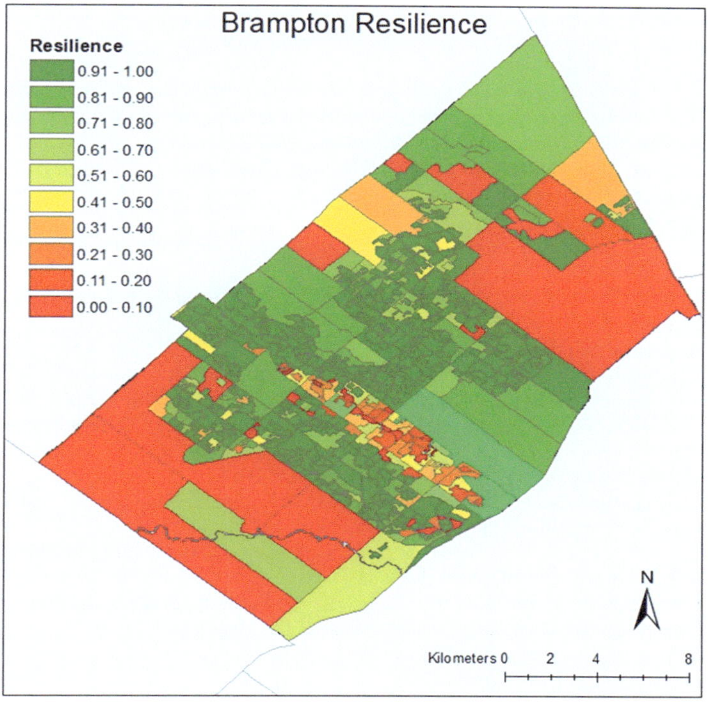

Fig. 4.27 Resilience is measured on a scale of 0 (no resilience) to 1 (highest resilience)

References

Adger WN (2006) Vulnerability. Glob Environ Chang 16(3):268–281

Agrawal N, Simonovic SP, Elliott M (2017) Disaster adaptiveness, resilience, and capacity building by accounting for people's perceptions: a Canadian case study, 7th international conference on flood management, University of Leeds, UK, Sep 5–7 2017

Annan K (2003) Annan calls for bridging digital divide in world's poorest countries. UN News Centre, 17 May 2003

Armenakis C, Nirupama N (2013) Estimating spatial disaster risk in urban environments, geomatics, natural hazards and risk. Taylor & Francis 4(4):289–298

Arnstein SR (1969) A ladder of citizen participation. JAIP 35(4):216–224

Ashmore P, Church M (2001) The impact of climate change on rivers and river processes in Canada. Geol Surv Can Bull 555

Becker JS, Johnston DM, Paton D, Ronan K (2009) Community resilience to earthquakes: understanding how individuals make meaning of hazard information, and how this relates to preparing for hazards. Proceedings of 9th New Zealand Society for Earthquake Engineering Conference 3–5 Apr 2009

Birkmann J (2006) Basic principles and theoretical concepts. In: Birkmann J (ed) Measuring vulnerability to natural hazards: towards disaster resilient societies. United Nations University Press, New York, 524p

Birkmann J, Cardona OD, Carreño ML, Barbat AH, Pelling M, Scheiderbauer S, Kienberger S, Keiler M, Alexandar D, Zeil P, Welle T (2013) Framing vulnerability, risk and societal responses: the MOVE framework. Nat Hazards 67:193–211

Boin A, Comfort LK, Demchak CC (2010) The rise of resilience. In: Comfort LK, Boin A, Demchak CC (eds) Designing resilience: preparing for extreme events. University of Pittsburgh Press, Pittsburgh, pp 1–12

Britton NR, Clark GJ (2000) From response to resilience: emergency management reform in New Zealand. Nat Hazard Rev 1(3):145–150

Brooks N (2003) Vulnerability, risk and adaptation: a conceptual framework. Tyndall Centre working paper

Bruneau M, Chang SE, Eguchi RT, Lee GC, O'Rourke TD, Reinhorn AM, Shinozuka M, Tierney K, Wallace WA, von Winterfeldt D (2003) A framework to quantitatively assess and enhance the seismic resilience of communities. Earthquake Spectra 19(4):733–752

Burton I, Kates RW, White GF (eds) (1978) The environment as hazard. Oxford University Press, New York

CanVec+ (2016) http://geogratis.gc.ca/. Last accessed Jan 2016

Carney D (ed) (1998) Sustainable rural livelihoods: what contribution can we make? 1st edn. Department for International Development, London

Carney D (2002) Sustainable livelihoods approaches: progress and possibilities for change. Department for International Development, London

Carney D, Drinkwater M, Rusinow T, Neefjes K, Wanmali S, Singh N (1999) Livelihoods approaches compared. Department for International Development, London

Carpenter S, Walker B, Anderies M, Abel N (2001) From metaphor to measurement: resilience of what to what. Ecosystems 4(8):765–781

Cassidy JF, Rogers GC, Lamontagne M, Adams J (2010) Canada's earthquakes: 'the good, the bad, and the ugly'. Geosci Can 37(1):1–16

Chambers R, Conway G (1992) sustainable rural livelihoods: practical concepts for the 21st century. IDS Discussion Paper 296. Institute of Development Studies, Brighton

Chang SE, Shinozuka M (2004) Measuring improvements in disaster resilience of communities. Earthquake Spectra 20(3):739–755

Chen A (1997) A Guide for Assessing the Impact of Microenterprise Services at the Individual Level. Washington, DC, AIMS/MSI

City of Toronto (2016) http://www1.toronto.ca/wps/portal/contentonly?vgnextoid=1a66e03bb8d1e310VgnVCM10000071d60f89RCRD. Accessed Mar 2016

Collins A (2009) Disasters and development. Routledge, New York, 304p

Crowley K, Elliott JR (2012) Earthquake disasters and resilience in the global North: lessons from New Zealand and Japan. Geogr J 178:208–215

Cutter S (2011) Preface. In: Disaster resilience: a national imperative, committee on increasing national resilience to hazards and disasters. The National Academies Press, Washington, DC, 244p

Cutter SL, Boruff BJ, Shirley WL (2003) Social vulnerability to environmental hazards. Soc Sci Q 84:242–261

Cutter SL, Barnes L, Berry M, Burton C, Evans E, Tate E (2008) A place-based model for understanding community resilience to natural disasters. Glob Environ Chang 18:598–606

Dhobal DP, Gupta AK, Mehta M, Khandelwal DD (2013) Kedarnath disaster: facts and plausible causes. Wadia Institute of Himalayan Geology, Dehradun

Downing TE, Ringius L, Hulme M, Waughray D (1997) Adapting to climate change in Africa. Mitig Adapt Strateg Glob Chang 2:19–44

EMDAT (2009) Natural disasters trends data access, The International Disaster Database, Centre for Research on the Epidemiology of Disasters. http://www.emdat.be/disaster_trends/index.html

Endfield GH (2012) The resilience and adaptive capacity of social-environmental systems in colonial Mexico. Proc Natl Acad Sci U S A 109:3628–3631

Folke C, Carpenter S, Elmqvist T, Gunderson L, Holling CS, Walker B (2002) Resilience and sustainable development: building adaptive capacity in a world of transformations. Ambio 31:437–440

Folke C, Carpenter SR, Walker B, Scheffer M, Chapin T, Rockstro MJ (2010) Resilience thinking: integrating resilience, adaptability and transformability. Ecol Soc 15(4):20

Gardner JS, Dekens J (2007) Mountain hazards and the resilience of social-ecological systems: lessons learned in India and Canada. Nat Hazards 41(2):317–336

Han GY (2011) Understanding regional dynamics of vulnerability: a historical approach to the flood problem in China. PhD Dissertation Graduate School of Geography, Clark University, USA

Holling CS (1973) Resilience and stability of ecological systems. Annu Rev Ecol Syst 4:1–23

Holling CS (1996) Engineering resilience versus ecological resilience. In: Schulze PC (ed) Engineering within ecological constraints. National Academy Press, Washington, DC, pp 31–44

IPCC (2001) Third assessment report (ar3), Cambridge University Press, Cambridge

IPCC (2007) Fourth assessment report (ar4). Cambridge University Press, Cambridge

IPCC (2012) In: Field CB, Barros V, Stocker TF, Qin D, Dokken DJ, Ebi KL, Mastrandrea MD, Mach KJ, Plattner GK, Allen SK, Tignor M, Midgley PM (eds) Managing the risks of extreme events and disasters to advance climate change adaptation. A special report of working groups I and II of the Intergovernmental Panel on Climate Change. Cambridge University Press, Cambridge/New York

Irwin S, Simonovic SP, Nirupama N (2016) Introduction to ResilSIM: a decision support tool for estimating disaster resilience to hydro meteorological events. Water Resources Research Report No. 094, Facility for Intelligent Decision Support, Department of Civil and Environmental Engineering, London, Canada, 66p. http://www.eng.uwo.ca/research/iclr/fids/publications/products/94.pdf

JMA (2011) The 2011 off the Pacific coast of Tohoku earthquake first report, Japan Meteorological Agency. http://www.jma.go.jp/jma/en/News/2011_Earthquake_01.html. Accessed 22 Feb 2013

Johnson S (2005) Gender relations, empowerment and microcredit: moving on from a lost decade. Eur J Dev Res 17(2):224–248

Kasperson JX, Kasperson RE (2001) Climate change, vulnerability and social justice. Stockholm Environment Institute, Stockholm, pp 1–18

Kasperson JX, Kasperson RE (2005) The social contours of risk, vol 1. Earthscan, London

Khalafzai AK, Nirupama N (2011) Building resilient communities through empowering women with information and communication technologies: a Pakistan case study. Sustainability 3 (1):82–96

Kimhi S, Shamai M (2004) Community resilience and the impact of stress: adult response to Israel's withdrawal from Lebanon. J Commun Psychol 32(4):439–451

Klein RJT, Nicholls RJ, Thomalla FT (2003) Resilience to natural hazards: how useful is this concept? Environ Hazard 5:35–45

Kovacs P, Kunreuther H (2001) Managing catastrophic risk: lessons from Canada http://www.iclr.org/images/Managing_Catastrophic_Risk.pdf. Accessed 15 June 2016

Lamontagne M (2008) Earthquakes in Eastern Canada: a threat that can be mitigated. In: Proceedings of the 4th Canadian conference on geohazards: from causes to management, Presse de l'Université Laval, Québec, 594p.

Lapen DR, Hayhoe HN (2003) Spatial analysis of seasonal and annual temperature and precipitation normals in Southern Ontario, Canada. J Great Lake Res 29:529–544

Lei Y, Wang J, Yue Y, Zhou H, Yin W (2014) Rethinking the relationships of vulnerability, resilience, and adaptation from a disaster risk perspective. Nat Hazards 70:609–627

Matanle P (2011) The Great East Japan earthquake, tsunami, and nuclear meltdown: towards the (re)construction of a safe, sustainable, and compassionate society in Japan's shrinking regions, local environment. Int J Just Sustain 16(9):823–847

McLaughlin P (2011) Climate change, adaptation, and vulnerability: reconceptualizing societal-environment interaction within a socially constructed adaptive landscape. Organ Environ 24:269–291

Mijumbi R (2002) ICTs as toll for economic empowerment of women: experience from the use of a CD ROM by rural women in Uganda. Proc. Division for the Advancement of Women, United Nations Conference, Seoul, Korea, 11–14 Nov 2002

Miller F, Osbahr H, Boyd E, Thomalla F, Bharwani S, Ziervogel G, Walker B, Birkmann J, Vander Leeuw S, Rockstro¨m J, Hinkel J, Downing T, Folke C, Nelson D (2010) Resilience and vulnerability: complementary or conflicting concepts? Ecol Soc 15(3):11

Mitsuyoshi M (2011) Resilience in the face of disaster: reflections on the Japanese earthquake and tsunami. The New Zealand Medical Student Journal. http://hedc.otago.ac.nz/ojs/index.php/NZMSJ/article/viewFile/482/214. Accessed 22 June 2016

Morgan S, Heeks R, Arun S (2004) Researching ICT-based enterprises for women in developing countries: a gender perspective, Institute for Development Policy and Management, University of Manchester, UK http://www.womenictenterprise.org

MPAC (2016). Municipal property assessment corporation. https://www.mpac.ca/about/corporate_overview/default.asp. Accessed Jan 2016

Murthy R (2002) Micro-credit, poverty and empowerment: linking the triad. Sage Publications, New Delhi, 372p

Nirupama N, Armenakis C (2013) Editorial in sociological aspects of natural hazards, special issue. Nat Hazards 66:1–2

Nirupama N, Maula A (2013) Engaging public for building resilient communities to reduce disaster impact. Nat Hazards 66:51–59

Nirupama N, Sharma R, Verma K, Panigrahi S (2014) A multi-tier hazard-Part I: description of the event. Nat Hazards 76(1):259–269

Nirupama N, Popper T, Quirke A (2015) Role of social resilience in mitigating disasters. Int J Disaster Resilience Built Environ 6(3):363–377

Paton G (2006) Disaster resilience: building capacity to co-exist with natural hazards and their consequences. In: In Disaster resilience. Charles C Thomas Publishers, Springfield, p 321

Rathgeber EM (2007) WID, WAD, GAD: trends in research and practice. J Dev Areas 23:407–419

Resilience Alliance (2009) Assessing and managing resilience in social-ecological systems: a practitioner's workbook, Version 1.0. http://www.resalliance.org/resilience-assessment

Sacred Yatra (2013) Kedarnath Pilgrim Statistics. Badarikedar.org. http://www.sacredyatra.com/kedarnath-pilgrim-stats.html. Accessed 12 June 2016

Seemann M, Onur T, Cloutier-Fisher D (2011) Earthquake shaking probabilities for communities on Vancouver Island, British Columbia, Canada. Nat Hazards 58(3):1253–1273

Sharma R, Verma K, Panigrahi S, Nirupama N (2014) A multi-tier hazard-Part II: meteorological analysis. Nat Hazards 76(1):271–281

Shaw R (2012) Incorporating resilience of rural communities for proactive risk reduction in Shikoku, Japan. Graduate School of Global Environmental Studies/Kyoto University, Kyoto

Shaw R, Takeuchi Y (2011) Role of Local institutions in climate related disasters in small and medium sized cities of Japan. Asian J Environ Disaster Manag 3(2):137–150

Simonovic S, Nirupama (2005) A spatial multi-objective decision making under uncertainty for water resources management. J Hydroinf 7(2):117–133

Simonovic SP, Peck A (2013) Dynamic resilience to climate change caused natural disasters in coastal megacities quantification framework. British J Environ Clim Chang 3(3):378–401

Smit B, Burton I, Klein RJT, Wandel J (2000) An anatomy of adaptation to climate change and variability. Clim Chang 45:223–451

Smith K (ed) (1996) Environmental hazards: assessing risk and reducing disaster, 2nd edn. London, Routledge

Sousounis PJ (2001) Lake effect storms. In: Holton J, Pyle J, Curry J (eds) Encyclopedia of atmospheric sciences. Academic Press, Amsterdam, pp 1104–1115

Stakhiv E (1993) Evaluation of IPCC adaptation strategies. Institute for Water Resources, U.S. Army Corps of Engineers, Fort Belvoir, Virginia, USA

Stirling MW, McVerry GH, Gerstenberger M (2012) New national seismic hazards model for New Zealand: changes to estimated long-term hazard. Proc. 12th New Zealand society for earthquake engineering conference 13–15 Apr 2012

Subrahmanian E (2003) Information and communications technology for sustainable development: defining a global research agenda, Workshop based reports. Carnegie Mellon University/ Pittsburgh and Indian Institute of Science, Bangalore. http://www.cs.cmu.edu/~rtongia/ICT4SD_Full_Book.pdf

Thomalla F, Downing T, Spanger-Siegfried E, Han G, Rockström J (2006) Reducing hazard vulnerability: towards a common approach between disaster risk reduction and climate adaptation. Disasters 30(1):39–48

Timmerman P (1981) Vulnerability, resilience and the collapse of society: a review of models and possible climatic applications, Environmental monograph no. 1. Institute for Environmental Studies/University of Toronto, Toronto

Turner BL (2010) Vulnerability and resilience: coalescing or paralleling approaches for sustainability science? Glob Environ Chang 20(4):570–576

Turner BL, Kasperson RE, Matson PA (2003) A framework for vulnerability analysis in sustainability science. Proc Natl Acad Sci USA 100:8074–8079

UN Millennium Summit (2000). Key proposals. http://www.un.org/millennium/sg/report/key.htm. Accessed 29 Dec 2010

UNISDR (2009) Terminology on disaster risk reduction. http://www.unisdr.org/eng/terminology/UNISDRterminology-2009-eng.pdf

USEPA (2012) The Great Lakes: an environmental atlas and resource book. US Environmental Protection Agency. https://www.epa.gov/greatlakes

USGS (2011) United States geological survey updates magnitude of Japan's 2011 Tōhoku Earthquake to 9.0. http://www.usgs.gov/newsroom/article.asp?ID=2727&from=rss_home. Accessed 22 Feb 2013

Vogel C, Moser SC, Kasperson RE, Dabelko G (2007) Linking vulnerability, adaptation, and resilience science to practice: pathways, players, and partnerships. Glob Environ Chang 17 (3–4):349–364

Walker B, Holling CS, Carpenter SR, Kinzig A (2004) Resilience, adaptability and transformability in social-ecological systems. Ecol Soc 9(2):5

Walker B, Abel N, Anderies JM, Ryan P (2009) Resilience, adaptability, and transformability in the Goulburn-Broken catchment Australia. Ecol Soc 14(1):12

Welle T, Birkmann J (2015) The world risk index – an approach to assess risk and vulnerability on a global scale. J Extreme Events 2(1):34 pages

Wisner B, Blaikie P, Cannon T, Davis I (2004) At risk: natural hazards, people's vulnerability and disasters, 2nd edn. Routledge, London, p 471

Young OR, Berkes F, Gallopin GC, Janssen MA, Ostrom E, Van der Leeuw S (2006) The globalization of socio-ecological systems: an agenda for scientific research. Glob Environ Chang 16(3):304–316

Zhou H, Wang J, Wan J, Jia H (2010) Resilience to natural hazards: a geographic perspective. Nat Hazards 53(1):21–41

Chapter 5
Disaster Perceptions

Generally speaking, perception includes individuals' subjectivity in terms of how they see or assess the characteristics of a phenomenon. Risk perception is vital to understanding what risks people consider to be acceptable, and what risk reduction programs have a better chance of being accepted. Risk perception is influenced by a variety of factors including the kind of information available and how that information is processed; the personality and emotional state of the perceiver; their personal experiences and prejudices; and socio-economic factors, to name but a few. Risk perception, risk tolerance, and high or low risk-taking behaviors are all interconnected. Livelihood opportunities (Chambers and Conway 1992) can drive people to take more risk. The nature and consequences of a potential threat, as well as its proximity, also contribute to how it is perceived by society. In this era of social media, the media is vital to ensuring that disaster news is covered more objectively. This chapter includes survey-based studies conducted in Canada as powerful testimonies to the importance of risk perception among various groups, including average citizens and emergency managers.

5.1 Perception of Risk

Risk perception is controlled by sets of dynamic social and psychological processes that result in some hazards becoming of increased concern within society, while others become less of a concern (Etkin 2016). Some processes include trust, blame, and prior attitudes. It is a subjective judgement of an individual's feeling towards the plausibility of experiencing a hazard when there is minimal objective information. Expert judgement uses a risk management approach, and as a result is more technical and narrow; for example, using annual fatalities as a measurement of risk.

Since the emergence of the species, humans have been exposed to risk, as it is believed to be engrained in human thinking as an integral part of the thought process

N. Agrawal, *Natural Disasters and Risk Management in Canada*, Advances in Natural and Technological Hazards Research 49,
https://doi.org/10.1007/978-94-024-1283-3_5

(Wahlberg and Sjoberg 2000). So, if there is no singular definition of risk, how can it be comprehended? To begin, one must become exposed to the debates that surround the controversial term. One such debate is whether risk is socially constructed. Is it objective and measureable, or subjective and immeasurable, and to what degree? To clarify, the objective perspective is referred to as the *rationalist approach*, and the subjective perspective is referred to as the *constructionist approach* (Etkin 2016). The rationalist approach puts emphasis on scientific management, statistics, and decision theory in order to control risk. The constructionist approach suggests that "nothing is a risk in itself, but rather that it is a product of cultural, political, social and historical ways of seeing". The above-mentioned debate is one of major standing in the field of risk; however, risk does in fact reflect upon both perspectives. If risk is socially constructed, society has made decisions which have inadvertently determined who is at risk and what the risks are. There are several factors contributing to socially constructed risk, such as minorities, education, disability, elderly and children, poverty, and health (Pine 2009). A well-developed example of a socially constructed risk faced by many in urbanized areas has been provided by Etkin (2016) in the quote:

> Allowing housing construction near hazardous chemical plants is a social/political decision that puts people who live there in harm's way. The proximity of residential areas to hazardous industrial ones has become increasingly important due to urban growth.

When a person is exposed to a risk, they do not respond to that risk directly; rather they respond to their own perception of that risk. Generally speaking, the average person (non-expert) relies on intuition to assess a risk; this concept is referred to as *risk perception*. Within social groups, acting powers "downplay certain risks and emphasize others as a means of maintaining and controlling the group" (Slovic 1987). A common perception found within many industrialized nations is the belief that people are presently exposed to a higher degree of risk than traditionally faced in the past, and risks to be faced in the future will be larger than present risk (Schneider et al. 2006; Etkin and Haque 2003). The above is a general statement, attempting to express the common beliefs of the population. There is, however, two viewpoints of risk not mentioned in the above-mentioned common perception: lay judgment of risk and expert judgement of risk. Lay judgement is generally a rights-based approach that focuses on justice, uncertainty, who benefits from the risk, who is at risk, and dread. It is important to note that expert judgement is prone to the same biases as laypeople, especially if the experts "are forced to go beyond the limits of available data and rely on intuition" (Slovic 1987). With that being said, members of the public sometimes do not possess all of the information relating to a certain hazard, and therefore can be misinformed. It is beneficial to embrace both the public and expert viewpoints in order to develop a well-rounded grasp on risk, as both views offer unique intelligence and insight (in Hébert 2016).

Research suggests that one of the greatest influences on risk perception is cultural factors causing a distortion of perception that can travel between social groups, potentially distorting the actual/realistic threats. Risk perception is fueled by people's experiences (or lack-of), emotions, and social and cultural factors of the community, along with numerous influencers. Each individual experiences and perceives risk differently and therefore makes it difficult to truly define the concept (Gierlach et al. 2010; ISDR 2004; GTZ 2004). Risk perception is influenced by direct or indirect experiences of activities, events, and/or technologies; for example, receiving information from news sources, or witnessing a natural disaster such as a severe flood. The characteristics of potential dangers associated with a risk also heavily influences risk perception. People tend to believe that rare, sensational events pose a higher level of risk than more conventional events. People's judgements of risk stem from social learning, peer influences and cultural practices, and are continuously exposed to media reports and other processes of communication. Similar to risk, risk perception is viewed differently by each individual depending on the following factors: the type of risk, the context of the risk, the social context, and the individual's personality. An individual's perception of risk is a motivator, urging community members to spring in to action mitigate, avoid, and adapt to risks (Wachinger et al. 2013).

Some risk analysts regard perception as invalid because they arise from subjective influences. But, to the general public, perception are the only relevant views because they incorporate the expert's analysis together with individual judgement based on individual experience, social context and other factors. The public also suspects that limits exist to what experts know a suspicion that is justified in certain cases (Sjöberg 2001). Table 5.1 (Smith 2004) analyses differences between risk assessment and risk perception.

Lay people perceive hazards differently from technical experts for a variety of reasons, including geographical location and aspects of their personality. For example, rural dwellers often perceive flood hazard perception closer to objectively derived estimates than urban dwellers (Smith 2004; Nirupama and Simonovic 2007; Nirupama et al. 2014). Group perceptions can easily be influenced by social or cultural factors as the influence of personality is exercised mainly through the so-called 'locus of control'. This classifies people according to the extent that they believe hazardous events are dependent on fate (external control) or within their own responsibility (internal control). In order to reduce the stress associated with uncertainty, hazard perceivers tend to adopt certain recognizable models of risk perception with which they are more comfortable. These can be grouped into three basic types, all of which conflict with more objective risk analysis (Smith 2004):

- *Determinate perception*: people having determinate perception believe that extreme events, such as earthquakes and flash flooding do exist but they occur in a certain pattern.
- *Dissonant perception*: people having dissonant perception believe that natural hazards are freak events that are unlikely to be repeated.
- *Probabilistic perception*: people having probabilistic perception accept that natural hazards exist and they maybe random events. Therefore, they do not see any benefit in doing anything about something that is an *Act of God*. If decision makers responsible for disaster mitigation happen to hold probabilistic perception, they may not support investing of resources in mitigation measures.

Table 5.1 Some difference between risk assessment and risk perception (Smith 2004)

Phase of analysis	Risk assessment process	Risk perception process
Risk identification	Event monitoring, Statistical inferences	Individual intuition, Personal awareness
Risk estimation	Magnitude/ frequency, Economic costs	Personal experiences, Intangible losses
Risk evaluation	Cost/benefit analysis, Community policy	Personality factors, Individual action

Table 5.2 Factors influencing public risk perception with examples of relative safety judgements

Factors tending to increase risk perception	Factors tending to decrease risk perception
Involuntary hazard (high risk vocations)	Voluntary hazard (mountaineering)
Immediate impact (wildfire – Fort McMurray fire example in following section)	Delayed impact (drought)
Direct impact (earthquake)	Indirect impact (drought)
Dreaded health hazard (cancer)	Common hazard (road accidents)
Many fatalities per event (air crash)	Few fatalities per event (car crash)
Death grouped in space/time (avalanche)	Deaths random in space/time (accidents)
Identifiable victims (chemical plants workers)	Statistical victims (smoking, drugs)
Processes not well understood (nuclear)	Processes well understood (snow storm)
Uncontrollable hazard (tropical cyclone)	Controllable hazard (ice on highways)
Unfamiliar hazard (tsunami)	Familiar hazard (river floods)
Lack of belief in authority (private industrialist)	Belief in authority (university scientist)
Much media attention (virus such as Ebola, Zika)	Little media attention (chemical plants)

Adapted from Whyte and Burton (1982), Smith (2004)

Social amplification of risk occurs when relatively minor threats elicit a disproportionately strong degree of public concern as demonstrated in Table 5.2 (Kasperson et al. 1988). Risks are taken more seriously if they are understood by people as life-threatening, immediate, and direct. This means that an earthquake, a rapid onset event, is normally rated more seriously than a drought, a slow onset hazard. Risk is also perceived to be higher if children are at risk in comparison with seniors. Additionally, lack of understanding of complex technologies associated with factories and chemical industries in the vicinity leads to fear and distrust in technical experts and the authorities. Currently ongoing environmental and climate change concerns are perceived differently by younger generation than older people whose priorities tend to be around health and safety issues (Fischer et al. 1991). Awareness is heightened when public health is at stake. For example, the city of Flint, Michigan is facing in the worst public health crisis seen to date in the United States. With Flint's struggling economy, in 2014 the local government officials made a decision to switch the water from being supplied by Lake Huron and pre-treated in Detroit, to be supplied by the Flint River without adding in the anti-corrosive agent in order to save the city money. The decision to not add in the anti-corrosive agent would cascade in to a major health disaster as entire region was exposed to lead poisoning, as well as exposure to Legionella bacteria for 18 months. The health effects this water crisis caused Flint was unimaginable to the community. The people were unaware that there was lead in the water that could be absorbed through the skin as well as being ingested (Flint Task Force 2016). The community was starting to see the side effects of the water in forms of rashes, eye irritations, and behavioural changes (Gupta et al. 2016).

People are also extremely fearful of nuclear accidents and nuclear power plants and have a great deal of skepticism in the industry. The world has seen accidents such as, the Three Miles Island in 1979, Chernobyl in 1986, and recently Fukushima meltdown in Japan in 2011. The Three Mile Island Unit 2 (TMI-2) reactor, near Middletown, Pennsylvania, USA, partially melted down on March 28, 1979. A combination of personnel error, design deficiencies, and component failures caused the Three Mile Island accident, which permanently changed both the nuclear industry and the US Nuclear Regulatory Commission (NRC). Public fear and distrust increased triggering nationwide debate (Fig. 5.1) NRC's regulations and oversight became broader and more robust, and management of the plants was scrutinized more carefully. Careful analysis of the accident's events identified problems and led to permanent and sweeping changes in how NRC regulates its licensees – which, in turn, has reduced the risk to public health and safety (USNRC 2016).

At Chernobyl, Ukraine (part of former Soviet Union – Fig. 5.2), on April 26, 1986, a sudden surge of power during a reactor systems test destroyed Unit 4 of the nuclear power station. The accident and the fire that followed released massive amounts of radioactive material into the environment. Emergency crews responding to the accident used helicopters to pour sand and boron on the reactor debris. The sand was to stop the fire and additional releases of radioactive material; the boron was to prevent additional nuclear reactions. After the accident, officials closed off the area within 30 km of the plant, except for persons with official business at the plant and those people evaluating and dealing with the consequences of the accident and operating the undamaged reactors. The government evacuated about 115,000 people from the most heavily contaminated areas in 1986, and another 220,000 people in subsequent years (USNRC 2016; UNSCEAR 2008).

The Great East Japan Earthquake of magnitude 9.0 on Friday 11 March 2011 generated a large tsunami that destroyed the Sendai region (Fig. 5.3). The earthquake and tsunami caused great loss of life and widespread devastation in Japan. More than 15,000 people were killed, over 6000 were injured and, thousands went missing. Considerable damage was caused to buildings and infrastructure, particularly along

Fig. 5.1 Time cover of April 29, 1991. Cover Credit: Steve Smith-Westlight http://content.time.com/time/covers/0,16641,19910429,00.html

Fig. 5.2 Chernobyl shown in map (source: World Nuclear Association http://www.world-nuclear.org/information-library/safety-and-security/safety-of-plants/chernobyl-accident.aspx)

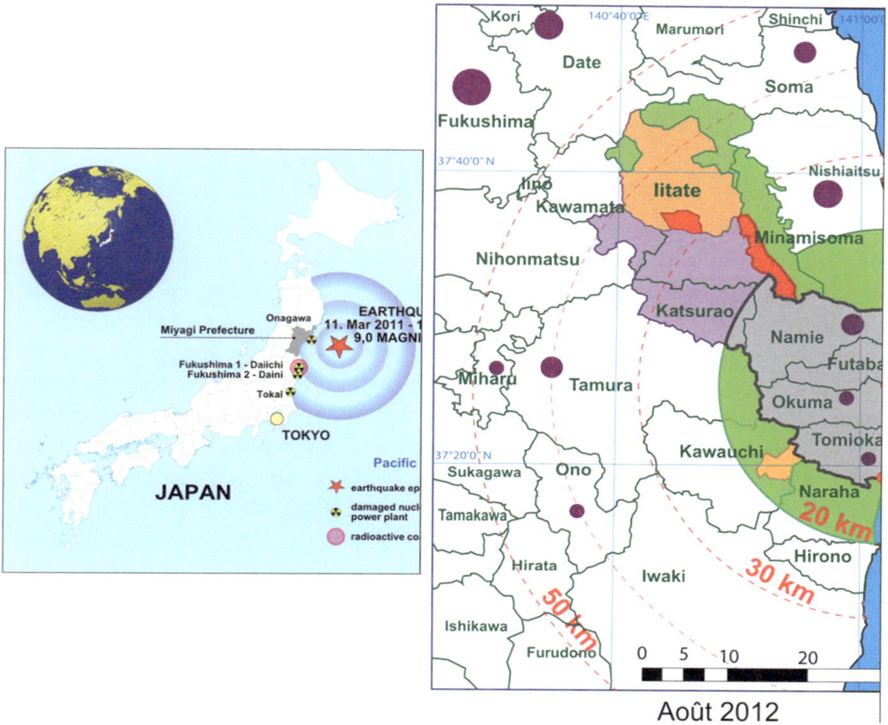

Fig. 5.3 Fukushima Daiichi nuclear plant location and impact zone. Source: By Japan_location_map_with_side_map_of_the_Ryukyu_Islands.svg: Maximilian Dörrbecker (Chumwa) File:Japan (orthographic projection).svg: (Connormah) derivative work: W.Rebel [CC - BY-SA 3.0 (http://creativecommons.org/licenses/by-sa/3.0) or GFDL (http://www.gnu.org/copyleft/fdl.html)], via Wikimedia Commons (left). By Roulex_45 [GFDL (http://www.gnu.org/copyleft/fdl.html) or CC BY-SA 4.0-3.0-2.5-2.0-1.0 (http://creativecommons.org/licenses/by-sa/4.0-3.0-2.5-2.0-1.0)], via Wikimedia Commons (right)

Japan's north-eastern coast. The tsunami caused meltdown of the Fukushima Daiichi nuclear power plant due to the loss of the cooling function at the operating reactor units2 as well as at the spent fuel pools. Despite the efforts of the operators at the Fukushima Daiichi nuclear power plant to maintain control, the reactor cores in Units 1–3 overheated, the nuclear fuel melted and the three containment vessels were breached. Hydrogen was released from the reactor pressure vessels, leading to explosions inside the reactor buildings in Units 1, 3 and 4 that damaged structures and equipment and injured personnel. Radionuclides were released from the plant to the atmosphere and were deposited on land and on the ocean. There were also direct releases into the sea. People within a radius of 20 km of the site and in other designated areas were evacuated, and those within a radius of 20–30 km were instructed to shelter before later being advised to voluntarily evacuate. Restrictions were placed on the distribution and consumption of food and the consumption of drinking water. The accident was rated 7 on the International Nuclear Events Scale (INES) due to high radioactive releases (IAEA 2013; Pletcher 2016).

5.1.1 Media's Influence on Risk Perception

The media has a fascination with disastrous events, influencing and inflating public anxiety and perceptions of danger (Borum et al. 2010). When events such as school or college shootings occur, public fear is heightened due to the fact that schools are portrayed as a safe place for children (Lindle 2008). Like the general public, teachers hear about school shootings occurring, exacerbating their existing fear of violence against children. Similarly, wild/forest fires that have potential to engulf communities and explosions caused by various reasons with a potential to hurt communities receive much and prolonged attention by the media. People are generally fearful of fires due to their destructive nature and stomach turning reaction to burn injuries or death. They provoke extreme passion in people's minds. There are also advantages to media attention in that it creates awareness, emotional engagement, and encourages donations and other forms of help by local, national, and international community. A recent example of a forest fire is Fort McMurray, Alberta fires in May 2016. Being an oil patch of Canadian Midwest, the community sits on the fringes of natural forest creating an easy access for a forest fire to spread to the community. This phenomenon is known as interface fire which is becoming far more common due to various reasons. The soaring 32 °C temperatures fueled the fire and tinder dry forest fire swept through the community destroying homes and buildings and forcing the largest wildfire evacuation in Albertan history. It continued to spread across northern Alberta and into Saskatchewan, consuming forested areas and impacting Athabasca oil sands operations until mid-June when rain helped firefighters to hold the fire (Ramsay and Shum 2016; Parsons and Graney 2016). It has become the costliest disaster in Canadian history. The fire destroyed 2400 structures, nearly 10% of the city, and forced more than 80,000 residents to flee. As a risk mitigation effort, Alberta instituted a province wide ban on open fires, including campfires and the use of charcoal briquettes (Globe and Mail 2016). Figure 5.4 shows the raging fire and fleeing residents and Fig. 5.5 illustrates the extent of loss in Fort McMurray.

5.2 Perception of Vulnerability

According to the International Disaster Database (www.emdat.be), vulnerability is degree of loss (from 0% to 100%) resulting from a potential damaging phenomenon. The Public Safety Canada (PSC 2012) describes vulnerability as, a condition or set of conditions determined by physical, social, economic and environmental factors or processes that increases the susceptibility of a community to the impact of hazards.

Fig. 5.4 A wall of fire rages outside of Fort McMurray on May 3, 2016 (Photo by Terry Reith/Canadian Broadcasting Corporation (CBC))

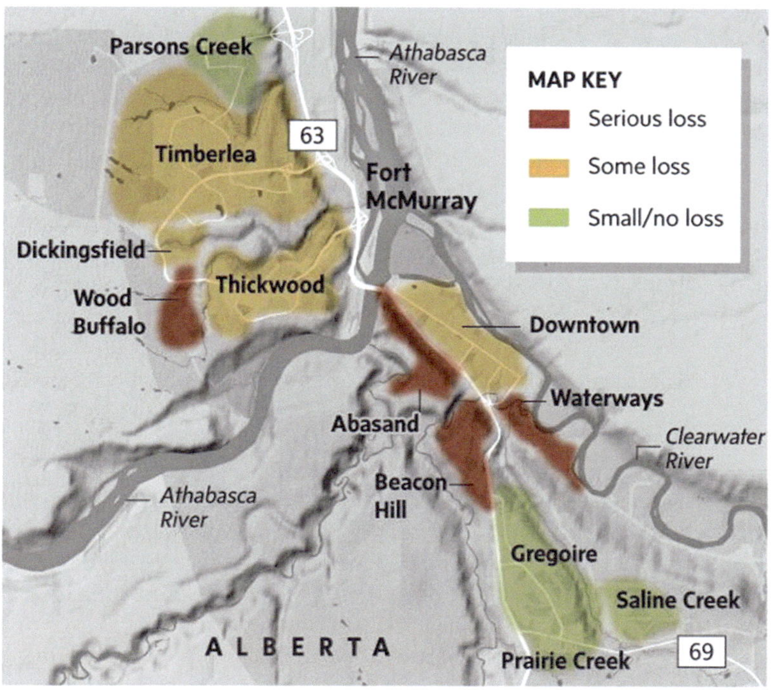

JOHN SOPINSKI/THE GLOBE AND MAIL | SOURCE: CMHC; GOOGLE MAPS

Fig. 5.5 Reports of loss shown in Fort McMurray fire affected region (Globe and Mail 2016)

Also defined in the PSC (2012) report is the vulnerability assessment, as the process of identifying and evaluating vulnerabilities, describing all protective measures in place to reduce them and estimating the likelihood of consequences.

Studies have shown (Etkin 2016; Scanlyn et al. 2013; Armenakis and Nirupama 2013, 2014a, b; Stewart 2007; Tierney 1999; Hewitt 1997; Whyte and Burton 1982) that certain people are more vulnerable than others due to various reasons such as lack of education and adequate income, age, poor health, physical disability, and living in hazardous locations. Many times, vulnerable people living in hazard-prone areas do not perceive their exposure to risk concerning enough to becoming their top priority (Nirupama 2015) as basic necessities of life remains their main focus. Perception about people's behaviour during emergencies defines, to a large extent, how authorities would plan resource allocation for community emergency response as well as develop and implement mitigation measures. During the past decade, a paradigm shift in the approach to disaster management has been apparent and community participation is being encouraged by policy makers. It is believed that community participation, not a top-down approach, will bring about a comprehensive and accurate appreciation of people's perception regarding hazard risk, vulnerability, and resilience. Experts (Wisner et al. 2004; Ferrier and Haque 2003; Twigg 2007; UNISDR 2001) have also delved in explaining the progression of people's vulnerability by employing various arguments given the social, physical, and political environments. Birkmann (2006) has developed indicators for identifying and assessing vulnerability. Emphasis on assessing people's vulnerability and potential risks they may be exposed to, in order to mitigate losses through knowledge based actions, is clearly noticeable (Cutter 2012; ICSU 2008; Pelling 2003; Jaeger et al. 2001; Tobin and Montz 1997) preference as a way to go forward.

In the flood risk mapping methodology developed by Armenakis and Nirupama (2014a), it is clearly demonstrated (Fig. 5.6) that accurate understanding and estimation of various types of vulnerabilities play vital role in the process of risk assessment.

5.3 Perception of People

According to Mileti and Fitzpatrick (1991), we process information in five different steps: hear, understand, believe, personalize, and decide/respond. In having others help us process information, we are able to legitimize the source of the information, assess its credibility, and confirm the best course of action based on the actions, those around us, wish to take. Knowing better leads to doing better which leads to action that would be based on sound information. For example, living near a railway track can be associated with potential risks such as train derailment, harmful emissions, toxic spills, fires, and explosions. This was the case in 2015 when a small engine fire from a train left oil and debris on various properties in Mississauga in the GTA, presenting a health risk to residents (City News 2015). A program with policies and procedures designed to consider cultures and community needs will not only benefit the community tremendously, but also allow for knowledge-based and well understood perceptions of people. Another major accident occurred in July 2013 when a train carrying inflammable petroleum crude oil derailed and exploded in the

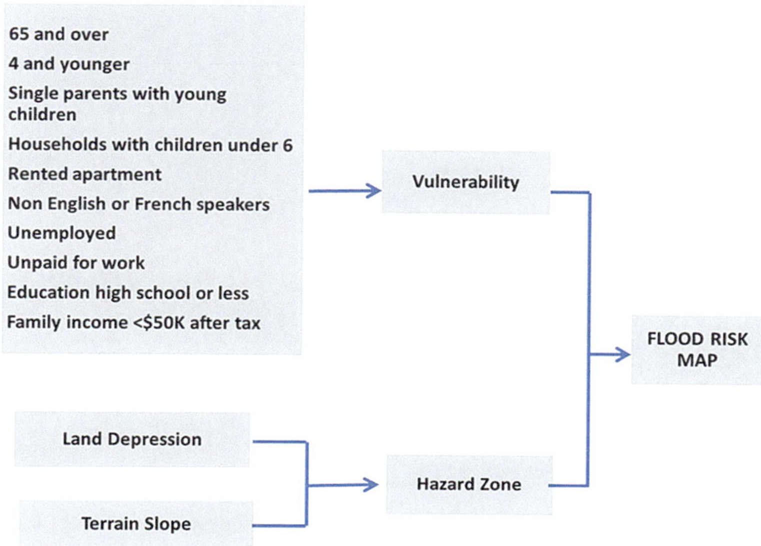

Fig. 5.6 Schematic of flood risk mapping, developed for the 2013 severe flooding in the City of Toronto, Canada

Fig. 5.7 Train route and direction of the travel at the time of the accident in Lac-Megantic, Quebec (Transport Safety Board of Canada www.tsb.gc.ca)

downtown area of in Lac-Mégantic, Quebec. Figure 5.7 shows the route of the train with Canadian cities on the track drawn on Google map, and Fig. 5.8 shows the extent of damage. The tragedy triggered a feeling of deep grief and sorrow in the close knit community, and a movement to move the rail track away from the town. Media reported conversations with impacted people in which they openly expressed

Fig. 5.8 Lac-Mégantic train derailment and explosion killing 47 people and forcing 2000 people from their homes (Transport Safety Board of Canada www.tsb.gc.ca)

their lack of trust with the authorities, including the Transport Safety Board of Canada and the Government of Canada in terms of how risks are being managed. People also perceived the situation as a case of companies putting their interest and convenience ahead of the safety of people and the environment, and impact on communities – both emotional and physical.

In Toronto, Canada, though the exponentiallyincreasing number of immigrants from around the world is a positive sign, new challenges arise from the standpoint of emergency management, institutional culture and practices. Proper governance is vital to creating an environment that would help new immigrants integrate in the society. According to the 2006 Census (Statistics Canada 2008), Toronto is one of the only four census divisions where more than 16% new immigrants (2000–2005) fall in the low income category – 5% higher than the national average. Ontario is the largest population centre; ten of the top twenty-five most populous Canadian municipalities are in Ontario, Toronto being number one at more than 2.5 million. Ontario is home to more than half of Canada's visible minorities, out of which more than 30% reside in the GTA, while the national average is only 16.2%. Though two-thirds of Toronto's adult population has completed postsecondary education, the percentage of allophones (persons whose first language is neither English nor French) is as high as 86% in several prominent municipalities. Employment numbers are discouraging, and newer dwellings are built further away from places of work – an additional contributing factor in the progression of vulnerability. In this scenario, it is prudent to pay attention to how vulnerabilities are perceived by policy makers with regards to emergency management and how they might impact potential disasters (Mileti 1999; Mitchell 2003; Tierney 2007).

Fig. 5.9 Participants at SAWC with Director, Kripa Sekhar (centre, in red dress)

5.3.1 Case Studies

Two studies are discussed in this section to illustrate the role and importance of people's perception of risk and vulnerability. The first study (Nirupama and Maula 2013) is based on a focus group session conducted at the South Asian Women's Centre (SAWC) in Toronto, Canada (Fig. 5.9). The participant women in the group were regular members of the Centre, using the resource because they were vulnerable in more than one way. They were mainly over 40 years old, the majority with little education, unemployed, facing language barriers, low income group, and reliant on public transit. Many participants identified that they lived in the vicinity of rivers, lakes, railways, or power plants (Fig. 5.10), but did not seem to be aware of their exposure to potential threats – indicating a lack of awareness and engagement with their surroundings. Figure 5.11 is an example that participants did not understand what was meant by level of safety, exposure to risk, and sense of belonging in their community as most of them chose not to respond to the question. Figure 5.12 illustrates group's interest and involvement in local government.

The second study was carried out in the Region of Peel in the GTA (Fig. 5.13) in Ontario, Canada (Nirupama and Jubril 2016; Jubril 2016). The Region of Peel has a Regional Emergency Management (REM) program in place. A survey conducted by the REM highlighted that among immigrants and visible minorities, presence of fire safety devices and other precautionary measures are less common; and there is a lack of social support for emergency situations. This research is based on the premise set by the REM survey to further explore people's knowledge, preferences, interests, priorities, and perceptions in order to identify strengths and weaknesses in the society. A questionnaire based survey was used to collect data from three different community locations in the City of Brampton, namely, South Fletcher's Sportsplex

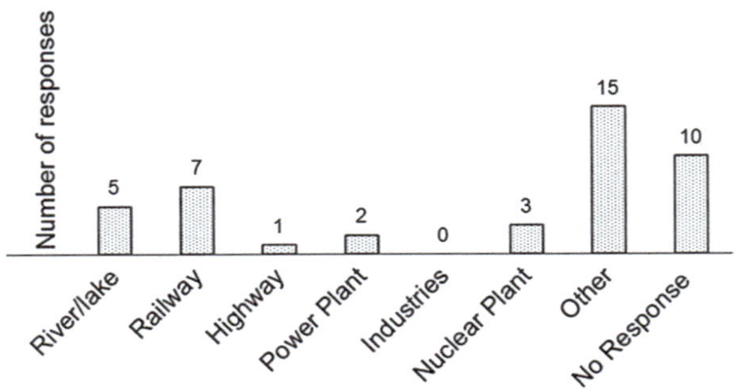

Fig. 5.10 Proximity to potential risks

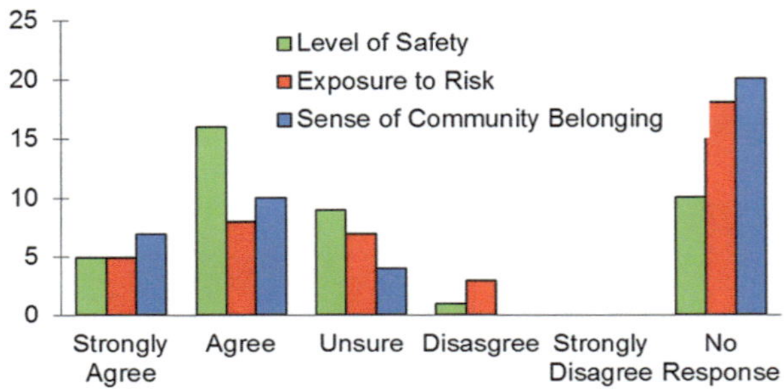

Fig. 5.11 People's perception of their safety, exposure to risk or threat, sense of belonging with their community, and preparedness to deal with emergencies

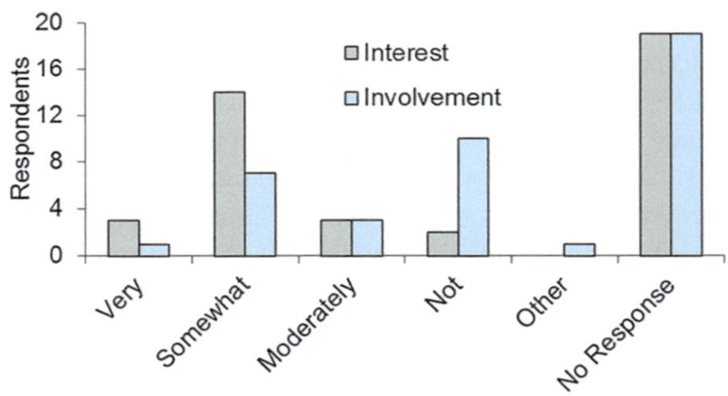

Fig. 5.12 Interest versus involvement in local government

community centre, New Birth Tabernacle (a local non-denominational faith centre), and a local restaurant (MJ'S BBQ & Suya). The questionnaire consisted of 29 questions intended to gather information from a sample of entities for the purpose of constructing attributes of the larger population of which the entities are members. Figs. 5.14 and 5.15 represent people's responses on questions such as, proximity to potential risks (Fig. 5.14), importance of social networks (Fig. 5.15), level of safety in their community (Fig. 5.16), and people's interest in the local government (Fig. 5.17).

It is apparent from the responses in the two case studies that to a certain degree, people understand the importance of social network but they do not see a need to

Fig. 5.13 The City of Brampton in Peel Region, Greater Toronto Area, Ontario, Canada (https://commons.wikimedia.org/wiki/File:Greater_toronto_area_map.svg Retrieved June 27, 2016)

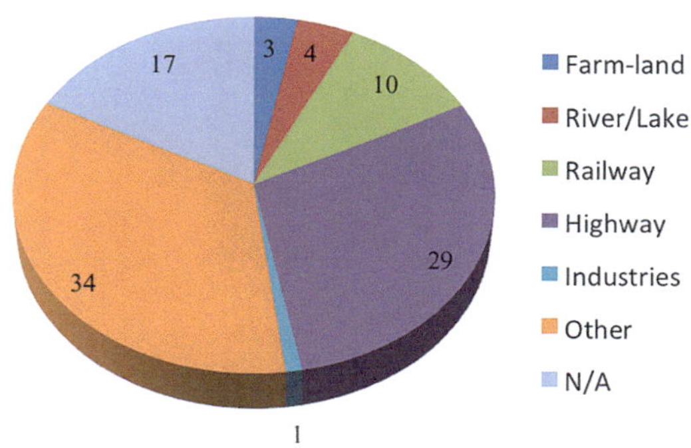

Fig. 5.14 Physical environment where people reside (Jubril 2016)

engage in the local government which is supposed to give them a sense of belonging. There is also a lack of awareness and understanding in regards with what is meant by potential risk, as a number of responses were 'other', 'not applicable' or 'no response'. People's perception on issues concerning assessment of their vulnerability in the context of their society must be observed after disseminating sufficient knowledge in addition to making efforts to attract their attention to the importance of such issues.

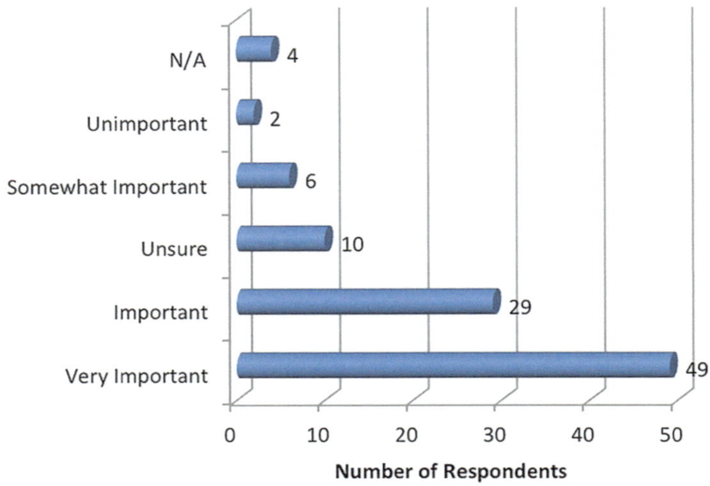

Fig. 5.15 Importance of social networks in people's mind (Jubril 2016)

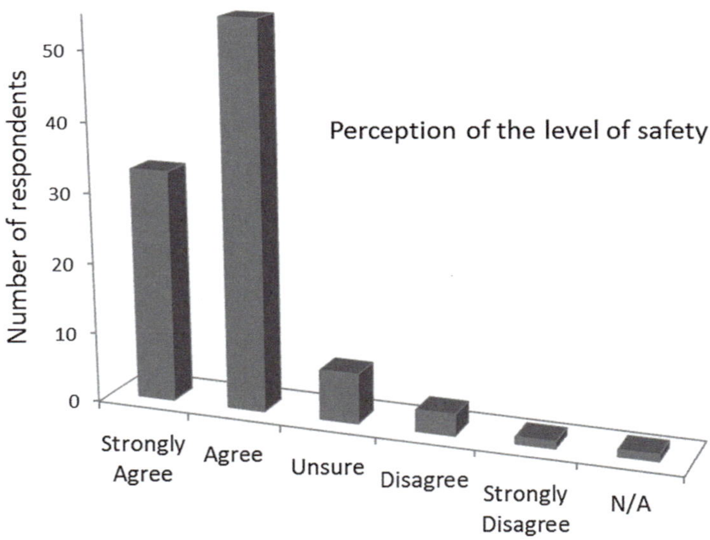

Fig. 5.16 People's perception of the level of safety (Jubril 2016)

5.4 Perspectives of Emergency Managers

Particularly with an emphasis on how cultural myths and false beliefs affect decision-making, various authors have discussed barriers to good disaster management (e.g. Der Heide 1989; Alexander 2002). These include: post-disaster recreation

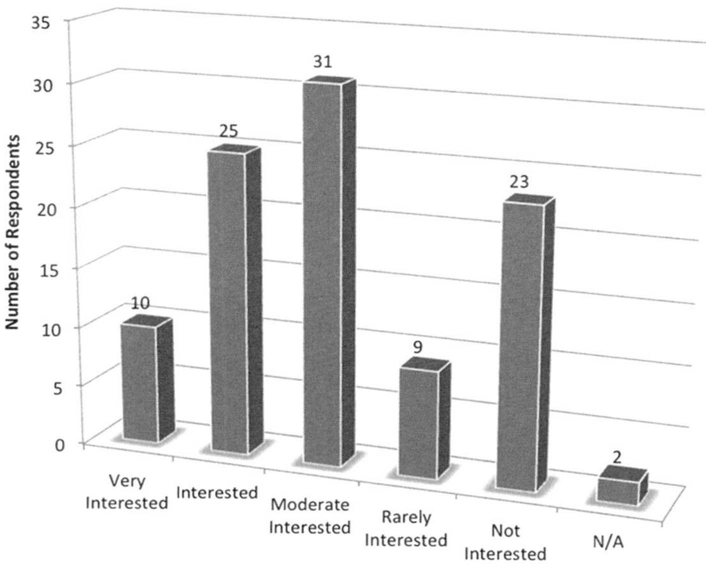

Fig. 5.17 Interest in the local government (Jubril 2016)

of vulnerability; removal of natural protective barriers; failure to learn from the mistakes of others; failure to correct existing but known deficiencies; overdependence upon technology; lack of recognition of system problems; the inter-governmental paradox; institutional ambiguities; apathy; underestimation of risk; overestimation of capacity; lack of resources; cultural attitudes, such as fatalism, defeatism etc.; social pressures; opposing special interest groups; and reliance upon myths/false beliefs in disaster planning, response and management. Research in the nascent field of disaster management suggests that it is often ineffective because of a large number of reasons. How the professional emergency management community perceives barriers that hinder effective emergency management and views itself may provide useful insights, and suggest strategies that might be used to help develop a culture of safety (Nirupama and Etkin 2009).

The way many disasters unfold can be attributed, in part, to a lack of institutional preparedness and a general perception of risk (Whyte and Burton 1982; Slovic 2000; White et al. 2001; Alexander 2002; Twigg 2007; Olanubi 2009). To better understand this issue, Nirupama and Etkin (2012) conducted a study to obtain insights into the minds and thoughts of emergency management professionals in Ontario, Canada. In order to assess how emergency management institutions perceive their importance in terms of Canadian society and the Emergency Management community the authors interviewed a number of experts positioned in policy-making and decision-making capacities. Based on these interviews, a questionnaire was prepared that highlighted some major concerns such as disaster myths/false beliefs, institutional barriers, knowledge limitations, cultural barriers, and resource limitations.

In total nine experts (three per sector) in emergency management were interviewed from (a) the public, (b) the private and (c) non-profit sectors in order to ascertain their opinions and perspectives on cultural and other barriers to risk reduction. Specific agencies targeted include (a) Emergency Management Ontario, Public Safety Canada, and Health Canada, (b) Bank of Montreal, IBM, and Ontario Hydro and (c) Red Cross, Salvation Army, and Canadian Centre for Emergency Preparedness. The interviews were semi-structured with open-ended questions to elicit rich details on the barriers to emergency management in Canada. Further, the open-ended format minimized the influence of researchers' biases on the issue. These interviews were taped and anonymous. Post interview analysis highlighted the attitudes and perceptions of the interviewees, with respect to themselves, their own organizations, their role in emergency management in Canada and their clients.

The questionnaire that was prepared based on these expert interviews was employed to conduct an anonymous survey of approximately 70 emergency managers in Ontario with the assistance of the *Ontario Association of Emergency Managers*, which is the provincial professional organization for emergency managers. Twenty-four or 34% of the surveyed people replied.

5.4.1 Highlights of the Findings

Generally speaking respondents scored emergency management institutions as having an average performance with the exception of their own, which they tended to rate as significantly higher than others. Most of them asserted that the amount of education and training they had received in their field was good or very good. This might not necessarily reflect a disinterest in improving job performance. The half of the respondents that were undecided might hesitate due to other concerns, such as being at the twilight of their career, potential costs or the difficulty of balancing further education with career and family. One respondent felt that it is better to use proven procedures and techniques rather than pursuing "new creative ways of doing things". For those that wanted further education/training, IMS (Incident Management System) was the training identified most as being needed. On deciding whether or not to take further education or training, two-thirds of the respondents said they would prefer short courses. A number of them also proposed online courses, as they are convenient.

Emergency managers' views were sought regarding whether a *Command & Control* hierarchical model would be preferred over a *Community-Based* one (Fig. 5.18). The question was in the context of the four phases of the Emergency Management Cycle, namely, mitigation and prevention; preparedness; response; and recovery. This question was placed within the following context:

> Two different models of emergency and disaster management are (a) one that is top down, command & control, and (b) another that is community based. Hierarchical command and control models based upon a pyramidal authority structure have been criticized as not being the most effective for handling complex disasters. In particular, it has been suggested that

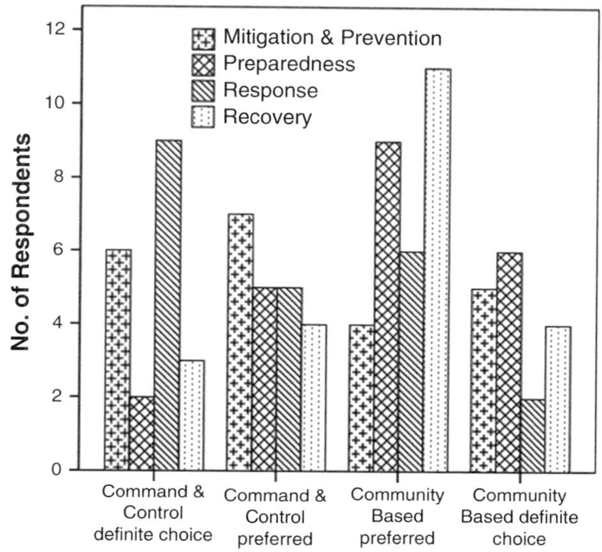

Fig. 5.18 Preferred approach – Command & Control or Community-based for handling the complex four phases of disaster management? (Nirupama and Etkin 2012)

> they insufficiently incorporate local concerns, authority, culture and expertise. Community based models that encourage such interactions can often be more effective. From an alternate perspective, President Bush after Hurricane Katrina said "It is now clear that a challenge on this scale requires greater federal authority and a broader role for the armed forces...

It is not surprising that command and control was emphasized more in the response phase. It is much easier to implement community based approaches during normal day-to-day non-emergency operations where there are few time constraints and decisions are not urgent. During crisis situations such approaches are far more challenging. What is surprising is that so many preferred a command and control approach during the mitigation and prevention phases; in the personal experience of the authors emergency managers often prefer community engagement in this part of the cycle. One of the most notable features of Fig. 5.18 is a large variance in terms of how the respondents viewed command-and-control versus community-based approaches. Recent academic literature has emphasized problems with top-down management structures and emphasized the importance of including community involvement (e.g. Mileti 1999; Canton 2007). In part this may reflect a tendency for emergency managers to come from military or first responder backgrounds, where this approach works well and is the basis for much of their training and work.

Information sharing came up as an important component of EM as can be seen in the select four quotes from the 'comments' section of the survey, as given below:

> How can you expect members of the community to become involved if they aren't aware of the risks they face?

> Agencies and jurisdictions are dependent upon one another, and all elements of the community are dependent upon a number of critical infrastructure sectors/organizations. Time and again at tabletop exercises, participants discover that there are greater interdependencies

than they had previously anticipated. More robust and systemic information sharing would minimize this kind of surprise, and often these moments of "Oh? We didn't know that!" are the most useful outcomes of multi-party and multi-sector exercises.

Emergency events don't respect geographical boundaries – neither planning nor response should be done in a vacuum.

Public has a right to know and to be informed. Public is responsible for themselves and must educate and prepare accordingly. I believe IS promotes resiliency.

On the contrary, few but striking arguments were presented against information sharing:

Restricting the information received on a community's vulnerabilities, would be in our best interest.

Encouraging information sharing can be a difficult proposition to sell because I have noticed that the general perception is that sharing the results of a HIRA would point out where all the vulnerabilities are. That is true to an extent, but that neglects the 'unknown' vulnerabilities that arise from not knowing how our partner agencies, jurisdictions, and critical sectors rely on us and on one another. The general thinking is still that each agency and jurisdiction looks out for itself, particularly at the senior management levels in my opinion, and there is little focus on the overall, coordinated EM and Disaster Response effort. There is information sharing, to be sure, often at multi-party exercises. It is just that I think there needs to be more, and it needs to be integrated into our processes and the way we think about the how and why of information sharing.

Figure 5.19 illustrates the responses concerning possible hindrances to effective and efficient emergency and disaster management. Even though 79% of the responses blame insufficient funds, some noted that effective programs and policies are also necessary that would make emergency management a greater part of Canadian culture. A large number (75%) also thought political factors are a major barrier to emergency and disaster management. Many felt that there is a lot of "politicking" when it comes to emergency management, thus action is not taken until forced (the 2003 SARS crisis was given as an example). By politicking we refer to the need to engage in the political process to obtain resources, conflicts between

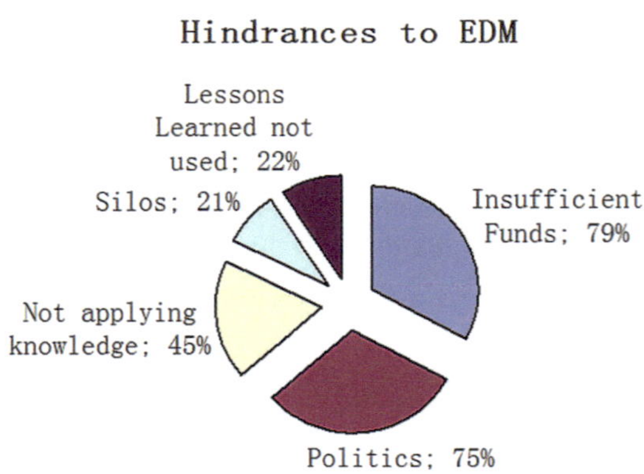

Fig. 5.19 Hindrances to disaster and emergency management (Nirupama and Etkin 2012). The question was designed to seek more than one response from respondents suggesting their preference, which is why the total percentage of all categories combined is more than one hundred percent

various organizations, and difficulties in convincing decision makers to invest in emergency management during non-emergency periods. Some expressed the opinion that politicians do not take EM seriously because they do not see how it would translate into votes.

While few professionals expressed concern that the institution is too bureaucratic and too slow to react, others wanted to see more advertising and awareness campaigns being initiated. Regarding the federal initiative, National Disaster Mitigation Strategy, it was felt that the document, despite being great, lacked a cost/benefit analysis and support for mitigation measures, and instead focussed on the costs of response and recovery. Furthermore, not so optimistic comments were also found in the interviews, suggesting that there is a lack of familiarity with the mitigation strategy, and that governments simply legislate and make more rules without actually implementing much.

Based upon the set of disaster myths listed in Alexander (2002) (see Appendix 4 for "reality"), the following statements were made, with which the practitioner was asked for their level of agreement (1 = agree and 10 = disagree):

- *After a disaster, survivors tend to be dazed and apathetic:* half the respondents did not disagree with the claim that disaster survivors are dazed and apathetic, though some noted that it depends on the type of disaster.
- *Looting is a common and serious problem after disasters:* 46% agree that looting is a common problem after disasters.
- *Disasters give rise to spontaneous displays of antisocial behavior*: fifty eight percent disagree with the assumption that disasters give rise to spontaneous displays of antisocial behaviour. In general, the survey respondents feel that people who survive a disaster display pro-social behaviour in the immediate aftermath.
- *Any kind of aid and relief is useful after disaster, provided that it is supplied quickly enough*: fifty four percent of the respondents said that not all aid or relief is useful after a disaster. This group felt that only solicited aid is useful, as uncoordinated supplies are often a hindrance. One noted that if relief is to be sent, it should be specific to the needs of the people. Another good point was that if the aid required a substantial amount of resources to manage, it would be more of a hindrance than a help, as EM teams might be put to better use the resources elsewhere.
- *People will flee in large numbers from a disaster:* thirty four percent of respondents were confident that people tend to flee.

The responses on disaster myths suggest a positive bias since a large proportion of the emergency managers subscribed to common disaster myths. This is a well known phenomenon – where people tend to view themselves and the world in a considerably more positive light than is objectively justified (Bazerman and Watkins 2004). This result is similar to the findings of Fischer (1998) who found that the frequency of belief in myths by emergency managers was independent of years in the job or experience, but only depended upon level of education.

To summarize, emergency managers felt that priority of the institution should be to mitigate impact on people and assets. They acknowledged that not enough time is allotted to fill gaps and consequently, occasionally same mistakes are repeated. Disaster scenario simulations reveal that there are greater interdependencies than anticipated. The role of Hazard Identification and Risk Assessment at the provincial level was felt vital. It was recognized that risk and vulnerability assessment and prioritization is pivotal for making connections with disaster mitigation strategies and resource allocation. In accordance with what Etkin (1999) presented regarding risk transference and related trends, we notice that the process of risk assessment lies more within the jurisdiction of municipalities and cities than at higher levels of government.

5.5 Exercise

Identify three recent disaster events and analyze them from different perception point of view – (i) how impacted people reacted to the event(s) (ii) what you think of the extent of vulnerability in the impact region(s) (iii) how first responders and the authorities perceived the event(s).

References

Alexander D (2002) Principles of emergency planning and management. Oxford University Press, Oxford

Armenakis C, Nirupama N (2013) Estimating spatial disaster risk in urban environments, geomatics, natural hazards and risk. Taylor & Francis 4(4):289–298

Armenakis C, Nirupama N (2014a) Flood risk mapping for the city of Toronto. Procedia Economics and Finance, Elsevier 18:320–326

Armenakis C, Nirupama N (2014b) Urban impacts of ice storm – Toronto 2013. Nat Hazards 74 (2):1291–1298

Bazerman MH, Watkins MD (2004) Predictable surprises: the disasters you should have seen coming and how to prevent them. Harvard Business School Press, Boston

Birkmann J (ed) (2006) Measuring vulnerability to natural hazards: towards disaster resilient societies. United Nations University Press, New York, 524p

Borum R, Cornell D, Modzeleski W, Jimerson S (2010) What can be done about school shootings? A review of the evidence. Mental Health Law & Policy Faculty Publications/University of South Florida. Paper 534. http://scholarcommons.usf.edu/mhlp_facpub/534

Canton LG (2007) Emergency management: concepts and strategies for effective programs. Wiley, Hoboken, 349p

Chambers R, Conway G (1992) Sustainable rural livelihoods: practical concepts for the 21st century, IDS discussions paper no. 296. University of Sussux, Brighton

City News (2015) Oil and debris spew over Mississauga due to train engine failure – CityNews. http://www.citynews.ca/2015/09/21/oil-and-debris-spew-over-mississauga-due-to-train-engine-failure/. Accessed 21 Sept 2015

Cutter S (2012) Preface. In: Disaster resilience – a national perspective. The National Academies, Washington, DC. www.nap.edu

Der Heide EA (1989) Disaster response: principles of preparation and coordination. Mosby-Year Book, St. Louis, 363p

Etkin D (1999) Risk transference and related trends: driving forces towards more mega disasters. Global Environ Change B Environ Hazard 1(2):51–92

Etkin D (2016) Disaster theory: an interdisciplinary approach to concepts and causes. Butterworth-Heinemann, Elsevier, 359p

Etkin D, Haque EC (2003) An assessment of natural hazards and disasters in Canada. Kluwer, Dordrecht

Ferrier N, Haque E (2003) Hazards risk assessment methodology for emergency managers: a standardized framework for application. Nat Hazards 28:271–290

Fischer HW (1998) Response to disaster: fact versus fiction and its perpetuation – the sociology of disaster, 2nd edn. University Press of America, Landham

Fischer GW, Morgan MG, Fischhoff B, Nair I, Lave LB (1991) What risks are people concerned about? Risk Anal 11(2):303–314

Flint Water Advisory Task Force (2016) Final report. The Office of Governor Rick Snyder State of Michigan. http://www.greatlakeslaw.org/Flint/FWATF_FINAL_REPORT_March_2016.pdf. Retrieved 3 June 2016

Gierlach E, Belsher BE, Beutler LE (2010) Cross-cultural differences in risk perceptions of disasters. Risk Anal 30(10):1539–1549

Globe and Mail (2016) The latest update on Fort McMurray fire, June 10, 2016. http://www.theglobeandmail.com/news/alberta/the-fort-mcmurray-disaster-read-the-latest-weekend/article29930041/. Accessed 30 June 2016

GTZ (2004) Guidelines risk analysis – a basis for disaster risk management. Federal Ministry for Economic Cooperation and Development, Eschborn

Gupta S, Tinker B, Hume T (2016) 'Our mouths were Ajar:' doctor's fight to expose flint's water crisis. http://www.cnn.com/2016/01/21/health/flint-water-mona-hanna-attish/. Posted by Cable News Network. 22 Jan 2016. Accessed 9 Jun 2016

Hébert LV (2016) The concerns of Ontario elementary school teachers on school-based emergencies and emergency preparedness, Major Research Paper, Master of Disaster & Emergency Management, York University, Canada

Hewitt K (1997) Regions of risk: a geographical introduction to disasters. Person Education, London, 388p

IAEA (2013) The Fukushima Daiichi accident, Report of the Director General, International Atomic Energy Agency. http://www-pub.iaea.org/

ICSU (2008) A science plan for integrated research on disaster risk – addressing the challenge of natural and human-induced environmental hazards. International Council for Science, France. ISBN 978-0-930357-66-5

ISDR (2004) Living with risk: a global review of disaster reduction initiatives. United Nations, New York, 429p. www.unisdr.org

Jaeger CC, Renn O, Rosa EA, Webler T (2001) Risk, uncertainty, and rational action. Earthscan Publications, London

Jubril JO (2016) Social capital: what impact do social capital, public education, and outreach initiatives on emergency preparedness have on immigrant residents of Brampton? Major Research Paper, Master of Disaster & Emergency Management, York University, Canada

Kasperson RE, Renn O, Slovic P, Brown HS, Emel J, Goble R, Kasperson JX, Ratick S (1988) The social amplification of risk: a conceptual framework. Risk Anal 8(2):177–187

Lindle JC (2008) School safety: real or imagined fear? Educ Policy 22(1):28–44

Mileti DS (1999) Disasters by design. Joseph Henry Press, Washington, DC

Mileti D, Fitzpatrick C (1991) Communication of public risk: its theory and its application. Sociological Practice Review 2(1):20–28

Mitchell JK (2003) The Fox and the Hedgehog: Myopia about homeland vulnerability in US Policies on terrorism. Terrorism and Disaster: New Threats, New Ideas – Research in Social Problems and Public Policy 11:53–72

Nirupama N (2015) Understanding risk from floods and landslides in the Himalayan region: a discussion to enhance resilience. planet@risk. Global Risk Forum 3(2):231–235
Nirupama N, Etkin D (2009) Emergency managers in Ontario: an exploratory study of their perspectives. J Homeland Security and Emergency Management 6(1):Article 38
Nirupama N, Etkin D (2012) Institutional perception and support in emergency management in Ontario, Canada. Disaster Prevention and Management, Emerald 21(5):599–607
Nirupama N, Jubril JO (2016) Social capital and disaster resilience: a Canadian case study. 6th international disaster and risk conference IDRC Davos, Switzerland, Global Risk Forum, 28 Aug–1 Sept 2016
Nirupama N, Maula A (2013) Engaging public for building resilient communities to reduce disaster impact, special issue on sociological aspects of natural disasters Springer. Nat Hazards 66:51–59
Nirupama N, Simonovic SP (2007) Increase of flood risk due to urbanization: a Canadian example. Natural Hazards, Springer 40:25–41
Nirupama N, Armenakis C, Montpetit M (2014) Is flooding in Toronto a concern? Nat Hazards 72 (2):1259–1264
Olanubi T (2009) Crisis management and psychosocial intervention: perception of immigrants, Major Research Paper, Master of Disaster & Emergency Management, York University, Toronto, Canada
Parsons P, Graney E (2016) Fort McMurray residents flee in the largest fire evacuation in Alberta's history, Edmonton Journal, May 4, 2016
Pelling M (2003) Tracing the roots of urban risk and vulnerability. In: The vulnerability of cities: natural disasters and social resilience. Routledge, Sterling, 212p
Pine JC (2009) Natural hazards analysis: reducing the impact of disasters. CRC Press, Boca Raton, 304p
Pletcher K (2016) Japan earthquake and tsunami of 2011. http://global.britannica.com/event/Japan-earthquake-and-tsunami-of-2011. Accessed 20 June 2016
PSC (2012) Public safety Canada. http://www.publicsafety.gc.ca/prg/em/cdd/srch-eng.aspx
Ramsay C, Shum D (2016) Ocean of fire destroys 2,400 structures but 85% of Fort McMurray still stands. Global News, 10 May 2016
Scanlyn J, Thomas D, Brett J (2013) Understanding social vulnerability. Theoretical framing of worldviews, values, and structural dimensions of disasters. In: Thomas DSK, Phillips BD, Lovekamp WE, Fothergill A (eds) Social vulnerability to disasters. CRC Press, Boca Raton, 513 p
Schneider TH, Basler E, Partner AG (2006) A delicate issue in risk assessment. In: Ammann, Dannenmann, Vulliet (eds) RISK21 -coping with risks due to natural hazards in the 21st century. Taylor & Francis Group, London. 255 pages
Sjöberg L (2001) Limits of knowledge and the limited importance of trust. Risk Anal 21 (1):189–198
Slovic P (1987) Perceptions of risk. Science 236(17 April):280–285
Slovic P (2000) The perception of risk. Earthscan/Routledge, London, 473p
Smith K (2004) Environmental hazards: assessing risk and reducing disaster. Routledge, New York, 306p
Statistics Canada (2008) Census Profile. Available at www.statcan.gc.ca/start-debut-eng.html
Stewart RM (2007) Community perspectives of flood risk and social vulnerability reduction: the case of the Red River Basin, Doctoral Thesis, Natural Resources Institute, University of Manitoba, Canada
Tierney KJ (1999) Toward a critical sociology of risk. Sociol Forum 14(2):215
Tierney KJ (2007) Testimony to house committee on oversight and government reform. 31 July 2007. http://www.iaem.com/committees/governmentaffairs/documents/tierney073107.pdf
Tobin GA, Montz BE (1997) Natural hazards: explanation and integration. Guilford Press, New York
Twigg J (2007) Characteristics of a disaster-resilient community. DFID Disaster Risk Reduction Interagency Coordination Group, 39p

UNISDR (2001) Targeting vulnerability: guidelines for local activities and events. United Nations International Strategy for Disaster Reduction, Geneva
UNSCEAR (2008) The Chernobyl accident. United Nations Scientific Committee on the Effects of Atomic Radiation. http://www.unscear.org/unscear/en/chernobyl.html. Accessed 30 June 2016
USNRC (2016) Fact sheets, United States nuclear regulatory commission report. http://www.nrc.gov/reading-rm/doc-collections/fact-sheets/. Accessed 30 June 2016
Wachinger G, Renn O, Begg C, Kuhlicke C (2013) The risk perception paradox – implications for governance and communication of natural hazards. Risk Anal 33(6):1049–1065
Wahlberg AA, Sjoberg L (2000) Risk perception and the media. J Risk Res 3(1):31–50
White GF, Kates RW, Burton I (2001) Knowing better and losing even more: the use of knowledge in hazards management. Environ Hazard 3(3):81–92
Whyte AV, Burton I (1982) Perception of risk in Canada. In: Burton I, Fowle CD, McCullough RS (eds) Living with risk. University of Toronto, Toronto, pp 39–69
Wisner B, Blaikie P, Cannon T, Davis I (2004) At risk: natural hazards, people's vulnerability and disasters, 2nd edn. Routledge, London, 471p. Accessed

Chapter 6
Disaster Risk Evaluation – Quantitative Methods in Canada

Understanding risk and vulnerabilities forms the basis for a comprehensive approach to addressing the potential impacts of disasters in populated areas. Risk is a function of hazard and people's vulnerabilities. It must be noted that there is a likelihood of specific hazards occurring in certain geographic areas and/or certain times of year. When they do occur, we can expect adverse impacts on people's health, lives, property and/or the environment. To keep the damage to a minimum, these impacts must be understood in advance. This chapter outlines three quantitative risk assessment methods developed by two of Canada's most populous provinces, Ontario and British Columbia, as well as the federal agency, Public Safety Canada.

6.1 Hazard Identification and Risk Assessment Method

The Ministry of Community Safety & Correctional Services of the Province of Ontario, Canada has developed a Hazard Identification and Risk Assessment (HIRA) process for preventing, mitigating, preparing for, and responding to hazards that may occur in the province. A basic conceptual diagram is found in Fig. 6.1. The process was first announced in 2004 with a short document and was revised with substantial changes in 2012. The revised HIRA methodology is risk based, assesses different types of hazards, incorporates both qualitative and quantitative information, includes scientific input, scalable as desired, and most importantly easily understood. The 2012 HIRA document available on the Ministry's website (EMO 2012) is intended to provide guidance on how to conduct a HIRA. The document outlines the methodology and instructions for reasons listed here:

N. Agrawal, *Natural Disasters and Risk Management in Canada*, Advances in Natural and Technological Hazards Research 49,
https://doi.org/10.1007/978-94-024-1283-3_6

Fig. 6.1 The basic steps in developing and maintaining a HIRA

- It helps emergency management professionals prepare for the worst and/or most likely risks.
- Allows for the creation of exercises, training programs, and plans based on the most likely scenarios.
- Saves time and resources by isolating hazards that cannot occur in the designated area.

6.1.1 Purpose

Emergency management programs in Ontario are required by the Emergency Management and Civil Protection Act (Government of Ontario 2009) to be risk-based. Section 4 of this Act requires that 'in developing its emergency management program, every municipality shall identify and assess the various hazards and risks to public safety that could give rise to emergencies and identify the facilities and other elements of the infrastructure that are at risk of being affected by emergencies'. Section 7 of this Act requires that 'in developing an emergency management program, every minister of the Crown and every designated agency, board, commission and other branch of government shall identify and assess the various hazards and risks to public safety that could give rise to emergencies and identify the facilities and other elements of the infrastructure for which the minister or agency, board, commission or branch is responsible that are at risk of being affected by emergencies'.

HIRA is an ongoing, ever evolving process as hazards may evolve and emergency management tools and processes may alter priorities over time.

6.1.2 Scope

The Ontario Provincial HIRA provides guidance for risk assessment for natural, technological and human-induced hazards in accordance with the definition of an emergency in the Emergency Management and Civil Protection Act. This document has been generated for use at a Provincial level; however, the process contained within can be adopted at a ministry, municipal or private sector level. The HIRA can provide all levels with guidance on how to undertake their own risk assessments which can lead to consistent assessments and improved information on risk.

6.1.3 Structure of the HIRA Process (Fig. 6.1)

Hazard Identification entails the following:

- Identify the hazards that have potential to occur in the area; this step may involve looking at past occurrences of hazards.
- A systematic review of all past hazards, their causes, and impacts in order to determine whether they may pose a threat to the area.

Risk Assessment involves the following:

- Likelihood of hazards along with their possible intensity and impact need to be examined.
- Current vulnerabilities (social, physical, economic, environmental) must be taken into account.
- Hazards may occur differently, causing varying degree of damages in different locations in the region, so, it is important to consult with subject matter experts, insurance industry, historical database, and government agencies for better understanding.

Various disaster databases are made available, maintained by a number of national and international organizations that can be used for evaluating hazards' frequency and their impact. The Canadian disaster database (Public Safety Canada 2015), Natural Resources Canada (2016), and International Disaster Database (EMDAT 2015), the National Weather Service (NWS 2009), the National Hurricane Centre (NHC 2016), National Oceanic and Atmospheric Administration (NOAA 2009, 2016), and the World Bank (2016).

Risk Analysis is based on analyzing the information collected according to hazard identification and risk assessment sections. This step is necessary to understand and determine frequency and potential impact of various hazards and prioritize them for the purpose of emergency management planning and resource allocation.

Monitor and Review is vital due to the nature of a HIRA that is an ongoing process. Changes are possible in hazards' frequencies, vulnerabilities of people and property, and mitigation practices therefore, monitoring and review process is an essential component in a HIRA process. Over time, with periodic review, all hazards will eventually get adequately examined and changing risks may be captured for better planning in future.

6.1.3.1 Risk Equation

$$Risk = Frequency \times Consequence \times Changing\ Risk \qquad (6.1)$$

6.1.3.2 Frequency Estimation

For the calculation of *Frequency*, Table 6.1 provides a guideline.

Frequency should be calculated whenever possible based on existing data from official and/or scientific sources. It should be remembered that some hazards do not have a long historical record and that their frequencies can be estimated based on the best sources available.

6.1.3.3 Consequence Estimation

Consequence of hazards is evaluated based on harm incurred to humans, property, critical infrastructure or the environment. Hazards that have the potential to occur frequently such as flooding tend not to take human lives, but cause infrastructural damage and disruption in essential services. The hazards that have lower likelihood of occurrence can be devastating to the area. For example, between 1900 and 2014, eastern Canada experienced 54 major winter storms and cold events, but only few caused fatalities as shown in Fig. 6.2 (Armenakis and Nirupama 2014).

It is proposed here that only damaging events are counted when determining the values for consequence, not non-damaging events. A hazardous event is considered

Table 6.1 Hazard frequency examples

Frequency (to be used in Eq. 6.1)	Category	Percent chance of occurrence (in any year)	Description
1	Rare	<1%	Hazards with return periods[a]>100 years
2	Very unlikely	Between 1–2%	Occurs in the province every 50–100 years and includes hazards that have not occurred in the province but are reported to be more likely to occur in the near future
3	Unlikely	Between 2–10%	Occurs in the province every 20–50 years
4	Probable	Between 10–50%	Occurs in the province every 5–20 years
5	Likely	Between 50–100%	Occurs in the province >5 years
6	Almost certain	100%	Hazards that occur annually

Also known as recurrence interval, a return period is the average length of time in years for a hazardous event of given magnitude to be equaled or exceeded. For example, the probability of the occurrence of a flood event of a 100 year return period would be 0.01 percent in any given time, any given year. It is a statistical measurement based on historic data over a long period of time

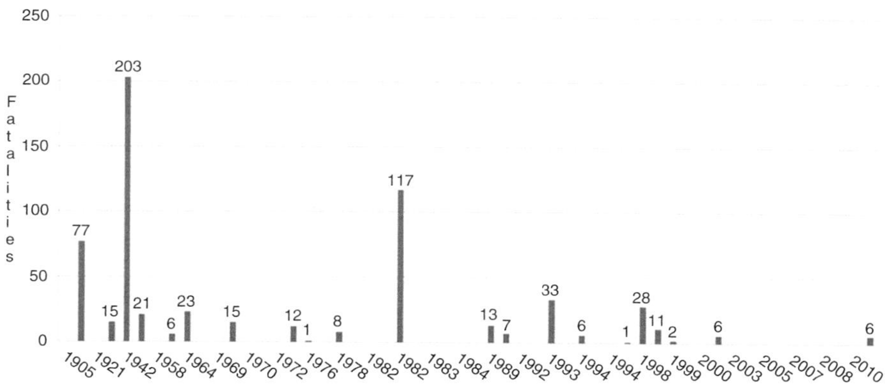

Fig. 6.2 Fatalities of historical winter storms in Eastern Canada during 1900 to 2014

to exceed the threshold for damage when any of the conditions summarized (adapted from the original document) below are met (Ferrier and Haque 2003; Jonkman et al. 2003; ADRC 2005):

- Assistance required from another community or province or country.
- Multiple fatalities and/or injuries reported resulting directly from the hazard or its immediate impact.
- Evacuations greater than 100 people.
- Severe damage to property or infrastructure reported. This would include partial or complete destruction of at least one building or widespread, less severe damage.
- Severe environmental damage deemed to be reported on by official sources requiring a form of response or monitoring. In particular, damage considered to be widespread, significantly affecting wild species, natural resources, or critical infrastructure or significant business/financial loss to an industry.
- Disruption/impact on critical infrastructure or essential service(s) for more than 10,000 people (this could be scaled to an entire town if the population is less than 10,000).

In the HIRA methodology, consequences/impacts are divided into six categories:

Social Impacts account for the physical health of people, and is further broken down into three groups; fatalities, injuries or evacuations.

Property Damage includes physical damage to buildings, structures and other forms of property, such as crops.

Critical Infrastructure Service Disruptions/Impact accounts for the interdependent, interactive, interconnected networks of institutions, services, systems and processes that meet vital human needs, sustain the economy, protect public safety and security, and maintain continuity of and confidence in government.

Environmental Damage accounts for the environment, including the soil, water, air and/or plants and animals.

Business/Financial Impact is about negative economic consequences of the occurrence of a hazard.

Psychosocial Impacts include the negative response of community to a hazard, including self-evacuation, mass hysteria, hoarding etc.

Tables 6.2, 6.3, 6.4, and 6.5 outlines the consequence scoring guideline for the six categories of consequences described above.

The magnitude categories in this HIRA methodology are a scale of impact, rather than a prioritization. The same value in two categories does not mean that the consequences of the two are equal and interchangeable.

After the consequences for the six categories are determined the consequence scores are summed up and reclassified using Table 6.6 for the final consequence evaluation.

Changing Risk – hazards and their risks do not remain static over time, and the frequency and consequence of future hazards can be affected by factors such as climate change and mitigation measures in place. Eq. (6.2) is used to calculate this variable with careful consideration given to the units of various variables.

$$\textit{Changing Risk} = \textit{Change in Frequency} + \textit{Change in Vulnerability} \tag{6.2}$$

Where, *Change in Frequency* is determined using Table 6.7 in which four questions are posed to be answered.

In Eq. (6.12), *Change in Vulnerability* is determined using Table 6.8.

The scores for the two components – Change in Frequency and the Change in Vulnerability are to be added together to get the value of Changing Risk.

Table 6.9 gives the guideline to determine hazard risk levels based on the Risk value determine using the HIRA approach of incorporating the *Changing Risk* component.

6.1.4 Application of HIRA for Hazards in Ontario, Canada

Table 6.10 shows qualitative assessment of Risk level of hazards using both conventional equation (Consequence × Frequency (C × F)) and the risk equation (Eq. 1) proposed in the HIRA process (Consequence × Frequency × Changing Risk (C × F × CR)). The values of Consequences and Frequency have been adopted from the Ontario Hazards database as given in Appendix 5.

Table 6.2 Social impact – guideline for quantitative scores (HIRA 2012)

Social impact – fatalities			Social impact – injuries			Social impact – evacuation		
Consequence	Category	Description	Consequence	Category	Description	Consequence	Category	Description
0	None	Not likely to result in fatalities within the province.	0	None	Not likely to result in injuries within the province.	0	None	Not likely to result in an evacuation shelter-in-place orders, or people stranded.
1	Minor	Could result in fewer than five fatalities within the province.	1	Minor	Could injure fewer than 25 people within the province.	1	Minor	Could result in fewer than 100 people being evacuated, sheltered-in-place or stranded.
2	Moderate	Could result in 5–10 fatalities within the province.	2	Moderate	Could injure 25–100 people within the province.	2	Moderate	Could result in 100–500 people being evacuated, sheltered-in-place or stranded.
3	Severe	Could result in 10–50 fatalities within the province.	3	Severe	Could injure +100 people within the province.	3	Severe	Could result in more than 500 people being evacuated, sheltered-in-place or stranded.
4	Catastrophic	Could result in +50 fatalities within the province.						

Table 6.3 Property damage – guideline for quantitative scores (HIRA 2012)

Property damage		
Consequence	Category	Description
0	None	Not likely to result in property damage within the province.
1	Minor	Could cause minor and mostly cosmetic damage.
2	Moderate	Localized severe damage (a few buildings destroyed).
3	Severe	Widespread severe damage (many buildings destroyed).

Table 6.4 Critical infrastructure service impact and environmental damage – guideline for quantitative scores (HIRA 2012)

Critical infrastructure service impact			Environmental damage		
Consequence	Category	Description	Consequence	Category	Description
0	None	Not likely to disrupt critical infrastructure services.	0	None	Not likely to result in environmental damage.
1	Minor	Could disrupt 1 critical infrastructure service.	1	Minor	Could cause localized and reversible damage. Quick clean up possible.
2	Moderate	Could disrupt 2–3 critical infrastructure services.	2	Moderate	Could cause major but reversible damage. Full clean up difficult.
3	Severe	Could disrupt more than 3 critical infrastructure services.	3	Severe	Could cause severe and irreversible environmental damage. Full clean up not possible.

6.2 Hazard Risk and Vulnerability Assessment Tool

The Ministry of Public Safety and Solicitor General, British Columbia developed an effective tool to assess hazard risk and vulnerability assessment for communities to help the province with risk reduction planning. Developed in 2004, the HRVA (2004) is still a critical part of every emergency program and is a requirement mandated by the Local Authority Emergency Management Regulation of the BC Emergency Program Act. Section 2(1) of this regulation required local authorities to prepare emergency plans that reflect the *local authority's assessment of the relative risk of occurrence and the potential impact on people and property of the emergencies or disasters that could affect all or any part of the jurisdictional area for which the local authority has responsibility.*

Table 6.5 Business/financial impact and psychosocial impact – guideline for quantitative scores (HIRA 2012)

Business/financial impact			Psychosocial impact		
Consequence	Category	Description	Consequence	Category	Description
0	None	Not likely to disrupt business/financial activities.	0	None	Not likely to result in significant psychosocial impacts.
1	Moderate	Could result in losses for a few businesses.	1	Moderate	Significant psychosocial impacts including limited panic, hoarding, self evacuation and long-term psychosocial impacts.
2	Severe	Could result in losses for an industry.	2	Severe	Widespread psychosocial impacts, e.g. mass panic, widespread hoarding and self-evacuation and long-term psychological impacts.

Table 6.6 Reclassification of consequence based on the six categories of consequences (HIRA 2012)

Sub variable total	Consequence	Description
1–4	1	Minor
5–6	2	Slight
7–8	3	Moderate
9–10	4	Severe
11–12	5	Very severe
+13	6	Catastrophic

Table 6.7 Guideline to determine the value of change in frequency

	Question	Change in frequency value
1	Is the number of reported non-emergency occurrences of the hazard increasing?	= **2** – if the answer to two or more questions is a YES
2	Is human activity (e.g. population expansion, altering of drainage flow patterns) likely to lead to more interaction with the hazard or an increase in frequency?	
3	Is there an environmental reason (e.g. climate change) why the frequency of this hazard may increase?	= **1** – if only one answer is a YES
4	Are human factors such as business, financial, international practices more likely to increase the risk?	

Table 6.8 Guideline to determine the value of change in vulnerability

	Question	Change in vulnerability value
1	Is a large percentage of the population vulnerable to this hazard or is the number of people vulnerable (see vulnerable groups) to this hazard increasing?	= **2** – if the answer to two or more questions is a YES
2	Does critical infrastructure reliance or our 'just-on-time' delivery system (e.g. stores not keeping a supply of food and relying on frequent shipments for restocking) make the population more vulnerable?	= **1-** if only one answer is a YES
3	Are response agencies not aware of, practiced and prepared to response to this hazard?	
4	Are no prevention/mitigation measures currently in use for this hazard?	

Table 6.9 Risk level of hazard using Eq. (6.11) proposed in the HIRA process

Risk level	Description	Hazards
>50	Extreme	Flood, Forest/Wildland Fire, Freezing Rain, Hazardous Materials Incident, Human Health Emergency, Snowstorm/Blizzard, Tornado
41–50	Very high	Drinking Water Emergency, Geomagnetic Storm, Oil/Natural Gas Emergency, Terrorism/CBRNE
31–40	High	Agricultural and Food Emergency, Critical Infrastructure Failure, Drought/Low Water, Nuclear Facility Emergency
21–30	Moderate	Civil Disorder, Cyber Attack, Earthquake, Human-Made Space Object Crash, Landslide, Transportation Emergency, Windstorm
11–20	Low	Building/Structural Collapse, Dam Failure, Explosion/Fire, Extreme Temperatures, Hurricane, Natural Space Object Crash, Radiological Emergency
<10	Very low	Energy Emergency (Supply), Erosion, Fog, Hail, Land Subsidence, Lightning, Mine Emergency, Sabotage, Special Event, War and International Emergency

6.2.1 Purpose

The purpose of Hazard, Risk and Vulnerability Analysis (HRVA) is: to help a community make risk-based choices to address vulnerabilities, mitigate hazards and prepare for response to and recovery from hazard events (HRVA 2004).

"Risk-based" means informed choices of alternative unwanted outcomes. In other words, communities make risk reduction choices based on the acceptability of consequences and the frequency of hazards.

Table 6.10 Risk levels using both conventional eq. (C × F) and the risk equation (Eq. 6.1) proposed in the HIRA (C × F × CR)

Hazard	Risk = C × F	C × F × CR	Hazard	Risk = C × F	C × F × CR
Agricultural and food emergency	Moderate	High	Human health emergency	Very high	Extreme
Building/structural collapse	Moderate	Low	Human-made space object crash	Moderate	Moderate
Civil disorder	Moderate	Moderate	Hurricane	Moderate	Low
Critical infrastructure failure	Moderate	High	Land subsidence	Very low	Very low
Cyber attack	Low	Moderate	Landslide	Moderate	Moderate
Dam failure	Moderate	Low	Lightning	Low	Very low
Drinking water emergency	Very high	Very high	Mine emergency	Low	Very low
Drought/low water	Moderate	High	Natural space object crash	Low	Low
Earthquake	Low	Moderate	Nuclear facility emergency	Moderate	High
Energy emergency (supply)	Very low	Very low	Oil/natural gas emergency	Very high	Very high
Erosion	Very low	Very low	Radiological emergency	Low	Low
Explosion/fire	High	Low	Sabotage	Very low	Very low
Extreme temperature	Moderate	Low	Snowstorm/blizzard	Extreme	Extreme
Flood	Extreme	Extreme	Special event	Low	Very low
Forest/wildland fire	Extreme	Extreme	Terrorism/CBRNE	Moderate	Very high
Freezing rain	Extreme	Extreme	Tornado	Extreme	Extreme
Geomagnetic storm	High	Very high	Transportation emergency	High	Moderate
Hail	Low	Low	War and international emergency	Very low	Very low
Hazardous materials incident	Extreme	Extreme	Windstorm	Moderate	Moderate

6.2.2 Objective

The purpose of hazard, risk and vulnerability analysis planning is to anticipate problems and possible solutions to help save lives and property, reduce damage, and speed a community's recovery. HRVA is designed to help us work towards establishing disaster-resilient communities.

To gain the support of the local Emergency Program Executive and Management committees, the HRVA committee chair must first become familiar with the HRVA process and the tool kit and explain it to the Executive and Management Committees in detail. The HRVA process has to be fully supported by Regional Managers and Emergency Management Analysts of the Provincial Emergency Program. The Executive committee must guide the local Emergency Program for the community, which usually consists of the Mayor/City Chief, a Counselor, and an appointed Officer. The Management committee's mandate is to manage the local emergency program on a daily basis, and usually consists of the Chair/Coordinator, an appointed officer, about 6 to 10 members from agencies that will have direct functional responsibilities during a major emergency. The combination of the two committees and subject matter experts should form an HRVA Advisory Committee.

The online HRVA Web tool at http://hrva.embc.gov.bc.ca/toolkit.html includes helpful tips and must be used for storing contact information for hazard subject matter experts (such as geologists or hazardous materials specialists).

HRVA should be conducted annually or each time there is a change in the hazards or vulnerabilities in your communities.

6.2.3 Steps Required for HRVA

It is an eight step process as shown in Fig. 6.3.

In this chapter, the focus is on hazard and vulnerability identification, risk analysis, and risk evaluation. For the complete HRVA tool kit, please visit http://hrva.embc.gov.bc.ca/toolkit.html.

After forming the required committees, the task force should gather risk information pertaining to all the communities in the jurisdiction. This would include:

- List of potential hazards in the region– natural, biophysical, technological, human induced, etc.
- List of communities, with data on their demography, and socio-economic details
- Land use type: residential and commercial properties, parks, water courses and bodies, spaces for public use, shopping centres, community centres, etc.
- List of critical facilities: hospitals, schools, community centres, police stations, fire stations, etc.
- List of critical infrastructure: telecommunication and transportation networks, water supply and electricity distribution networks, natural gas lines, etc.
- List of major employers in the area

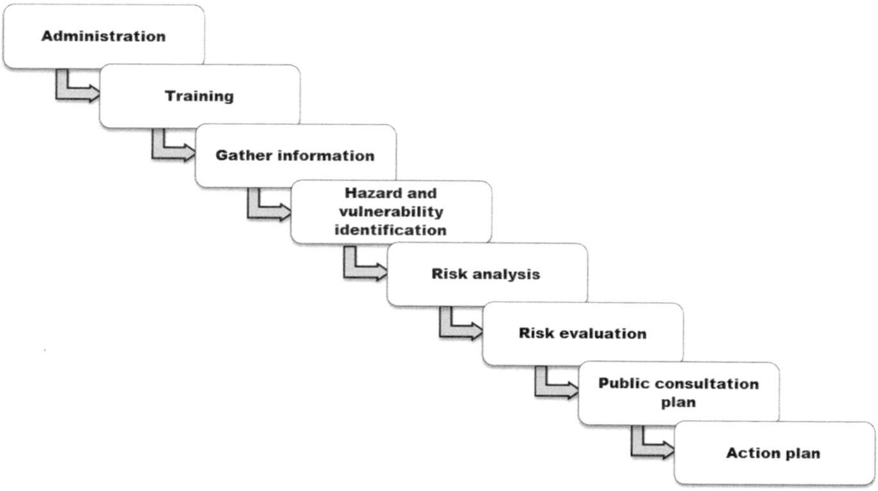

Fig. 6.3 Eight steps involved in HRVA

The above information should also be made available in maps for visualization in spatial format. In most cases, maps are already being used in municipalities and cities for planning purposes.

The next step is to organize a workshop with the HRVA Advisory Committee to review the information gathered and identify potential hazards and their potential impacts. A description of hazards' scenarios and their impacts enables the process of ranking of the severity of consequences and allows for identification of vulnerabilities, which leads to the development of risk reduction measures. This should follow by cost-benefit analysis of potential risk reduction options and actions required for the best and most feasible measures to be implemented.

It is important to identify the likelihood of a hazard scenario and describe the severity of impact because the more severe a scenario is, the less likely it is to occur. Also, mitigation and response capabilities and capacities need to be taken into account at this point in order to accurately assess the impact of potential scenarios considered.

Following the workshop, provisions of a hazard list based on historical data for the region must be made available. For example, in Canada, Public Safety Canada is the federal agency that provides a list of natural hazards of Canada. The search engine also allows the user to filter their query based on specific parameters, such as, province, dates, hazard type etc. The database also gives a general understanding and definitions of hazards as well as what constitutes a disaster that is the aftermath of the hazard. The Canadian Disaster Database (CDD) (PSC 2015) is an extension of the list of hazards, and provides an interactive map (Fig. 6.4) to users to select various variables to extract specific information.

Fig. 6.4 Canadian disaster database interactive map (PSC 2015)

The definition of disaster according to the Emergency Management Framework of Canada (Public Safety Canada PSC 2015) is an event that meets one or more of the following criteria:

- 10 or more people killed
- 100 or more people affected/injured/infected/evacuated or homeless
- an appeal for national/international assistance
- historical significance
- significant damage/interruption of normal processes such that the community affected cannot recover on its own

The database is also available in the classic format, not requiring JavaScript to access it. Each data field has been defined in the database as shown in Table 6.11.

For details, it is best to visit the website of Public Safety Canada at https://www.publicsafety.gc.ca/cnt/rsrcs/cndn-dsstr-dtbs/index-eng.aspx

Major categories of hazards are defined in Table 6.12. It is imperative to define hazards for better understanding, clarity, and consistency.

Table 6.11 data fields in the database with their description

Data field	Description
Disaster type	The type of disaster (e.g. flood, earthquake, etc.) that occurred.
Date of event	The date a specific event took place.
Specific location	The city, town or region where a specific event took place.
Description of event	A brief description of a specific event, including pertinent details that may not be captured in other data fields (e.g. amount of precipitation, temperatures, neighbourhoods, etc.)
Fatalities	The number of people killed due to a specific event.
Injured/infected	The number of people injured or infected due to a specific event.
Evacuees	The number of individuals evacuated by the government of Canada due to a specific event.
Latitude & longitude	The exact geographic location of a specific event.
Province/territory	The province or territory where a specific event took place.
Estimated total cost	A roll-up of all the costs listed within the financial data fields for a specific event.
DFAA payments	The amount, in dollars, paid out by Disaster Financial Assistance Arrangements (Public Safety Canada) due to a specific event.
Insurance payments	The amount, in dollars, paid out by insurance companies due to a specific event.
Provincial/territorial costs/payments	The amount, in dollars, paid out by a Province or Territory due to a specific event.
Utility costs/losses	The amount of people whose utility services (power, water, etc.) were interrupted/affected by a specific event.
Magnitude	A measure of the size of an earthquake, related to the amount of energy released.
Other federal institution costs	The amount, in dollars, paid out by other federal institutions.

6.2.4 Understanding Hazards

In Ontario, between 1900 and 2013 (113 years), a total of 160 events have been recorded, which have included hydrological, geological, and biophysical events. This averages to roughly 1.5 hazardous events per year. Figure 6.5 demonstrates hazard type and their frequency.

6.2.5 Identification of Vulnerabilities

Vulnerabilities can be classified into four major categories:

1. Social vulnerabilities
2. Physical vulnerabilities
3. Economic vulnerabilities
4. Environmental vulnerabilities

Table 6.12 Major categories of hazards

Hazard categories	Hazard
Geological	Earthquakes
	Avalanches
	Landslides caused by earthquake
	Tsunamis
Hydrometeorological	Storms (snow, ice, hail, blizzards, lightening, wind)
	Hurricanes
	Tornadoes
	Heat waves, cold wave
	Floods, droughts
	Landslides due to heavy rainfall
Fire	Fire (urban, rural, interface)
Epidemics or pandemics	Human diseases
	Plant diseases
	Pest infestations
Dam failure	Dam failure (including foundations and abutments)
Technological (including human error and explosions)	Transportation (road, air, marine)
	Industrial
	Nuclear
Terrorism	Explosions, suicide bomb, mass shooting, etc.

Social vulnerabilities

- Elderly – over 65 years of age, retirement homes
- Young children – under 6 years of age, primary and middle schools, daycare centres
- Infirm – hospitals, rehabilitation facilities
- Disability issues – physical (visible), physical (non-visible such as vision, hearing, autism, etc.), mental
- Gender – women with young children, victims of domestic violence or abuse
- Minority groups – visible, aboriginal, LGBTQ
- High density – shopping malls, community centres, places of worship, entertainment
- Other – incarcerated

Physical vulnerabilities

- Critical structures – bridges, transportation networks (roadways, railways, airways, waterways)
- Telecommunications systems and electrical distribution network
- Water reservoirs, dams, supply systems
- Oil and gas storage, distribution, and transportation

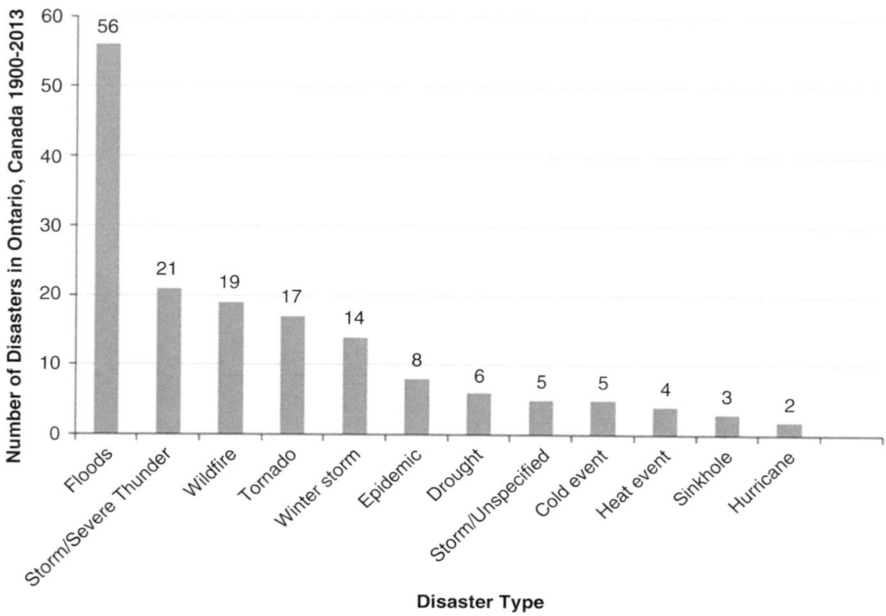

Fig. 6.5 Hazard types with frequency in the Province of Ontario, Canada during 1900–2013. PSC database was used for this plot (Nirupama et al. 2014b)

- Hazardous waste sites – mine tailing ponds, large garbage collection and recycling plants
- Heritage and historic sites
- Neighbourhoods in close proximity to hazardous locations

Economic vulnerabilities

- Lack of employment opportunity and diversity
- Farms and livestock – inherent uncertainty
- Limited access to credit
- Minimal access to critical services
- Insufficient or no insurance
- Low income

Environmental vulnerabilities

- Natural resources, ecologically valuable, sensitive areas – forests, grasslands, coasts, and wetlands
- Water systems – rivers, canals, lakes, ground water aquifers
- Parks, marinas, fisheries, estuaries

Fig. 6.6 Critical facility –a community centre near propane explosion site in Toronto, 2008

6.2.6 Impact Assessment and Ranking

According to the conventional approach, disaster risk is evaluated as:

$$Risk = Likelihood \times Consequences \tag{6.3}$$

Where, Likelihood for a hazard is based on the frequency of occurrence in the past. According to the Environment Canada recommendation at least 30 years of data is essential to determine the frequency of occurrence. However, longer time series, preferably a minimum of 100 years, should be used to ensure the quality and accuracy of the results. The Canadian Disaster Database has been available since the 1900s making it a reliable source to help evaluate Likelihood. The International Disaster Database, EMDAT (2015 www.emdat.be) is also a useful resource for this purpose.

Consequences can be assessed using the following 7 categories. Some photo examples from the 2008 Toronto propane explosion, near Downsview area in Toronto, are shown in Figs. 6.6, 6.7, 6.8, and 6.9.

- Fatality
- Injury
- Critical facilities
- Lifelines (critical infrastructure)
- Property
- Environment
- Economic & social

Fig. 6.7 Economic impact – Bombardier facility, the largest employer in the area

Fig. 6.8 Critical facility –a school close to the explosion site

Fig. 6.9 Explosion as seen from midtown Toronto (public domain Wikipedia.org)

Each category is further explained in terms of ranking criteria in Tables 6.13 and 6.14.

Risk profile is obtained by inputting the data into the online tool at (HRVA 2004).

An example of risk profile is shown in Fig. 6.10.

Two screen shots of the online tool are shown in Figs. 6.11 and 6.12 for better understanding how the information on hazards likelihood and their impact is fed into the tool. It is a well thought, comprehensive application that is user friendly.

6.2.7 Case Studies on the Application of the HRVA

(i) Natural hazard example – the 1998 Ice Storm in Montreal, Canada
(ii) Technological hazard example – The 2008 Propane Explosion in Toronto, Canada

6.2.7.1 The 1998 Ice Strom with Focus on the City of Montreal, Canada

Information gathering: in January 1998, the most devastating and catastrophic ice storm in Canadian history occurred. Many regions were caught in the storm

Table 6.13 Ranking criteria for the impact assessment categories

	Category	Examples	Criteria	Description	Rank
1	Fatality		0–4	Very low	1
			4–10	Low	2
			10–50	High	3
			50 +	Very high	4
2	Injury	Includes homeless, missing	0–4	Very low	1
			4–50	Low	2
			50–2000	High	3
			2000 +	Very high	4
3	Critical Facilities	Hospitals, emergency services	Temporary relocation	Very low	1
			Closure of few days	Low	2
			Loss of 50% of capability	High	3
			Long term disruption	Very high	4
4	Lifeline	Water, gas, electricity	Temporary situation	Very low	1
			Interruption: few days	Low	2
			Interruption: week	High	3
			Long term disruption	Very high	4
5	Property damage	Public, commercial, private	Minimal damage	Very low	1
			Localized damage	Low	2
			Localized & severe	High	3
			Widespread & severe damage	Very high	4
6	Environmental impact	Green/park, asbestos exposure, toxic releases	Minimal damage	Very low	1
			Localized damage	Low	2
			Localized & severe	High	3
			Widespread & severe damage	Very high	4
7	Economic and social impact	Industries, businesses and employers	Temporary impact	Very low	1
			Temporary & widespread	Low	2
			Extended & widespread	High	3
			Permanent impact	Very high	4

Table 6.14 Likelihood ranking

Return period in years	Measure of likelihood	Rank
1–3	Very likely	6
3–10	Likely	5
10–30	Slight chance	4
30–100	Unlikely	3
100–200	Highly unlikely	2
200–300	Very rare	1

Modified from HRVA (2004)

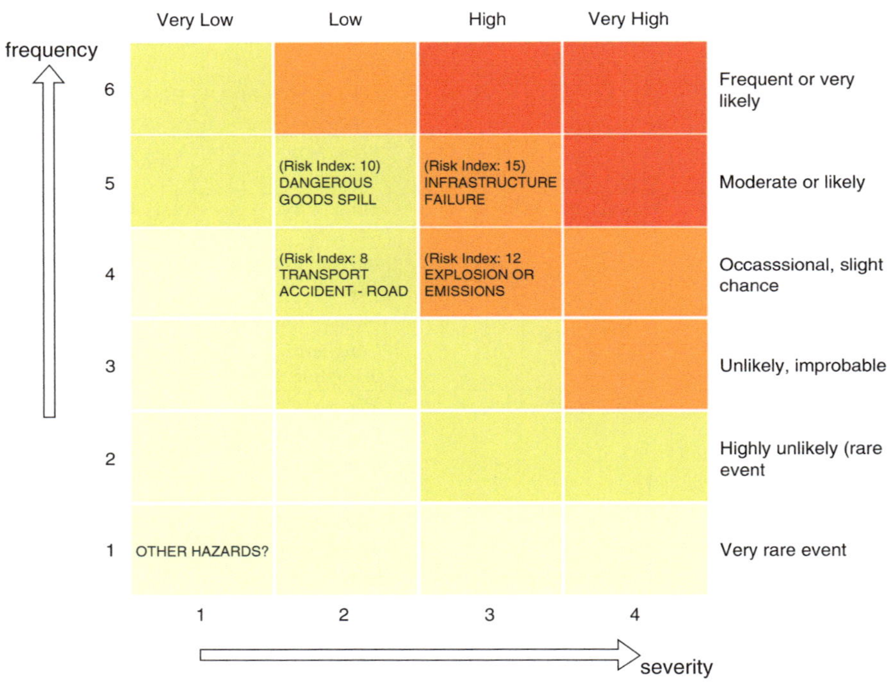

Fig. 6.10 An example of risk profile obtained using the HRVA online tool http://hrva.embc.gov.bc.ca/toolkit.html

including, Atlantic Canada, Eastern Ontario and Southern Quebec. While many regions were badly hit, the city of Montreal was hit quite severely. The rapid urbanization, growing population, and inadequate preparedness of the city made the region extremely vulnerable to the event. The City of Montreal is the second-largest city in Canada and the largest city in the province of Quebec. The City‘s population is 1.8 million people, with a population density of 3,629 inhabitants/km^2. The industrial sector is made up of approximately 3,000 large, midsized, and small

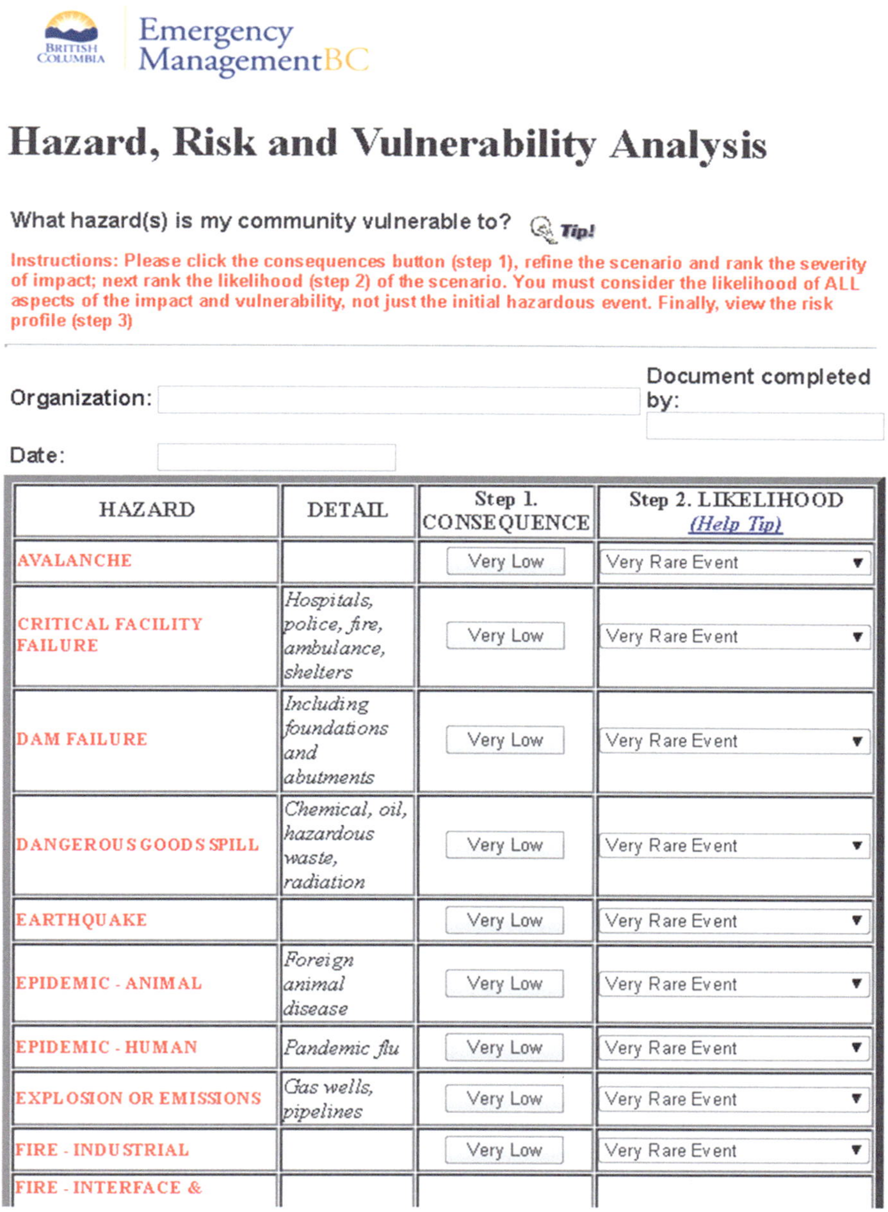

British Columbia Emergency ManagementBC

Hazard, Risk and Vulnerability Analysis

What hazard(s) is my community vulnerable to? Tip!

Instructions: Please click the consequences button (step 1), refine the scenario and rank the severity of impact; next rank the likelihood (step 2) of the scenario. You must consider the likelihood of ALL aspects of the impact and vulnerability, not just the initial hazardous event. Finally, view the risk profile (step 3)

Organization:

Document completed by:

Date:

HAZARD	DETAIL	Step 1. CONSEQUENCE	Step 2. LIKELIHOOD *(Help Tip)*
AVALANCHE		Very Low	Very Rare Event
CRITICAL FACILITY FAILURE	*Hospitals, police, fire, ambulance, shelters*	Very Low	Very Rare Event
DAM FAILURE	*Including foundations and abutments*	Very Low	Very Rare Event
DANGEROUS GOODS SPILL	*Chemical, oil, hazardous waste, radiation*	Very Low	Very Rare Event
EARTHQUAKE		Very Low	Very Rare Event
EPIDEMIC - ANIMAL	*Foreign animal disease*	Very Low	Very Rare Event
EPIDEMIC - HUMAN	*Pandemic flu*	Very Low	Very Rare Event
EXPLOSION OR EMISSIONS	*Gas wells, pipelines*	Very Low	Very Rare Event
FIRE - INDUSTRIAL		Very Low	Very Rare Event
FIRE - INTERFACE &			

Fig. 6.11 Screen shot of the online tool hazard likelihood page http://hrva.embc.gov.bc.ca/toolkit.html

British Columbia Emergency Management BC

Consequence Analysis: AVALANCHE

Close and Return to Hazard Analysis Tip!

Organization:

Document completed by:

Date:

VULNERABILITY & CAPABILITY ISSUES	DETAIL	ASSESSMENT
VULNERABLE POPULATION	*Density, age, gender, ethnicity, social-economic status*	False ▼
VULNERABLE AREAS CLOSE TO HAZARD (AVALANCHE)	*People, buildings, infrastructure, property, environment*	False ▼
INADEQUATE ALERT OR EVACUATION PLANS	*Dated notification list, inadequate maps, routes or shelters*	False ▼
LIMITED CAPABILITY TO RESPOND OR RECOVER	*Inadequate resources, training or equipment*	False ▼
DATED RISK ANALYSIS, RESPONSE OR RECOVERY PLANS		False ▼
INADEQUATE HAZARD-SPECIFIC CONTINGENCY PLANS	*If not covered by the general emergency response plan*	False ▼

Fig. 6.12 Screen shot of the online tool vulnerability and impact evaluation page front http://hrva.embc.gov.bc.ca/toolkit.html

IMPACT	DETAIL	ASSESSMENT
POTENTIAL EXTENT OF DEATHS		Very Low: 0 - 4 ▼
POTENTIAL EXTENT OF INJURY	*Injury, health effects, suffering*	Very Low: 0 - 4 ▼
POTENTIAL EXTENT OF DAMAGE OR LOSSES TO CRITICAL FACILITIES	*Hospitals, police/fire/ambulance, schools, shelters*	Very Low: Temporary relocation ▼
POTENTIAL EXTENT OF DAMAGE OR LOSSES TO LIFELINES	*Transportation, power, water, gas, telephone*	Very Low: Temporary interruption ▼
POTENTIAL EXTENT OF DAMAGE OR LOSSES TO PROPERTY	*Public, commercial, private*	Very Low: Minimal damage ▼
POTENTIAL EXTENT OF DAMAGE OR LOSSES TO ENVIRONMENT		Very Low: Minimal damage ▼
POTENTIAL EXTENT OF ECONOMIC OR SOCIAL IMPACT		Very Low: Temporary impact ▼

Close and Return to Hazard Analysis

Fig. 6.12 (continued)

companies. About 180 of these firms process, produce, or store hazardous materials. In case of a hazardous event these industrial firms become prone to the possibility of industrial accidents, exposing neighbourhoods to serious threat. In following sections, gathered information on vulnerabilities and impact is presented for the purpose of carrying out risk evaluation using the tool. Montreal is also at the heart of Canada's French population, with the majority of communication in the French language, accessibility was low for those who spoke English, also approximately 13% of the area's population is defined as visible minority, which again meant many elderly might not have been able to communicate and obtain required help which they might have been able to otherwise. There are 10 senior homes in the city and 5 major hospitals, these facilities need to be taken care of extra carefully as the hospitals became the place for refuge. Also, it was important for the hospitals to maintain a high supply of non-perishable food items which were during the ice storm used as food aid. The city of Montréal is also home to 6 Universities, 8 Junior colleges, and over 200 schools. During school hours children become prone to physical damage caused by the ice storm, power outage etc. Schools were closed for a few days. There are approximately 30 shopping centres in the city alone, all of which became exposed to the disastrous events. The business owners of these shopping malls faced significant economic loss due to power outages and also because many could not get to work due to road conditions. People, in times of crisis, move towards staple goods and discretionary spending drops therefore causing a decline in sales for weeks to follow for many industries. Montreal is a well-developed city and has enormous network of physical assets and critical infrastructure. The lengthened freezing rain and snow storm damaged about 120,000 km of telephone and powerlines. Approximately another 1000 steel electrical posts and another 35,000 wooden utility poles had fallen down further hindering the return of electricity in the city. The ice storm left many industries handicapped for days to follow which resulted in a huge loss for the area's economic wellbeing. Nine out of the 22 major industry groups which accounted for almost 80% of *manufacturing* declined in the area. Quebec's *maple syrup* industry accounts for 70% of the world's maple syrup supply. Many Quebec maple syrup producers were ruined with much of their sugar bush permanently destroyed. The domestic sugar maples are more vulnerable to ice damage than natural ones, and the falling branches and heavy ice buildup also caused a lot of damage to many of the pipelines. As estimated by the Ontario Maple Syrup Producers Association, the loss incurred by the industry would take up to 40 years to recover and for the area to restore its high level of production on maple syrup. The storm covered nearly 30% of the area where the *dairy* cows reside. The storm hit areas suffered massive electricity shortages, resulting in no electricity to operate milking machines, since the cows were not milked for days this made them vulnerable to inflammation in their udder. As compared to declines in other sectors the fall down in the dairy industry was comparatively small, but there was a concern that the cows which faced and survived the storm were never able to regain their pre-storm productivity levels.

The Insurance Bureau of Canada reported a soaring high insurance claims (535,200) following the ice storm. These claims added up to approximately $790 million. The nation's economic output fell by almost 0.7% in January alone. The decline in GDP was largely due to the fall off of many industries which were down due to the loss in electricity and the electric power industry itself resulted in a decline of 14.2%. Additionally, the goods-producing industries fell 1.4%. According to the Conference Board of Canada the manufacturing, transportation, communications and retail sectors sustained a short-term loss of $1.6 billion to Canada's economic output. Over 5000 **trees** in Montreal's famous Mount Royal Park were cut or trimmed due to damages. Visual information adds value to the information gathered, as shown in Fig. 6.13.

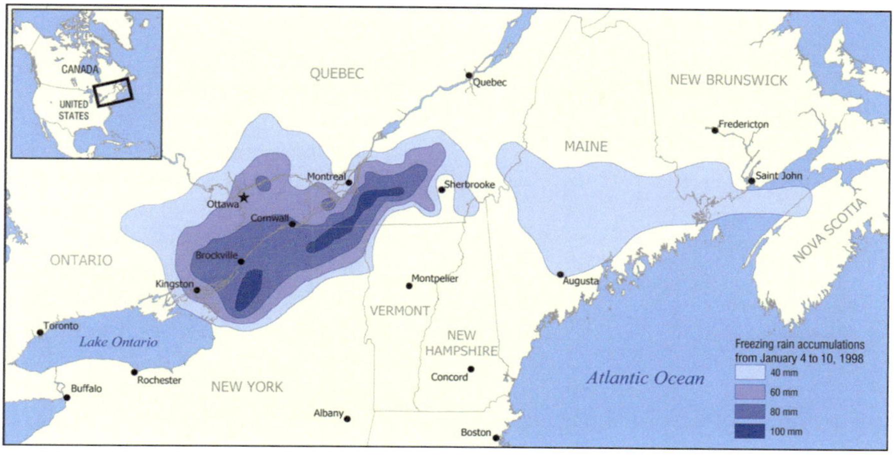

Fig. 6.13 The ice storm affected a large part of eastern Ontario, southwest Quebec, southwest Quebec, New Brunswick, and parts of New York, Vermont, New Hampshire, and Maine. The map shows the accumulation of ice in millimeter from January 4 to January 10, 1998. Increments shown are 40 mm, 60 mm, 80 mm, and 100 mm (Map created by Norman Einstein, January 30, 2006 – based on data from Environment Canada)

Objective of the HRVA process: to help a community make risk-based choices to address vulnerabilities, mitigate hazards and prepare for response to and recovery from hazard events.

Next step is to identify vulnerabilities in four defined categories as: social, physical, economic, and environmental as shown in Table 6.15.

Risk Analysis

Consequence Severity

Seven categories to assess the impact are: Fatality, Injury, Critical facilities, Lifelines, Property damage, Environmental impact, Economic and social. Table 6.16 shows the compiled data.

Next step is to estimate hazard likelihood (Table 6.17). The likelihood rank for a hazard is the occurrence frequency for the hazard scenario of similar intensity. The Public Safety Canada Database is used to estimate the likelihood of ice storms in Quebec and eastern Ontario, however, not necessarily of exactly the same intensity.

Risk Profile: Fig. 6.14 shows the risk profile obtained using the online HRVA tool, with a Risk Index value as 20.

Sources used for information:

https://www.publicsafety.gc.ca/cnt/rsrcs/cndn-dsstr-dtbs/index-eng.aspx
http://www.unisdr.org/disaster-statistics/introduction.htm http://www.usgs.gov/

Table 6.15 Identification of vulnerabilities in the 1998 Ice Storm in Montreal, Canada

Social vulnerability:	Physical vulnerability:
Elderly and very young	Communications systems
High density places	Transportation systems
Minority groups	Critical infrastructure
Low income groups	
Language barrier	
Economic vulnerability:	Environmental vulnerability:
Agriculture	Areas of biodiversity and ecological value
Major employers	Resource degradation or depletion
Economic activities	Parks, forests, wetlands

Table 6.16 Consequences severity assessment using HRVA method for the 1998 Ice Storm event in Montreal

Category	Rank	Description	Criteria
Fatality	3	High	28 deaths reported
Injury	3	High	945 injured
Critical facility	3	High	**Loss of 50% of capability**: bridges and tunnels linking Montreal with the South Shore were closed
Lifelines	4	Very high	**Long term disruption**: many power lines snapped and over 1000 pylons collapsed; city's water pumping stations disabled
Property damage	4	Very high	**Widespread & severe**: 2 billion worth damage
Environmental impact	3	High	**Localized & severe**: extensive damage to trees: 5000 trees cut in Mt. Royal Park; 80% of the rest severely damaged. Toxic spill
Economic and social impact	4	Very high	**Long term disruption**: power loss, barns collapsed under weight of ice, loss of cattle

6.2.7.2 Case Study – The 2008 Propane Explosion in Toronto, Canada

Information gathered: on August 10, 2008 at 3:50 ET in Downsview, Toronto was the site of one of the worst explosions to have ever occurred in Toronto. The disaster involved a large explosion equivalent to over 70 tons of TNT. The site of the explosion was a propane storage facility called Sunrise Propane, an industrial facility that has had a history of unlawful practices. Sunrise Propane has had a history of unsafe storage issues including concerns dating back to 2002 when over 117 commercial sized propane cylinders were left within an unsafe distance from combustible material.

Table 6.17 Likelihood of the occurrence (highlighted in red) of a similar ice storm

Measure of likelihood	Return period in years	Rank
very likely	Every 1 – 3 years	6
Likely	**Every 3 – 10 years**	**5**
Slight chance	Every 10 – 30 years	4
Unlikely	Every 30 – 100 years	3
Highly unlikely	Every 100 – 200 years	2
Very rare event	Every 200 – 300 years	1

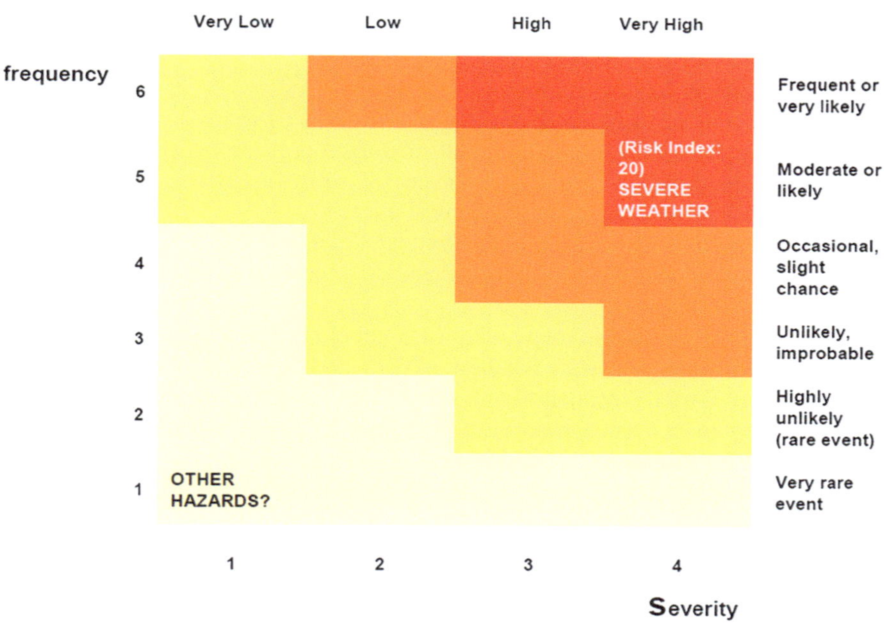

Fig. 6.14 Risk profile showing the value of Risk Index using the online tool http://hrva.embc.gov.bc.ca/toolkit.html

Fig. 6.15 Map showing (left) the location of propane explosion site (Notable features are: the Downsview airport, dense residential area, and two major highways 401 and 400. On the right (by Floydian via Wikimedia Commons), the entire Highway 401 is shown; red solid circle is approximate location of the explosion)

- The explosion caused a large fireball, with clouds of smoke and flying metal debris into the air. This was the result of storage propane cylinders exploding. Flames could be felt more than 10 km away and seen 30 km away and continually burned for 6 hours after the initial explosion.
- The incident resulted in 2 deaths: a Sunrise employee and a Toronto fireman. 6 additional people were hospitalized and 18 others were reported to needing medical attention.
- Twelve thousands people living within 1.6 km radius had to be evacuated in the middle of the night and were not allowed back into their home for 24 hours. The main areas affected were densely populated and homed to low income residents.
- Extensive property damage occurred to homes and local business costing $1.8 Million to the Province of Ontario.
- This devastating fire was rated as a level 5 alarm incident and was upgraded to a 6 alarm incident which took over 200 firemen and 34 Emergency vehicles to battle the massive inferno.
- The effects of the explosion extended to regions of Toronto and York Region. Local transportation was impacted as parts of Highway 401 (shown in Fig. 6.15) had to close for over 12 hours as well as interruptions to subway lines.
- Yorkdale Mall (a popular local shopping centre) was evacuated and closed for part of the day and the area was considered a 'no fly zone' by Pearson International Airport for small aircraft.
- The cause has been identified as a propane leak that resulted from a hose failure during an illegal "tank-to- tank" transfer from one cargo truck to another that is prohibited by Safety Standards.
- Bombardier Aerospace has a plant located within the blast zone which was also forced to evacuate their facility and sustained extensive external damage to facility where over 4500 employees worked.

Fig. 6.16 A cleanup crew combing the grass for evidence of asbestos at Ancaster Park on August 13, 2008, near propane explosion site in Toronto (in Toronto Toronto Star/Aaron Vincent Elkaim)

North York's population is estimated at 640,000. The area is predominantly occupied by residential housing and home to York University staff and students. Over 45% of housing in this area is categorized as high density population dwellings. The east side of Dufferin Street is primarily residential, while the west side is industrial. This street has been relegated to a quiet service road in the adjacent neighbourhood of Bathurst Manor. Immediately beside Dufferin Street S, William R. Allen Road brings large traffic volumes from the Ontario Highway 401 exit just a minute south.

Vulnerability Assessment

Social vulnerability and impact

The main demographics of people within are of Italian, African American and Latin American decent. Many residents were not allowed to return to their homes until 3 days after the explosion until they were deemed safe to occupy by police and hazardous materials specialists due to asbestos concerns (Fig. 6.16). According to the police, about 50 of the houses near the blast site were deemed uninhabitable. Many of the residents were low income providers and do not have accessibility to insurance and protection to handle the effects of damage to their homes and property. The elderly, single women with children were mainly affected by explosion that resulted in the closure and relocation of many essential services including transportation and community daycare such as Ancaster Child Care Center until the facility was absolutely safe to resume to regular operations. In addition, parks of the child care facility were also closed until cleanup was completed. Just north of the explosion radius is York University campus which is home to faculty and students and was used as an evacuation centre along with Yorkdale Mall. As a result of explosion people who were evacuated to York University resulted in both classes and athletic activities to be cancelled.

Physical vulnerability and damage

The explosion of Sunrise Propane had a severe effect on the city's critical infrastructure. The explosion left community members without electricity or gas. The large explosion of the propane tanks left the nearby area surrounded by debris and asbestos chemical. Many homes and offices were damaged by shattered windows, doors ripped from their hinges, debris from the explosion and other things in the surrounding that were blown up. Closure of many roads and streets within the vicinity was common. Cars were damaged or melted due to the impact of the explosion. Estimated 6 homes had structural damage beyond repair and 50 homes were deemed uninhabitable. Local businesses were looted during the chaos after the explosion.

The explosion also caused the closure of part of highway 401 in both directions due to its close proximity. The 401 is a major artery leading north out of Toronto for travelers as well as leading into the Don Valley Parkway (DVP). Regular commercial air traffic were allowed to continue in and out of Pearson International Airport while smaller privately owned aircraft were restricted from a no fly-zone issued due to the thick smoke above the area. Downsview is most known for its post WWII subdivisions and historical heritage. It is the home to one of Toronto's the oldest Jewish cemeteries where 20 graves were destroyed. Within the area is Downsview Airport, the former site of Canadian Forces Base Downsview, which has since been largely converted following the end of the Cold War into an urban park known as Downsview Park. the airport is still used as a manufacturing and testing facility for Bombardier Aerospace.

Economic vulnerability and damage

Residents were generally low income providers and do not have accessibility to insurance and protection of any kind to handle the effects of damage to their homes and property. Class action lawsuits were filed against Sunrise Propane by both homeowners as well as local business who suffered financial losses due to the explosion. Many businesses were forced to close their stores or were not accessible due to road closures and rerouting of streets. Local businesses were looted during the chaos after the explosion. Yorkdale mall was also closed for 6 hours due to the evacuation of community members and the closure of major roads and highways that connected the mall. Furthermore this was attributed to the rerouting of local transportation facilities such as Toronto and York Region Transit that were required to change their bus routes resulting in losses to business. Bombardier Aerospace located south of Downsview Airport was forced to evacuate their plant which consisted of over 4500 employees. This facility also sustained extensive external damage to their building. Total cost for cleanup and rebuilding was estimated to be over $1.8 million.

Risk Analysis

Consequence severity: Table 6.18 gives an account of the severity of consequences in the case of the 2008 propane explosion in Toronto.

Table 6.18 Severity of consequences, summarized for the propane explosion in Toronto

Category	Rank	Description	Criteria
Fatality	1	very low	2 deaths reported
Injury	2	Low	24 injuries attributed to the explosion
Critical facility	2	High	12,500 Residents Evacuated to local shelter, over 34 fire trucks with 200 firemen caused decrease in Emergency Services capacity.
Lifelines	4	Very high	Local Gas & Electricity within 2 km's cut off for up to 48 hours
Property damage	3	High	Localized severe damage: debris field approx. 2 km's; over 580 homes damaged; over 20 homes waited more than 48 hours due to concerns for airborne asbestos. 20 Graves at Toronto's oldest cemetery Mount Sinai Memorial Park were damaged
Environmental impact	3	High	Localized severe damage: smoke, asbestos, burning metal and ensuing land degradation was contained to 2 km radius
Economic and social impact	3	High	Extended & Widespread: cleanup costs exceeded 1.8 Million and took over a year. There was also a noted reduction in local property values. This incident had a major impact on the community's long term mental and physical health with high concern for safety

Propane Hazard Likelihood

Records indicate there are over 330 Propane Storage and handling sites within GTA, including larger facilities similar to Sunrise located within high density population areas. Therefore, likelihood is estimated to be Slight Chance as shown in Table 6.19.

Risk Profile: Fig. 6.17 shows the risk profile obtained using the online HRVA tool, with a Risk Index value as 20.

Sources used for information on this case study:

http://en.wikipedia.org/wiki/Toronto_propane_explosion
http://www.thestar.com/topic/propane
http://www.sunrisepropaneclassaction.com/
http://www.ctv.ca/CTVNews/WinnipegHome/20080810/to_explo_080810/
http://www.cbc.ca/news/canada/toronto/story/2008/08/11/mount-sinai.html
http://www.citytv.com/toronto/citynews/news/local/article/10690--sunrise-propane-explosion-victim-claims-city-charging-outrageous-fees-as-he-rebuilds-his-home
http://www.ofm.gov.on.ca/en/Media%20Relations%20and%20Resources/News/2010/08-04-10.asp
http://www.ofa.gov.on.ca/docs/emergency.pdf
http://www.bloomberg.com/apps/news?pid=newsarchive&sid=a7e0CSl8EsOs&refer=canada
http://www.claimscanada.ca/issues/article.aspx?aid=1000224912
http://www.thestar.com/generic/article/476267

Table 6.19 Likelihood of the occurrence of a similar event shown in red highlight

Measure of likelihood	Return period in years	Rank
very likely	Every 1 – 3 years	6
Likely	Every 3 – 10 years	5
Slight chance	**Every 10– 30 years**	**4**
Unlikely	Every 30 – 100 years	3
Highly unlikely	Every 100 – 200 years	2
Very rare event	Every 200 – 300 years	1

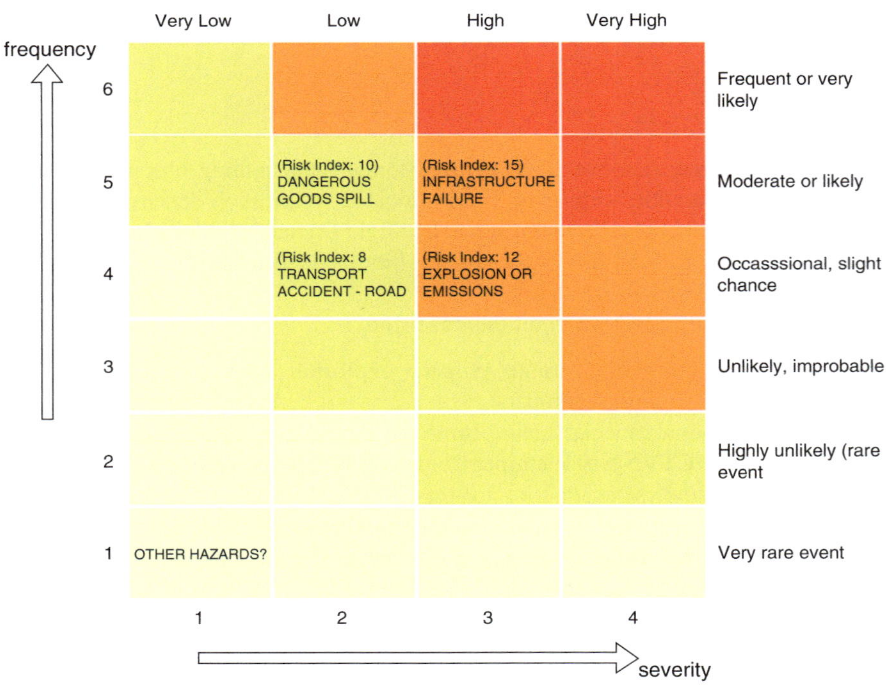

Fig. 6.17 Risk profile showing the value of Risk Index values for various technological hazards using the online HRVA tool

http://digitaljournal.com/article/295552
http://www.thestar.com/News/GTA/article/477410
http://airfresh-society.blogspot.com/2010/02/2008-toronto-propane-explosion.html

6.3 All Hazards Risk Assessment Method

The federal agency responsible for public safety regulations and guidelines, Public Safety Canada has developed an All Hazard Risk Assessment (AHRA) methodology guideline (PSC 2016) for the purpose of conducting risk assessment, primarily, for federal government institutions in Canada. The guidelines document has been developed by Public Safety Canada, in close partnership with Defense Research and Development Canada (DRDC) – Centre for Security Science (CSS), as part of the federal All Hazards Risk Assessment initiative endorsed by the Assistant Deputy Minister Emergency Management Committee in October 2009 (PSC 2016). International government partners such as the United States, the United Kingdom, and the Netherlands have also been consulted during the process.

6.3.1 Purpose

The intention of the process is to produce a holistic government risk picture to support emergency management (EM) planning across federal government institutions and to ensure that interdependencies are recorded and managed. It also provides a venue for the creation of a federal AHRA community of practice, and a forum for sharing risk information, tools, and methodologies.

More specifically, the AHRA's objectives are to:

- Enable federal government institutions to perform AHRA consistently and efficiently as part of their risk management responsibilities under the EMA and other relevant legislation and policies.
- Address the interconnected nature of Canada's risk environment and provide a means to produce a collective judgment of risk assessments currently being carried out by different federal government institutions into a holistic government picture to inform future actions and initiatives.
- Support the relative ordering of risk events based on their ratings at a federal level, while enhancing decision-making processes within the Government of Canada (GC).
- Capture risks that are significant and of federal interest.
- Raise awareness of risks that may presently not be of federal concern, but likely to be elevated to risk status in the future.
- Raise awareness of risks that are not of federal concern, but ensure that these risks are monitored.

- Capture changes in risks over time.
- Help to foster and nurture an AHRA community of practice for the federal community.

Scope Risk assessment specific to the critical infrastructure (CI) sectors is beyond the scope of the federal AHRA methodology and falls under the *National Strategy and Action Plan for Critical Infrastructure*.

6.3.2 Overview of the AHRA Process

The federal AHRA process employs a scenario-based risk assessment approach, focusing on following five steps as shown in Fig. 6.18:

6.3.3 AHRA Business Cycle (Fig. 6.19)

In order to ensure a coordinated approach of the AHRA process, an Interdepartmental Risk Assessment Working Group (IRAWG) has been created to represent federal institutions participating in the AHRA process. During the summer quarter, the IRAWG is responsible to choose, amongst the entire list of departmental priority threats and hazards, key risks that will be further assessed during each AHRA cycle. The working group is also responsible for providing PS with on-going and timely strategic safety and security advice related to the AHRA methodology and process.

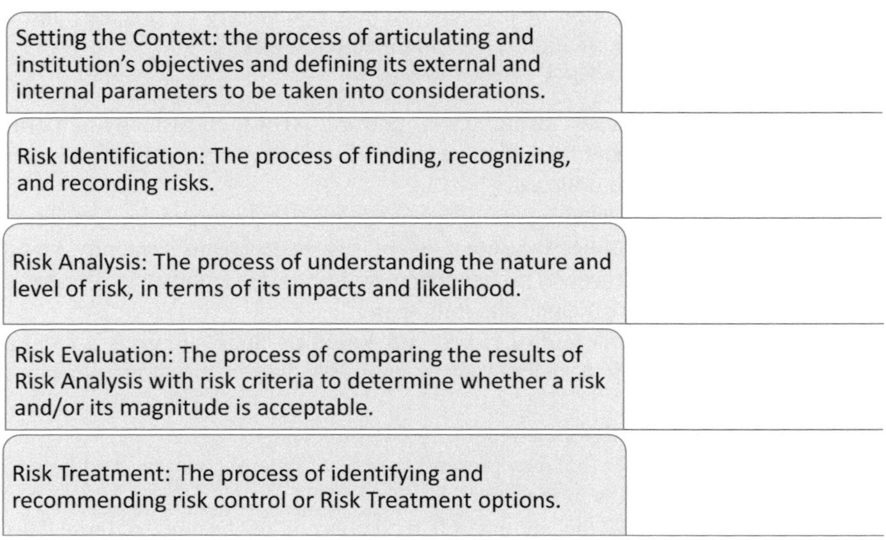

Fig. 6.18 The five steps employed in AHRA process

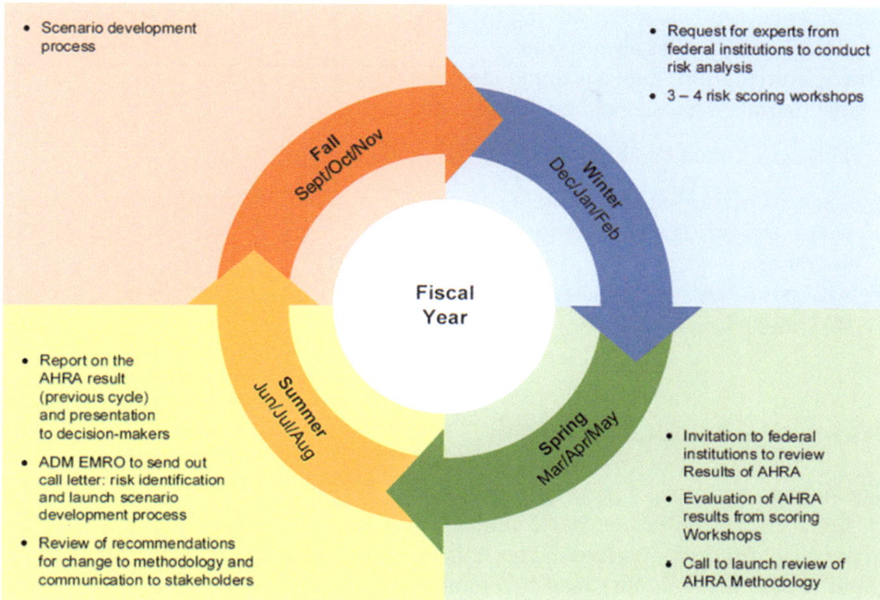

Fig. 6.19 AHRA business cycle (PSC 2016)

6.3.3.1 Step 1: Setting the Context

Each federal government institution is responsible to research, review and gather relevant data such as applicable legislation, reports on plans and priorities, departmental performance reports, etc. in order to enhance its understanding of its mandate, responsibilities, areas of interest, and various information sources that would help inform the next AHRA process step: risk identification.

Inputs required in this step are:

Departmental planning and reporting documentation: this would include the budgetary information, departmental priorities and strategic plans, performance indicators, and resource requirements on a 3 year basis.

Environmental scans: involves a process of gathering and analyzing information and typically considers both internal and external factors such as policies, capabilities, and societal indicators (e.g. demography, economy, technology) at a local, national, and international level.

Historical records: include information on past risk assessment and associated risk database. The Canadian Disaster Database (CDD) contains detailed disaster information of over 900 natural, technological, and conflicts events that have impacted Canadians over the past century. Additionally, census information (Statistics Canada 2015) and other economic and infrastructural details can be obtained from research institutions such as National Research Council of Canada (NRC 2015).

Intelligence reports: these are highly relevant to a range of risk types including intentionally malicious activities.

Other information: relevant to the identification of risks within the domain of the institution's mandate, responsibilities, and interests.

This step should be able to provide following as outputs:

- Analysis of short-term (within the next 5 years) threats and hazards, accompanied with a certain level of comprehension regarding an institution's level of risk tolerance.
- Analysis of emerging and future (in 5 to 25 years) threats and hazards.
- Risk themes.

6.3.3.2 Step 2: Risk Identification

The outputs from Step 1 are expected to serve as inputs to Step 2.

Initial risk identification based on their priority should be identified by June of every year. Appendix 6 gives the hazards risk taxonomy to assist with this process. Some of the methods that may be employed for environmental scanning include brainstorming, risk source analysis, checklists, scenario analysis, surveys and questionnaires, and interviews and focus groups.

Risk event scenario development should be done using the template shown in Appendix 7.

- Salient features of the risk event scenario development process are listed below:
- the risk event scenario development process should take place in the fall
- the development of scenarios should continue until end of November
- in the early phase, federal government institutions are to provide their highest ranking risks

The risk event scenario should be realistic. It should indicate the capacity to respond in place and in time. Roles and responsibilities for the scenario Leads include – Identify and contact relevant federal government institutions; provide consultation if necessary; organize regular working group meetings; if necessary, establish terms of reference; develop a work plan and timelines; revisions and meetings with subject matter experts to discuss inconsistencies; and finalize the risk event scenario for the risk scoring workshop.

Figure 6.20 provides an illustration of the risk event scenario development process and mandatory fields for which information must be provided when developing a risk event scenario.

These scenarios should also provide adequate flexibility so that different consequences associated with different location, weather conditions, population, etc. can be modeled. One very basic approach would be to model variations of each developed scenario, for example, nominal impact, reduced impact, or elevated impact variations.

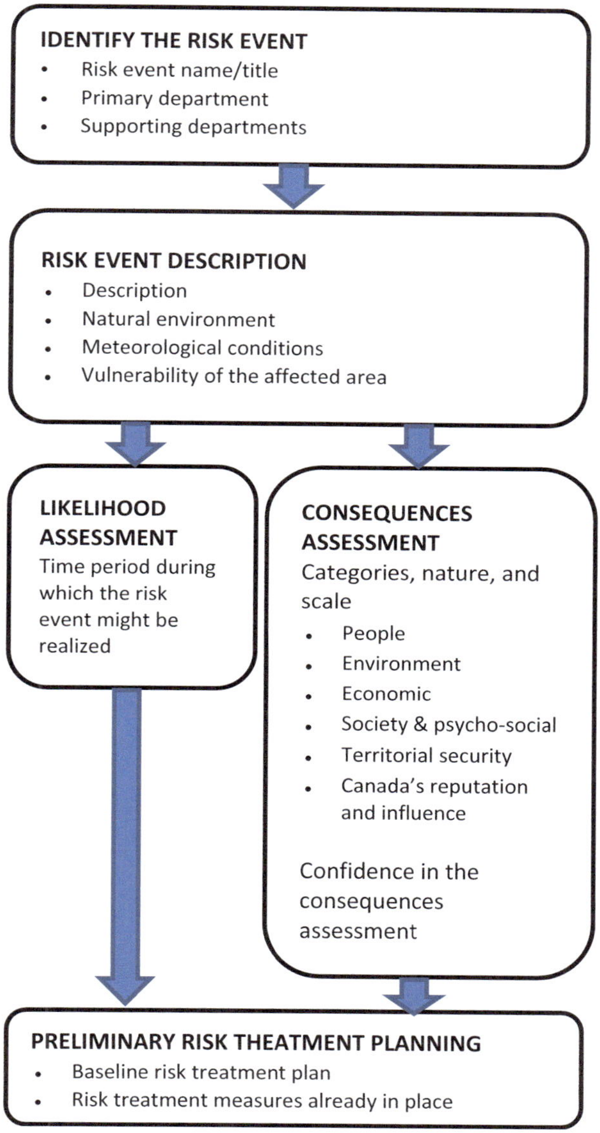

Fig. 6.20 Process for developing a risk event scenario (Adapted from PSC 2015)

Table 6.20 Impact ratings and associated adult fatalities

Magnitude of impact rating	No impact	0	0.5	1	1.5	2	2.5	3	3.5	4	4.5	5
Number of adult fatalities	0	1	3	10	30	100	300	1000	3 K	10 K	30 K	100 K

Adapted from PS (2015)

6.3.3.3 Step 3: Risk Analysis

Historical data on past events should be used for establishing an understanding on potential events' frequency. Related sequential events and associated consequences must also be taken into account. These impacts should be classified into six categories as can be seen in Fig. 6.20 (consequences assessment box).

Likelihood analysis: likelihood is an estimate of the chance of an event happening. It can be defined objectively, as a probability or a frequency over a time period, or subjectively (i.e. rare, less likely, likely, more likely or almost certain). For quantitative estimation of likelihood, various approaches and methods are available in the literature. For qualitative estimation, interviewing experts is a widely used approach.

Impact analysis: impact or consequences from hazard risks can also be estimated objectively and subjectively. Objective or quantitative estimates are derived based on real facts from past events. When information is gathered from interviewing experts or impacted population, consequences are presented in descriptive form. Impacts are usually expressed in terms of financial losses, technical failures, operational disruptions, and human traumas. The primary six categories of impact are given in Fig. 6.20 (consequences assessment box), and a detailed account of each of the category is as follows:

- People – fatalities and injuries, displacement, chronic diseases, and emotional stress. Ratings are determined according to the World Health Organization (WHO) Disability-Adjusted Life Year (DALY) measurement (WHO 2015). An example is shown in Table 6.20.
- Economy – direct and indirect losses.
- Environment – the geographical extent, magnitude, and duration of damage.
- Territorial Security – the disruption in the effective functioning of an area or a border, including the area affected, combined with duration and population density.
- Canada's Reputation and Influence – situations that would result in a shift in views towards the reputation and influence of Canada and actions taken by citizens and/or stakeholders as a consequence.
- Society and Psycho-Social – the impact of widespread public anxiety and outrage.

Table 6.21 Economic impact category rating guideline

Magnitude of impact rating	No impact	0	0.5	1	1.5	2	2.5	3	3.5	4	4.5	5
Economic Loss	No impact	10 M	30 M	100 M	300 M	1B	3B	10B	30B	100B	300B	1000B

People impact category ratings are determined according to the WHO DALY measurement (WHO 2015). For the purpose of simplicity, however, only equivalent number of adult fatalities is shown in Table 6.20.

Economy Impact Category

Direct economic loss is measured by the repair or replacement cost of damaged buildings (commercial, institutional, recreational, etc.); critical infrastructure such as roads, water systems, oil and gas facilities, telecommunication network; agriculture industry; raw materials; and residential assets etc.

Indirect economic loss includes flows of goods and services that may result in direct economic loss. Examples of indirect losses include loss of business reputation or consumer confidence or reputation; broken supply chain resulting in lost or delayed production; etc. Indirect losses could also include the loss of income resulting from the non-provision of goods and services or from the destruction of previously used means of production. However, these are difficult to adequately quantify. Appendix 8 outlines examples of direct and indirect loss situations. For detailed description, the original document should be consulted at the Public Safety Canada website, following the Emergency Management link.

Table 6.21 shows the recommended rating guideline to estimate the magnitude of economic impact (PSC 2015).

Environment Impact Category

Canada's natural environment shapes national identity, health, and prosperity. Therefore, in this category, factors that are accounted for are – the preservation of air, water, and soil ecosystems. The environment rating scale focuses on environmental damage caused by a hazardous event. Tables 6.22 and 6.23 provide related rating guidelines.

A baseline is recommended for an environmentally affected area, which could be reflected as the typical geographical extent for the respective response rating. A modifier is applied only if the size of the environmentally affected area is more than what is typically expected for that type of response. For example, Table 6.23 gives the guideline of one such modifier indicating that the baseline (modifier value of 0) for the size of the environmentally affected area for local response with no federal

Table 6.22 Environment impact category response magnitude

Base rating	Type of response and magnitude
0	Some local general response, but no specialized response
1	Some local specialized response, and surveillance and monitoring from federal authorities
2	Multi-regional general response, and notification from federal authorities
3	Multi-functional, multi-regional specialized response, and notification from federal authorities
4	Multi-functional, multi-jurisdictional specialized response and federal mobilization
5	Multi-functional, national & international, specialized response and rapid federal mobilization

Table 6.23 Environmental impact category – modifier chart for geographical extent of damage assessment

Geographical extent modifier	Size of damage (km^2)
0	Up to 50
+0.5	150
+1	500
+1.5	1500
+2	5000
+2.5	15,000

Note: The Greater Toronto Area (GTA) is about 7200 km^2

monitoring (response rating of 0) would be 50 km^2 (below which no action/interest is taken). If the size is greater, a modifier is added in increments of 0.5.

Similarly, for response levels of 1–5, proposed modifiers are recommended that can be accessed on the Public Safety Canada website.

The magnitude of damage based on environmental impact refers to the severity which is applied as a modifier, and determined using the following guideline. The total value of the modifier is arrived at by adding all that applies to a risk event. Table 6.24 gives criteria and associated values of modifiers needed to be applied to assess the magnitude of environmental damage.

Modifier for duration of damage is also proposed in a similar fashion. A modifier to increase or decrease the rating is recommended for the assessment of the duration of damage. A baseline is recommended (Table 6.25) for the duration of environmental damage, which could be deliberated as the typical duration for the respective response rating. A modifier is applied only if the duration is more than what is typically expected for that type of response. If the baseline (modifier value of 0) for the duration of environmental damage for local response with no federal monitoring (response rating of 0) would be 3 weeks (anything below that would barely raise any interest, considering the scope of AHRA). If the duration is longer (duration increasing threefold at each rating step, to be consistent with the overall approach), a modifier is added (in increments of 0.5):

Table 6.24 modifier chart for magnitude assessment of environmental damage

Value	Criteria
2.0	Loss of rare or endangered species; and/or loss of critical/productive habitat; and/or loss of water resources
1.0	Reductions in species diversity; and/or loss of current use of lands resources
0.5	Transformation of natural landscapes; and/or environmental losses from air pollution

Note: these modifiers should only be used in exceptional cases, where the likely impacts of an event are beyond what would be captured by the primary environmental response

Table 6.25 The Baseline for the assessment of the duration of environmental damage

Duration modifier	Duration of environmental disruption
0	Up to 3 weeks (approx. 1 month)
0.5	10 weeks (approx. 2 months)
1	8 months
1.5	2 years
2	6 years
2.5	20 years

Similar to the case of geographical extent assessment modifier approach, based on the response level (1–5), modifiers are proposed to accurately assess the duration of environmental damage. Table 6.26 gives duration modifiers for a response level 1 scenario. For the other response levels, duration modifier tables can be found at the Public Safety Canada webpages.

A cumulative factor will affect the final score for the geographical extent of the damage as the magnitude of damage and the duration of the damage are to be added to the base rating score as assessed by the magnitude of response, without exceeding 5.

Territorial Security

This category captures conditions in which there is a loss in the ability of the Government of Canada to secure the territory or the border (land and marine inclusive area = 9,984,670 km^2) and to secure the safety of citizens. The baseline rating system (Table 6.27) for this category is rooted in geographical area of the country at risk or affected. The final score is determined by the area affected, with factors including the duration of disruption and population density as shown in Eq. (6.4).

$$\begin{aligned} \textit{Impact score} &= \textit{Area Affected} + \textit{Duration Score} \\ &\quad + \textit{Population Density Modifier} \end{aligned} \tag{6.4}$$

Table 6.26 For a response level of 1 (local specialized with federal monitoring), the duration baseline is considered to be 10 weeks, and a modifier is applied for a longer duration

Duration modifier	Duration of environmental disruption
0	10 weeks (approx. 2 months)
0.5	8 months
1	2 years
1.5	6 years
2	20 years

Table 6.27 Territory security – size of impact (PSC 2015)

Base rating for area affected	No impact	0	0.5	1	1.5	2	2.5	3	3.5	4	4.5	5
Size of impact (Thousand km^2)	No impact	0.1	0.3	1	3	10	30	100	300	1000	3000	10,000

Modifiers for duration of disruption and population density are to be applied according to the guideline given in Table 6.28. For the final and cumulative assessment of territorial security, the duration of disruption and population density modifiers should be added to the base rating score, with the total not exceeding 5.

Canada's Reputation and Influence

This category represents an expert assessment of the potential international reaction to an emergency event occurring in Canada, or involving Canadians abroad. This assessment should be made based on reactions to similar emergency events previously experienced within Canada and/or in other countries. Table 6.29 explains the rating systems proposed toward damage assessment in this category. Appendix 9 lists possible examples of changes in international positions towards Canada, and ways the GC and/or Canadians may be impacted by a risk event that has international implications.

Society and Psycho-social

Social events such as civil disturbances that can be provoked by a risk event and can impact response and recovery efforts are accounted for in this category. People's perception plays vital role here. At the same time, even if people do not engage in social action following an event, they can nevertheless experience the psycho-social effects of disaster that can lead to changes in their individual pattern of behavior over the short or long term. Psycho-social effects can also impact the effectiveness of the

Table 6.28 Modifiers for duration of disruption

Duration modifier	Duration of disruption	Density of affected area (people/km^2)
−2	1 hour	
−1.5	3 hours	
−1	10 hours (≈ ½ day)	0.1
−0.5	1 day	0.3
0	3 days (≈ ½ week)	1
+0.5	10 days	3
+1	1 month	10
+1.5	3 months	30
+2	1 year	100
+2.5	3 years	
+3	≥ 10 years, not permanent	
+3.5	Permanent	

Table 6.29 Canada's reputation and influence – repercussions assessment (PSC 2015)

Rating level	Repercussions/damage to Canada or Canadians
0	No damage
1	Insignificant damage – Minor, short term and localized reaction that is limited to small groups of individuals and has no repercussions
2	Minor damage – Minor, medium- to long-term, international reaction by groups of individuals that has a minor effect
3	Significant damage – Significant, short to medium-term, international reaction by groups of individuals, foreign governments and/or organizations that has a medium term effect
4	Major damage – Major, short- to medium-term, widespread reaction by large groups, foreign governments and/or organizations that has a long lasting effect
5	Severe damage – Major, long term, widespread reaction by large groups, foreign governments and/or organizations that has a lasting effect

overall response and recovery efforts if they are not appropriately managed. Over the longer term, continued exposure to the source of stress or lack of support for the population may result in secondary disorders.

The dimension of public mood should be considered when scoring for this impact category. The scoring of public mood (Table 6.30) is based on a subjective assessment that focuses on two criteria: public outrage and public anxiety. The descriptors for each of these attempt to capture how people's behavior might be affected by an event and the score, although subjective, points to the possibility of short to long-term psycho-social impacts.

Table 6.30 Society and psycho-social impacts (PSC 2015)

impact score	Public outrage descriptor	Public anxiety descriptor
No impact	No impact	No impact
0	Insignificant	Insignificant: no changes in people's normal routine
0.5–1	Minor	Minor anxiety but no change in people's behaviours
	No authority or person perceived to be culpable or incompetent	Short term avoidance of transport modes
	Impact targeted on one particular group associated with the government (rather than being indiscriminate)	Risk to children or future generations limited
		Strong public familiarity with/understanding of the risk and its consequences
	Little symbolic value	Less than a thousand people feel more vulnerable
1.5/2	Significant but localized and temporary	Minor, localized and temporary changes in people's normal routines
	Consequences are largely one-off	
	Public acceptance that the risk was a natural disaster or avoidable and largely not caused by human failure	Short to medium-term avoidance of transport modes
		Good public understanding of the risk
	Little symbolic value of site or target	Thousands of people feel more vulnerable (but less than 10,000 people)
2.5/3	Serious, widespread	Moderate anxiety leading to medium to short-term changes in peoples' routines
		Medium term avoidance of some modes of transport
		Shortage of essential supplies due to panic-buying
		Conceivable that the event could occur again
		Impact was indiscriminate
		Lack of control or helplessness
		Some concern about potential health risk to future generations
		Limited public understanding of the risk
	Consequences will not just be one-off, but still short-term	Tens of thousands of people feel more vulnerable (<100,000 people)
	High impact on those perceived as vulnerable (i.e. elderly, women, children)	
	Public perception that the disruptive outcome was a result of someone/the government's failure	
	High symbolic value	

(continued)

Table 6.30 (continued)

impact score	Public outrage descriptor	Public anxiety descriptor
3.5/4	Serious national-wide concern, with strong calls for government action	High levels of anxiety leading to sustained changes in people's normal routines
	The adverse impact was intentional/ malicious	Intense and widespread information seeking by the public
	Domestic pressure for resignations; public perception that government/ person significantly failed	High levels of concern about risks to children or future generations
	Risk results from human action rather than natural causes	High levels of concern that catastrophic event could occur again
	Consequences will be medium-term rather than just one-off	Impact was indiscriminate and affected large (< 1000,000) number of people
		Significant sense of lack of control/ helplessness
	Indiscriminate and very significant impact	Lack of informed public knowledge or understanding of the risk
	Significant impact on those perceived as vulnerable	Millions of people feel more vulnerable
	Very high symbolic value	
4.5/5	Extreme, nation-wide, sustained	Extreme, widespread, prolonged
	Widespread calls for severe governmental reprisal (i.e. the adverse impact was intentional/malicious)	Widespread avoidance of an area
		Social conflict and community tensions resulting from fear-induced behavior
	Persistent domestic pressure for resignations at national/CEO level	Severe loss of confidence in government's ability to protect citizens
	Consequences will be long-term rather than one-off	Severe and prolonged loss of confidence in the financial markets
	Risk results from human action rather than natural causes	Significant concern about risks to children or future generations
		Significant concern that catastrophic event could occur again
	Indiscriminate and catastrophic impact	Severe sense of lack of control/ helplessness
	Severe impact on those perceived as vulnerable (i.e. elderly, women, children)	Impact was indiscriminate and directly affected very large (i.e., 10,000,000 or more) number of people perception that the adverse consequences could happen to anyone
	Very high symbolic value	Very little informed public knowledge or understanding of the risk
		Significant proportions of people feel more vulnerable

Table 6.31 Likelihood – frequency relationship

Estimated frequency, once every X years, where X is (years):	100,000	30,000	10,000	3000	1000	300	100	30	10	3	1

Likelihood Assessment

Likelihood assessment for various hazardous scenarios has been discussed in the AHRA guideline document including malicious scenarios and associated intent. However, in this chapter, the focus is on non-malicious hazards. Likelihood estimates for natural or technological hazards would be based on historical frequencies, predictive models, or expert judgement. Various scenarios can be developed based on past events in order to project future possibilities. Table 6.31 shows the likelihood-frequency correlation chart.

6.3.3.4 Step 4 Risk Evaluation

Risk evaluation is the process of comparing the results of risk analysis with risk criteria to determine whether a risk and its magnitude are acceptable. The purpose of this process is to support the development of sound recommendations about prioritization of risks in order to see which ones may need mitigation measures right away, which ones can wait, and which ones do not face any pressing challenges. The risk evaluation entails following steps:

1. Determination of the risk magnitude
2. Aggregation of risk assessment results for all federal government risks into a government-wide AHRA.
3. Production of selected AHRA information products and/or graphical representations of results.

Risks are ranked by comparing them in terms of their magnitude. Among various approaches to do so, risk matrix (Fig. 6.21) is one of the most commonly used, which normally plots the likelihood and impact on the x- and y-axes. Another method is to plot different factors used in impact assessment on a bar chart as shown in Fig. 6.22.

This step gives risk evaluation results in the form of a report with graphs and charts of risk ratings. This information is used by the government and relevant authorities for prioritization purposes and resource allocation strategies.

6.3.3.5 Step 5: Risk Treatment

Risk treatment is the process of developing, selecting, and implementing risk control options and measures. The scope of risk treatment can include, but are not limited, to:

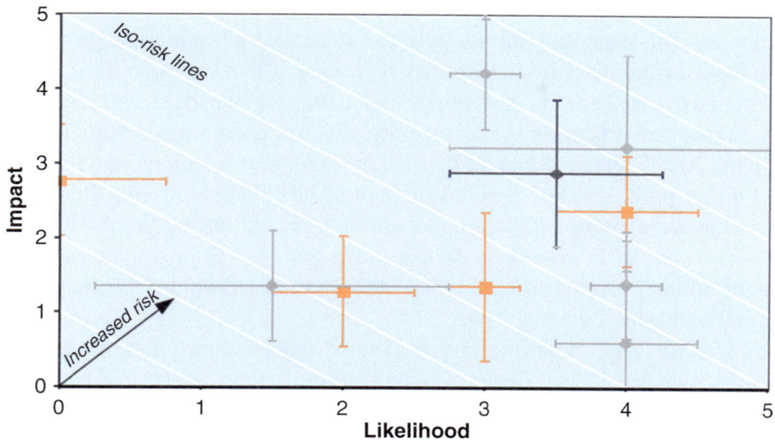

Fig. 6.21 Example of a risk rating scatter plot

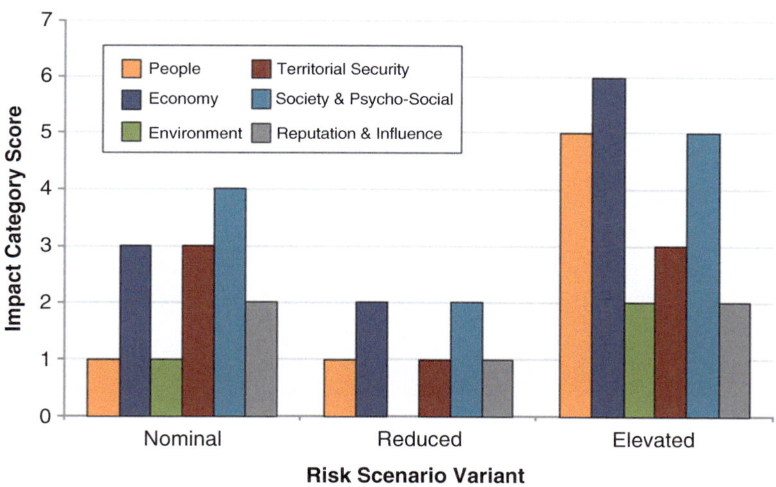

Fig. 6.22 Example of an impact based scenario chart

- Avoiding the risk by deciding not to continue with the activity that gives rise to the risk.
- Removing the source of the risk.
- Changing the nature or magnitude of the likelihood.
- Changing the consequences.
- Reducing exposures or vulnerabilities.
- Sharing the risk with another party.
- Retaining the risk by choice.

Risk Treatment options are prioritized by considering a number of factors, including institutional obligations, political impetus, humanitarian grounds, cost, etc. and by considering risk severity, risk tolerance, effectiveness of Risk Treatment measures, cost and benefits, the horizontal nature of the risk, and existing constraints. These treatment options, forming recommendations, would be used to develop the Risk Treatment step in the risk management or emergency management cycle. Furthermore, residual risk levels should be understood and their tolerance levels within an institution should be taken into account during the AHRA process.

Outputs from Risk Treatment would be a set of recommendations for Risk Treatment options, from a risk analyst perspective, based on the results of the Risk Evaluation and other considerations.

Data Management capability and associated resources are important aspects of any risk assessment process. Seamless data continuity must be ensured to account for technology advancements. The AHRA results are to be compiled by Public Safety (PS) in a report and inventoried in an electronic risk register that would be maintained by PS with data from each successive cycle of risk assessment. The register will permit easy access to the risk data for analysts and decision-makers. The AHRA register will catalogue the data relevant to the current AHRA cycle as well as previous AHRA data that can be used for tracking the progress over time, analysis of trends in emerging risks, reflections on mistakes made, and guidance on best practice.

6.4 Exercise

Apply each of the methods discussed in this chapter to the best of your ability using any recent disaster. Suggest modification and new ideas as you see fit.

References

ADRC (2005) Major disasters in FY2005. http://www.adrc.asia/publications/stats/MajorDisasters2005.html. Accessed 10 Jan 2016

Armenakis C, Nirupama N (2014) Flood risk mapping for the City of Toronto. Proc Econ Fin, Elsevier 18:320–326

EMDAT (2015) International disaster database. Centre for Research on the Epidemiology of Disasters. www.emdat.be. Accessed 17 Jan 2016

EMO (2012) Hazard identification and risk assessment for the Province of Ontario, Emergency Management Ontario. Ministry of Community Safety and Correctional Services, Ontario

Ferrier N, Haque CE (2003) Hazards risk assessment methodology for emergency managers: a standardized framework for application. Nat Hazards 28(2):271–290

Government of Ontario (2009) Emergency management and Civil Protection Act, RSO 1990, cE9, December 15, 2009, Ministry of Community Safety & Correctional Services, Ontario

HIRA (2012) The Ontario provincial hazard identification and risk assessment. Ministry of Community Safety and Correctional Services. https://www.emergencymanagementontario.ca/english/emcommunity/ProvincialPrograms/hira/hira_2012.html?_ga=1.142318386.161534779.1449636202. Accessed 27 Jan 2016

HRVA (2004) Hazard, risk, and vulnerability assessment toolkit. Emergency Management British Columbia, Canada. http://hrva.embc.gov.bc.ca/toolkit.html

Jonkman SN, van Gelder PHAJM, Vrijling JK (2003) An overview of quantitative risk measures for loss of life and economic damage. J Hazard Mater A99:1–30

National Hurricane Center (NHC) (2016) http://www.nhc.noaa.gov/. Accessed 11 Mar 2016

Natural Resources Canada (2016) Natural Resource Canada, Government of Canada. https://www.nrcan.gc.ca/hazards/natural-hazards

Nirupama N, Armenakis C, Montpetit M (2014a) Is flooding in Toronto a concern? Nat Hazards 72 (2):1259–1264

Nirupama N, Adhikari I, Sheybani A (2014b) Natural hazards in Ontario Canada – taking stock for mitigating disasters. Procedia Economics and Finance, Elsevier 18:55–61

NOAA (2009) Weather, national oceanic and atmospheric association. http://www.noaa.gov/wx.html. Accessed 14 Jan 2016

NOAA (2016) National oceanic and atmospheric administration, http://www.noaa.gov/climate-data-and-reports

NWS (2009) National weather service. http://www.nws.noaa.gov/. Accessed 14 Jan 2016

PSC (2015) The Canadian disaster database, Public Safety Canada., http://www.publicsafety.gc.ca/cnt/rsrcs/cndn-dsstr-dtbs/index-eng.aspx. Accessed 11 Mar 2016

PSC (2016) All hazards risk assessment (AHRA) methodology guidelines 2012–2013. Public Safety Canada https://www.publicsafety.gc.ca/cnt/rsrcs/pblctns/ll-hzrds-ssssmnt/index-eng.aspx. Accessed 17 Jan 2016

Statistics Canada (2015) http://www.statcan.gc.ca/eng/start. Accessed 17 Jan 2016

World Bank (2016) Data, World Bank. http://data.worldbank.org/. Accessed 11 Mar 2016

Chapter 7
Disaster Risk Evaluation – Other Quantitative Methods

Building on the previous chapter on quantitative risk evaluation approaches that are currently practiced in Canada, this chapter examines methods used in other developed nations. In the USA, the Federal Emergency Management Agency (FEMA) has developed a method to help the insurance industry assess flood damage and subsequent compensation as part of the federal government's mandate. The SMUG method was developed in New Zealand; and the World Risk Index was developed by two scientists from the University of Stuttgart, offering a new approach to assessing risk and vulnerability at the national scale and allowing countries to be compared.

7.1 Federal Emergency Management Agency Method

Federal Emergency Management Agency (FEMA) is the main federal agency responsible for planning, preparing, and responding to disasters in the United States. According to FEMA, a risk assessment is a process to identify potential hazards and analyze what could happen if a hazard occurs. A business impact analysis (BIA) is the process for determining the potential impacts resulting from the interruption of time sensitive or critical business processes (FEMA 2015a). Appendix 10 gives a sample of risk assessment table that should be used to collect information on risk.

FEMA rightly emphasizes on educating public as a vital and first step toward disaster mitigation. Their website is comprehensive and universally applicable with modification to suit local and regional facts. Topics such as basic protective measures for all hazards and disaster specific preparedness are insightful on the website. Facts about natural disasters are also given (FEMA 2015b). Pandemic related information can be found at (FEMA 2015c). The information is available in a number of languages, divided into following categories:

N. Agrawal, *Natural Disasters and Risk Management in Canada*, Advances in Natural and Technological Hazards Research 49,
https://doi.org/10.1007/978-94-024-1283-3_7

- *Be informed* – includes understanding of disasters, community plans, shelters, emergency alerts, evacuation details, recovering from disasters etc.
- *Make a plan* – includes plans for your risks, access and functional needs, infants and children, seniors, campuses, animal care, military families, first responders, First Nations etc.
- *Build a kit* – includes basic disaster supply kit and how to maintain and store it, food, car safety, water management etc.
- *Get involved* – discusses how to get involved before a disaster occurs, preparing your community, national preparedness community, volunteer opportunities etc.
- *Business* – includes guidelines on program management, planning, implementation, testing and exercising, program improvement etc.
- *Kids* – this section is the most important as not only do kids learn and adapt quickly, they make their parents pay attention as well.

Similar to other risk assessment tools used around the world, people are given the first consideration of the risk assessment process. Figure 7.1 shows the risk assessment process. Hazard scenarios that could cause significant injuries should be highlighted to ensure that appropriate emergency plans (FEMA 2015d) are in place. Physical assets such as buildings, communication networks, other utility systems, machinery, raw materials, and finished goods are accounted. The potential for environmental impact should also be considered. As part of the risk assessment process, special attention is required to finding vulnerabilities in social and physical structures. In social sense, special groups of people can be more vulnerable than others. In physical sense, a building without a fire sprinkler system would be more likely to burn to the ground than a building with a properly designed, installed and maintained fire sprinkler system.

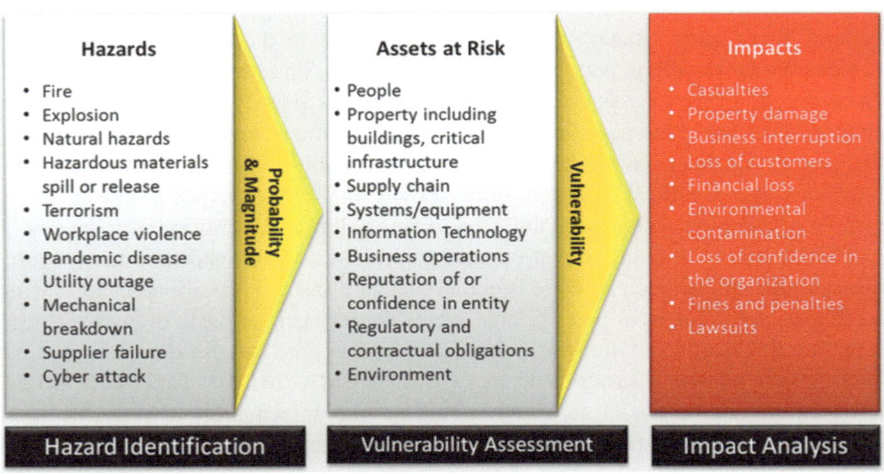

Fig. 7.1 Risk assessment process diagram developed by FEMA (www.ready.gov/risk-assessment)

The impacts from hazards can be reduced by investing in mitigation (http://www.ready.gov/risk-mitigation). If there is a potential for significant impacts, then creating a mitigation strategy should be a high priority.

The next step toward risk assessment is the Business Impact Analysis (BIA) that is meant to predict consequences of disruption of a business function and process and gathers information needed to develop recovery strategies. Potential operational and financial loss scenarios are to be identified during a risk assessment, such as disruption in supply chain processes or services etc. The following is a list of possible business disruption scenarios:

- Physical damage to a building
- Damage to or breakdown of machinery, systems or equipment
- Restricted access to a site or building
- Interruption of the supply chain including failure of a supplier or disruption of transportation of goods from the supplier.
- Utility outage (e.g., electrical power outage)
- Damage to, loss, or corruption of information technology including voice and data communications, servers, computers, operating systems, applications, and data
- Absenteeism of essential employees

Identifying and evaluating the impact of disasters on business provides the basis for investment in recovery strategies as well as investment in prevention and mitigation strategies. Impacts to consider include:

- Lost sales and income
- Delayed sales or income
- Increased expenses (e.g., overtime labor, outsourcing, expediting costs, etc.)
- Regulatory fines
- Contractual penalties or loss of contractual bonuses
- Customer dissatisfaction or defection
- Delay of new business plans

7.1.1 Timing and Duration of Disruption

The point in time when a business function or process is disrupted can have a significant bearing on the loss sustained. A store damaged in the weeks prior to the holiday shopping season may lose a substantial amount of its yearly sales. A power outage lasting a few minutes would be a minor inconvenience for most businesses but one lasting for hours could result in significant business losses. A short duration disruption of production may be overcome by shipping finished goods from a warehouse but disruption of a product in high demand could have a significant impact.

7.1.2 Conducting the BIA

Using a BIA questionnaire to survey managers and others within the business is critical. Survey those with detailed knowledge of how the business manufactures its products or provides its services. Ask them to identify the potential impacts if the business function or process that they are responsible for is interrupted. The BIA (FEMA 2015e) should also identify the critical business processes and resources needed for the business to continue to function at different levels as shown in Fig. 7.2.

7.1.2.1 BIA Report

The BIA report should document the potential impacts resulting from disruption of business functions and processes. Scenarios resulting in significant business interruption should be assessed in terms of financial impact, if possible. These costs should be compared with the costs for possible recovery strategies. The report should prioritize the order of events for restoration of the business. Business processes with the greatest operational and financial impacts should be restored first.

Businesses use information technology to quickly and effectively process information. Employees use electronic mail etc. to communicate. Data must be backed up on a regular basis in order to be able to retrieve it when needed. An information technology disaster recovery plan (IT DRP) should be developed in conjunction with the business continuity plan. Priorities and recovery time objectives for information technology should be developed during the business impact analysis. Technology recovery strategies should be developed to restore hardware, applications and data in time to meet the needs of the business recovery. Businesses large and small create and manage large volumes of electronic information or data. Much of that data is important. Some data is vital to the survival and continued operation of the business. The impact of data loss or corruption from hardware failure, human error, hacking or malware could be significant. A plan for data backup and restoration of electronic information is essential. A number of resources (CSRC 2015; Swanson et al. 2015; Grance et al. 2015; IIBHF 2015) are available in the United States for this purpose.

Fig. 7.2 Business continuity plan (www.ready.gov/business/implementation/continuity)

(a) Computer security resource center: http://csrc.nist.gov/publications/PubsSPs.html
(b) Contingency Planning Guide for Federal Information Systems: http://csrc.nist.gov/publications/nistpubs/800-34-rev1/sp800-34-rev1_errata-Nov11-2010.pdf
(c) Guide to Test, Training, and Exercise Programs for IT Plans and Capabilities: http://csrc.nist.gov/publications/nistpubs/800-84/SP800-84.pdf
(d) Insurance Institute for Business & Home Safety: http://www.disastersafety.org/

The Operational & Financial Impacts worksheet as given in Appendix 11 can be used to capture this information as discussed in Business Impact Analysis. The worksheet should be completed by business function and process managers with sufficient knowledge of the business. Once all worksheets are completed, the worksheets can be tabulated to summarize:

- The operational and financial impacts resulting from the loss of individual business functions and process
- The point in time when loss of a function or process would result in the identified business impacts

Those functions or processes with the highest potential operational and financial impacts become priorities for restoration. The point in time when a function or process must be recovered, before unacceptable consequences could occur, is often referred to as the "Recovery Time Objective."

7.1.2.2 Conducting the Business Continuity Impact Analysis

The worksheets given in Appendices 11 and 12 are used for collecting information from business process managers along with instructions about the process and how the information will be used. After all managers have completed their worksheets, information should be reviewed. Gaps or inconsistencies should be identified. Meetings with individual managers should be held to clarify information and obtain missing information.

After all worksheets have been completed and validated, the priorities for restoration of business processes should be identified. Primary and dependent resource requirements should also be identified. This information will be used to develop recovery strategies.

There are multiple strategies for recovery of manufacturing operations. Many of these strategies include use of existing owned or leased facilities.

Manufacturing strategies include:

- Shifting production from one facility to another
- Increasing manufacturing output at operational facilities
- Retooling production from one item to another
- Prioritization of production—by profit margin or customer relationship
- Maintaining higher raw materials or finished goods inventory
- Reallocating existing inventory, repurchase or buyback of inventory

- Limiting orders (e.g., maximum order size or unit quantity)
- Contracting with third parties
- Purchasing business interruption insurance

There are many factors to consider in manufacturing recovery strategies:

- Will a facility be available when needed?
- How much time will it take to shift production from one product to another?
- How much will it cost to shift production from one product to another?
- How much revenue would be lost when displacing other production?
- How much extra time will it take to receive raw materials or ship finished goods to customers? Will the extra time impact customer relationships?
- Are there any regulations that would restrict shifting production?
- What quality issues could arise if production is shifted or outsourced?
- Are there any long-term consequences associated with a strategy?

7.1.3 Resources Required to Supporting Recovery Strategies

Recovery of a critical or time-sensitive process requires resources. The Business Continuity Resource Requirements worksheet as given below should be completed by business function and process managers. Completed worksheets are used to determine the resource requirements for recovery strategies. Following an incident that disrupts business operations, resources will be needed to carry out recovery strategies and to restore normal business operations. Resources can come from within the business or be provided by third parties. Resources include:

- Employees
- Office space, furniture and equipment
- Technology (computers, peripherals, communication equipment, software and data)
- Vital records (electronic and hard copy)
- Production facilities, machinery and equipment
- Inventory including raw materials, finished goods, and goods in production.
- Utilities (power, natural gas, water, sewer, telephone, internet, wireless)
- Third party services

Since all resources cannot be replaced immediately following a loss, managers should estimate the resources that will be needed in the hours, days, and weeks following an incident.

7.1.4 Testing & Exercises

You should conduct testing and exercises to evaluate the effectiveness of your preparedness program, make sure employees know what to do and find any missing parts. There are many benefits to testing and exercises:

- Train personnel; clarify roles and responsibilities
- Reinforce knowledge of procedures, facilities, systems and equipment
- Improve individual performance as well as organizational coordination and communications
- Evaluate policies, plans, procedures and the knowledge and skills of team members
- Reveal weaknesses and resource gaps
- Comply with local laws, codes and regulations
- Gain recognition for the emergency management and business continuity program

7.1.4.1 Testing the Plan

When you hear the word "testing," you probably think about a pass/fail evaluation. You may find that there are parts of your preparedness program that will not work in practice. Consider a recovery strategy that requires relocating to another facility and configuring equipment at that facility. Can equipment at the alternate facility be configured in time to meet the planned recovery time objective? Can alarm systems be heard and understood throughout the building to warn all employees to take protective action? Can members of emergency response or business continuity teams be alerted to respond in the middle of the night? Testing is necessary to determine whether or not the various parts of the preparedness program will work.

7.1.4.2 Exercises

When you think about exercises, physical fitness to improve strength, flexibility and overall health comes to mind. Exercising the preparedness program helps to improve the overall strength of the preparedness program and the ability of team members to perform their roles and to carry out their responsibilities. There are several different types of exercises that can help you to evaluate your program and its capability to protect your employees, facilities, business operations, and the environment.

7.2 SMUG Model

This method has been developed by the Chatham Islands Civil Defence Emergency Management (CDEM 2015) Group for risk evaluation and hazard prioritization purposes. The Chatham Islands (Fig. 7.3) form an archipelago in the Pacific Ocean about 680 km southeast of mainland New Zealand. It consists of about ten islands within a 40-kilometre radius, the largest of which are Chatham Island and Pitt Island. Chatham Island is the largest island of the Chatham Islands group and said to be halfway between the equator and the pole and right on the International Date Line (CIC 2015). The CDEM Group is a partnership of agencies and organizations including emergency services. The primary partners are those with membership on the Coordinating Executive Group (CEG). Together the CEG and CDEM Group ensure the effective delivery of civil defence emergency management in the Chatham Islands, New Zealand.

The Ministry of Civil Defence and Emergency Management of the Island recommends the SMUG method for Prioritizing risks associated with natural and technological hazards in order to manage the risks posed by these hazards by developing strategies along the 4R's, namely reduction, readiness, response, and recovery.

The model is based on four main components – **Seriousness, Manageability, Urgency and Growth** (SMUG). Each of the four components is defined as follows.

7.2.1 *Seriousness*

The relative impact in terms of people and or dollars, including the number of lives lost and potential for injury, and the physical, social, and economic consequences of a hazardous event.

Fig. 7.3 Geographical location of the Chatham Islands (Google Map)

7.2.2 *Manageability*

The relative ability to mitigate or reduce the hazard (through managing the hazard, or the community, or both). Manageability refers to how well a hazard could be managed in the future. If a hazard has the potential to be mitigated by putting more emphasis on risk reduction initiatives, it would be rated high.

7.2.3 *Urgency*

The measure of how critical it is to address the hazard (associated with the probability of the risk of the hazard).

7.2.4 *Growth*

The rate at which the risk from the hazard will increase through either an increase in the probability of the extreme event occurring, an increase in the exposure of the community, or combination of the two.

The numeric ratings assigned to each of the four-prioritization criteria for the Chatham Islands CDEMG Plan (seriousness, manageability, urgency and growth) is provided in Table 7.1.

7.2.5 *Application of the SMUG Model*

The first step would be to review hazard reports for the area and existing mitigation plans. Having discussions with local people who have had significant knowledge or experience of those hazards is also vital for accurate assessment of the risk. A summary would be prepared and provided to a range of CDEMG member organizations, who will rate the hazards in terms of Seriousness, Manageability, Urgency, and Growth.

7.2.5.1 Method for Rating Seriousness

The number of lives lost and potential for injury and the physical, social, and economic consequences of a hazard event are considered in rating seriousness.

For each of the potential hazards in the area, a seriousness score of 0–5 is assigned to each vulnerable element (lives lost and injuries, physical, social, and economic components of the community). These are added together to attain a total seriousness

Table 7.1 The numeric score assigned to the high, medium, and low ratings for the four criteria used in the SMUG model

Criteria component	Numeric score		
Seriousness	High = 4–5	Medium = 2–3	Low = 0–1
Manageability	High = 7+	Medium = 5–7	Low = 0–4
Urgency	High $\geq$20 yrs	Medium $\leq$20	Low = 100 yrs
Growth	High = 3	Medium = 2	Low = 1

score. The top third of the range of seriousness scores are assigned a High rating; the middle third of the range of seriousness scores are assigned a Low rating. *Note: when rating seriousness, current risk treatment measures that are in place are taken into account.*

7.2.5.2 Method for Rating Manageability

The manageability rating is an estimate of how much extra effort is required for each hazard, across each of the 4R's (reduction, readiness, response, and recovery), to reach the desired level of preparedness for each R.

A manageability rating for each hazard is determined by using the following 5-step process:

1. For each hazard, the ideal amount of effort that should be spent treating the risk across each of the 4R's is estimated. A total of 12 points is spread across the 4R's to represent the relative amount of effort that should ideally be spent on each R.
2. The actual amount of effort that is or has been spent by the authorities on each of the 4R's is estimated for each hazard. No more than 12 points is spread across the 4R's to represent the relative amount of effort that is actually spent on each R.
3. The difference between the ideal and actual amounts of effort that is spent on risk treatment across each of the 4R's is then calculated for each hazard.
4. The difference between ideal and actual values for Reduction is added to the ideal-actual differences for Readiness, Response, and Recovery components.
5. The hazards with a total difference greater than 7 received a High rating, and those with a total difference less than 7 received a Low rating.

High = total score of future effort required is >7
Medium = total score of future effort required is >5 and <7
Low = total score of future effort required is <4

7.2.5.3 Example for Rating Manageability

If steps 1–5 above were undertaken for a flooding hazard, the results may look something like Table 7.2.

Table 7.2 Example for rating manageability using the SMUG model

Hazard	4R's	Ideal	Actual	Ideal-Actual	Total difference
Flooding	Reduction	4	3	1	
	Readiness	3	2	1	
	Response	3	2	1	
	Recovery	2	1	1	
Total		12			4 low

Method for Rating Urgency

It is a measure of criticality of addressing the hazard (associated with the probability of the risk of the hazards). Assign a High, Medium, or Low rating for the Urgency criteria, based upon the hazard return period following the guideline below. Please note that factors such as climate change or population growth etc. should be accounted for when evaluating return periods of hazards.

Low Urgency = > 200 year return period
Moderate Urgency = 20–200 year return period
High Urgency = < 20 year return period

Method for Rating Growth

It is the rate at which the risk from the hazard will increase through either an increase in the probability of the extreme event occurring, an increase in the exposure of the community, or a combination of the two. Assign a High, Medium, or Low rating for the Growth criteria as follows.

Low: risk increases from either an increase in the probability of an extreme event occurring or an increase in the exposure of the community.
Moderate: risk increases from both the probability of an extreme event occurring and the exposure of the community at a low moderate rate.
High: risk increases from both increase in the probability of an extreme event occurring and an increase in the exposure of the community at a high rate.

7.2.6 Challenges in Prioritizing Hazards

There are several difficulties associated with assigning a High, Medium, or Low rating to each of the criteria:

- The seriousness rating assigned to a given hazard depends upon the magnitude of the hazardous event under consideration.

- Lack of sufficient quantitative data that can be used to compare hazard against hazard for accuracy.

The severity and consequences of a hazard are variable and depend upon the magnitude and location of an event. In order to rate a specific type of hazard, scenarios should be developed to provide a threshold for assessing the consequences of a given hazardous event and therefore prioritizing hazards. These scenarios are usually based upon severe events that are on the threshold of causing major disruption. For example, the seriousness of a flood event may be described for a 100 year return period flood; and the seriousness of an earthquake may be based on a magnitude 9 event (430 year return period). An earthquake of this return period has a magnitude so high that it could be just around or even slightly over the earthquake code standards in place. Due to a lack of quantitative data, many of the hazard consequences or risks described may be qualitative, which may compromise the prioritization process. Table 7.3 shows a sample of SMUG Model results for the Chatham Islands.

It should be noted here that in the recent version (2011–2016) of the model, the Urgency component is missing, and risk assessment conducted on Chatham Islands is based on the three components – Seriousness, Manageability, and Growth.

7.2.7 *Case Study: Risk Profile for Chatham Islands*

Risk is measured in terms of consequences and likelihood.

Table 7.3 SMUG results

Hazard	S	M	U	G	Total
Tsunami – Local	3	7	2	3	15
Utility failure – Communications	3	6	3	3	15
Transportation – Air	3	5	3	3	14
Flooding	3	4	3	3	13
Wind storm	3	5	3	2	13
Fire – Rural	1	8	2	2	13
Public health emergency	3	5	2	2	12
Tsunami – Distant	3	4	2	3	12
Utility failure – Power	2	4	3	3	12
Erosion	2	5	2	2	11
Transportation – Marine	2	5	2	2	11
Bio-security emergency	3	2	2	3	10
Earthquake	1	5	2	1	9
Storm surge	2	3	2	2	9
Hazards substances	1	4	2	2	9
Transportation – Road	2	1	3	2	8
Utility failure – Water	2	1	2	2	7
Fire – Urban	3	1	2	1	7
Criminal damage	1	3	1	1	6
Civil unrest	1	3	1	1	6

$$\text{Risk} = \text{Likelihood} \times \text{Consequences}$$

The hazardous effects on Chatham Islands' social, built, economic and natural environments, will determine the consequences of hazards. Understanding the diversity of hazardous events, and how they interact with these environments is vital to establishing a risk profile. To achieve a risk profile, the hazards are identified, their risks assessed, and each hazard evaluated and assigned a level of priority. This prioritization provides guidance for directing resources and effort for the treatment of risk across the 4Rs. The approach is based on a premise that hazards are best managed through a degree of complexity and co-ordination.

Risk Profile Assessment

- A compressive summary of the natural, social, built, and economic environments
- Description of all hazards impacts, likelihoods, and consequences
- Assessment of Risks in the Chatham Islands including Pitt Island
- Prioritization of the Risks

Background information collection

Social Environment: include social structures such hospitals, schools, town hall, and sports clubs. Consideration is given to public safety and accessibility during an emergency.

Vulnerable Groups: this includes seniors with physical and mental disabilities, the young including early childhood locations, and primary schools when they are operating.

Ethnic Diversity: 65.6% of people living in the Chatham Islands belong to the European ethnic group and 64.2% belong to the Maori ethnic group (2006 Census).

Population: as identified in the 2006 Census is for the Chatham Islands Territory, which includes Pitt Island. There is not a breakdown of the individual settlements on main Chatham or Pitt Island but a holistic overview. Dwellings identified as private and non-private dwellings. Table 7.4 gives an overview of the population.

Next, the natural, built, and economic environments were examined and information recorded. Within the natural environment, the main categories of geology, geography, and climate were reviewed. In the built environment, the following details were recorded – commercial and industrial units, agricultural infrastructure, residential areas, lifeline utilities (water, gas, electricity, telecommunication,

Table 7.4 An overview of the population for the Chatham Islands based on 2006 census

Area	Population	Dwellings
Male	345	
Female	267	
Private dwellings		249
Non-private dwellings		6
Unoccupied dwellings		51
Chatham Island/Pitt Island	612	306

transportation network, waste water and storm water). The economic environment would include regional economy, growth, employment rate, income (comparison with national average), and tourism.

Risk Assessment

Some hazards may pose more threat to the communities than others in terms of their frequencies and impacts. The modified version, SMG model ANZS4360 was then applied for an in-depth evaluation of each hazard based on the following:

Seriousness – the relative impact of human, economic, social, infrastructural, and geographic factors.
Manageability – how difficult hazard risks are to manage and level of effort currently applied.
Growth – the rate at which the risk is likely to increase through heightened probability of the event occurring, an increase in the exposure of the community to the hazard, or a combination of both.

Appendix 13 gives the overall prioritization of hazards for the Chatham Islands.

7.3 World Risk Index Tool

The need for a widely acceptable and applicable concept and method for risk assessment led to the development of the concept of the World Risk Index (WRI) (see http://www.uni-stuttgart.de/ireus/Internationales/WorldRiskIndex/) by Welle and Birkmann (2015) from the University of Stuttgart. The WRI (Fig. 7.4) provides a new approach to assess risk and vulnerability towards natural hazards on country scale and allows the comparison of countries at global scale. The WRI evaluates the exposure to natural hazards faced by 171 countries and assesses the inherent vulnerability in the countries towards suffering from impacts when facing these

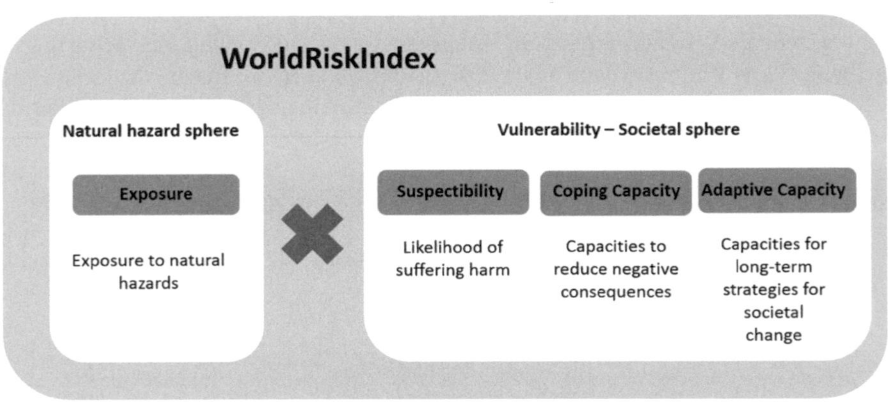

Fig. 7.4 World Risk Index (Welle and Birkmann 2015; Birkmann and Welle 2015)

hazards. The index shows that Vanuatu is the country with the highest disaster risk (Index value: 36.72), Tonga ranks 2nd (Index value: 28.45) and the Philippines, ranks 3rd (Index value: 27.98).

A detailed description of the WRI and method to calculate parameters (exposure, susceptibility, coping capacity and adaptive capacity) involved can be accessed at http://www.uni-stuttgart.de/ireus/Internationales/WorldRiskIndex/

7.4 Role and Assessment of Return Period

The age of our planet is more than 4.5 billion years and it is evolving continuously. Natural hazards such as volcano eruptions, earthquakes, floods etc. have been occurring for millennia. From theoretical facts and practical experience, it is understood that a number of different mutually exclusive events may occur at the same time. For example, flooding and landslides can occur at the same time during one single event of a heavy rainfall causing both landslides and floods in the region. Probability of occurrence of naturally occurring events can be derived from historical data based on past similar events. The quality and quantity of data would determine the quality of future probability predictions. For this reason, frequent events often have a better chance of accurate future predictions, compared with rare events, such as the 2004 Boxing Day tsunami in the Indian Ocean. Probability of potential loss (life, property, and the environment) is also essential to determine in this context, in order to estimate potential risk. So, what does this all mean? It means that in order to get a sense of what may come in future in terms of natural hazards and how often, and also how much damage the event may cause, time series of past events would be required. Each hazard type (e.g. earthquakes, floods, ice storms, hurricanes, forest fires, etc.) is measured using specific scales developed by scientists.

Let us consider an example of flooding in the Greater Toronto Area (GTA). Rainfall is measured by a network of rain gauge stations (Fig. 7.5) (Armenakis and Nirupama 2014) in which precipitation is measured at various intervals – every 5 minutes, 10 minutes, 15 minutes, 30 minutes, and hourly. Based on this data and the objective of the study, daily maximum and minimum, monthly maximum and minimum, or annual maximum and minimum values of precipitation can be acquired in millimeters or centimeters. In addition to this, any particular sever storm can be easily delineated from this data series. Figure 7.6 is an illustration of extraction of the July 2013 flooding event in the GTA (Armenakis and Nirupama 2014). Figure 7.7 (Nirupama et al. 2014) shows the four major flooding events in the GTA. Major players in the collection, storage, processing, and analysis of these datasets are Environment Canada and Toronto and Region Conservation Authority. Utility agencies, such as Hydro One (federal agency in Canada) or Toronto Hydro (provincial agency in Ontario, Canada) may also have some jurisdiction to these datasets for the purpose of maintaining continuity of their operations in the event of a major event.

Fig. 7.5 Rain gauge network in the GTA (Armenakis and Nirupama 2014)

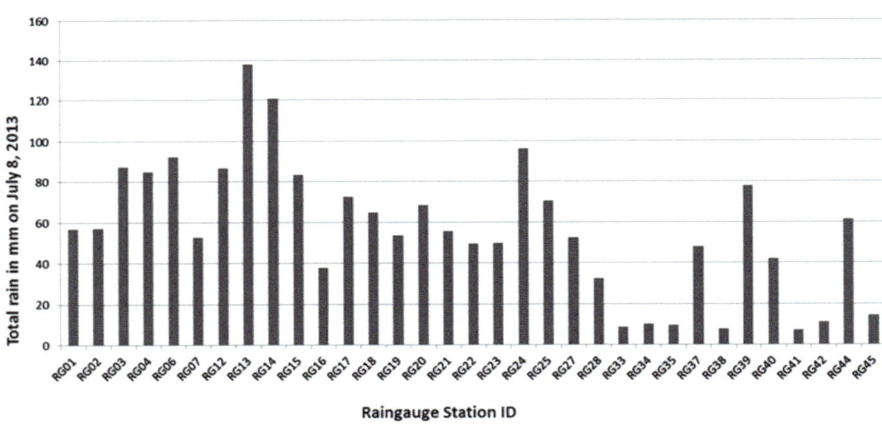

Fig. 7.6 Rainfall recorded at various rain gauge stations in the GTA (Armenakis and Nirupama 2014)

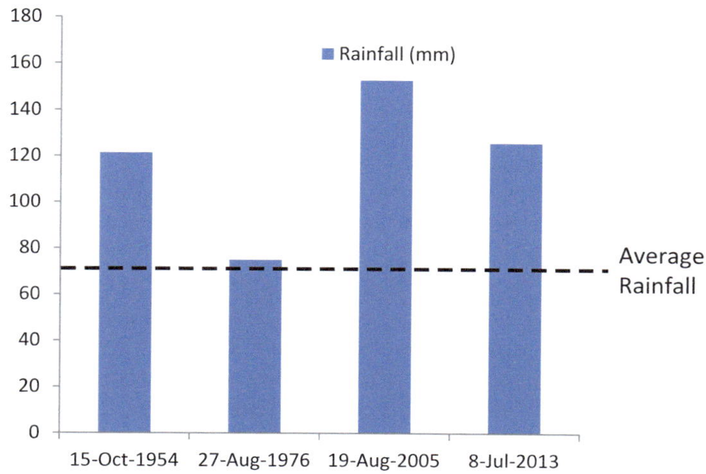

Fig. 7.7 Toronto rainfall amounts during major flooding events in the GTA. Data from Toronto and Region Conservation Authority has been used to extract the four major events. (Nirupama et al. 2014)

Most weather related extreme events are analyzed based on their annual maximum or minimum values at a given site. Hydrological and meteorological events are measured and analyzed this way. The other natural disaster types, such as earthquakes, tsunami generating earthquakes, and volcanic eruptions related predications are entirely based on past events and anticipation of underground built-up pressures and stresses. For example, as given in Smith (2004) annual maximum wind gusts recorded at Tiree in western Scotland, over a 59-year period, from 1927 to 1985 are available to determine the potential for windstorm damage. These events can be given a rank, m, starting with m = 1 for the highest value, m = 2 for the next highest and so on in increasing order. The Return Period or Recurrence Interval, T (in years) can then be computed from,

$$T = \frac{n + 1}{m} \tag{7.1}$$

Where m is event ranking and n is number of events in the period of record. The percentage probability for each event may then be obtained from,

$$P(\%) = \frac{100}{T} \tag{7.2}$$

And, the Annual Frequency (AF) can be given by,

$$AF = \frac{1}{T} \tag{7.3}$$

Major floods are referred to in terms of a certain return period event. For example, a 100-year flood is a flood with a 0.01 chance of occurrence in any given year, at any given time. A 200-year flood is a flood with a 0.005 chance of occurrence in any given year, and a 500-year flood would be a flood with a 0.002 chance of occurrence in any given year. It is important to note that by definition, any of these events that are used as examples can occur more than once in a given year or in two consecutive years. The idea behind explaining the real meaning of return period is to remove the misconception that a 100-year flood event can occur only once in 100 years. Instead, it is the probability or chance of occurrence that should be taken into account in determining the possible occurrence of a certain magnitude event. This is true for any hazard type that there is a c% chance of a t-year return period event happening in a usual life time of, say T years.

Smith and Petley (2009) also provides evidence of the relationship of flood magnitudes with probabilities (Fig. 7.8).

It is worth noting here that Environment Canada requires a minimum of 30 years of data records for probability calculations using Eq. (7.2) to be considered reliable.

7.4.1 Working Exercise to Calculate Return Period

Table 7.5 lists the top ten floods in the Mississippi at St. Louis from 1861–2002. In Table 7.6, the floods have been arranged in descending order according to their magnitude measured as discharge (flow rate). What would be the return period for a flood that would have a discharge of about 25,000 cubic meters per second (cumecs)?

Steps to follow to determine the return period:

- Find out the total number of data records (n) given (from 1861 to 2002).
- Arrange all the floods in descending order according to their given discharge in Table 7.5.
- Rank all the floods by their magnitude as shown in Table 7.6.
- Look for the desired magnitude (discharge of about 25,000 cubic m/s) and note the corresponding rank (m = 4 according to Table 7.6).
- Now, apply Eq. (7.1) to calculate the Return Period:

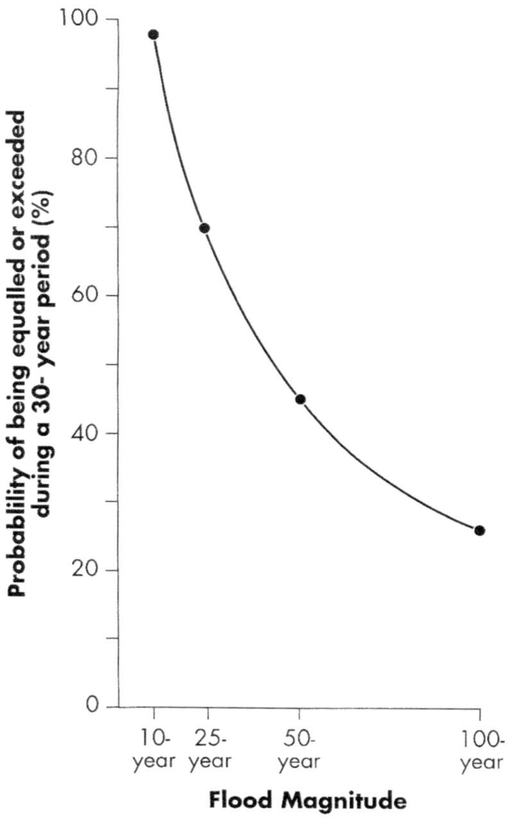

Fig. 7.8 The probability of occurrence of floods of various magnitudes during a period of 30 years, the length of a standard property mortgage (Smith 2004)

Table 7.5 Top ten Mississippi river floods at St. Louis during 1861–2002

Date	Discharge or flow rate (cumecs)
June 1883	24,432
May 1892	26,236
June 1903	28,855
June 1908	24,069
July 1909	24,369
April 1927	25,182
May 1943	23,786
April 1944	23,899
April 1973	24,126
August 1993	29,166

Table 7.6 Top ten Mississippi river floods at St. Louis

Date	Discharge or flow rate in cumecs arranged in descending order	Rank by magnitude
August 1993	29,166	1
June 1903	28,855	2
May 1892	26,236	3
April 1927	25,182	4
June 1883	24,432	5
July 1909	24,369	6
April 1973	24,126	7
June 1908	24,069	8
April 1944	23,899	9
May 1943	23,786	10

$$T = \frac{141 + 1}{4}$$
$$= 35.5 \text{ years}$$

- Next, the probability of the occurrence of this 35.5 year return period event can be calculated using Eq. (7.2) as,

$$P(\%) = \frac{100}{35.5}$$
$$= 2.8\%$$

If more than one event has been recorded as having the same magnitude in a data series, the ranking needs to be conducted carefully. Table 7.7 shows a dataset on lowest central pressure of notable recorded tropical cyclones around the world. It is known that the lower the central pressure, the deadlier the storm. Therefore, when ranks are assigned, the lowest central pressure event gets the top rank followed by the second lowest, and so on. The last event in this series will be the one with the least powerful storm or, in other words, highest value of central pressure. The same value of central pressure events are assigned the same rank, however, the next event after them will skip to the consecutive ranking, and receive the next rank instead. For example, in Table 7.7, Typhoons Nora and Ida are assigned rank number 3 based on their pressure values, and Typhoon Rita is assigned a rank of 5, skipping ranking value 4. If three events happen to have the same magnitude like Cyclone Cossack, Hurricane Janet, and Cyclone Mahina, the event falling next in the list will have a rank of 20 (skipping rank values 18 and 19).

Once ranks are determined, calculation of return period, probabilities, and annual frequencies will be determined the same as the example given above when each

Table 7.7 Lowest central pressures recorded in tropical cyclones/hurricanes/typhoons (adapted from Brayant 2005: 49)

	Event	Location	Date	Pressure (hPa)	Rank
1	Typhoon Tip	Philippines	October 1979	870	1
2	Typhoon June	Guam	November 1975	876	2
3	Typhoon Nora	Philippines	October 1973	877	3
4	Typhoon Ida	Philippines	September 1958	877	3
5	Typhoon Rita	Philippines	October 1978	878	5
6	Hurricane Gilbert	Caribbean	September 1988	902	11
7	Typhoon Nancy	NW Pacific	September 1961	888	6
8	Labour Day storm	Florida	September 1935	892	7
9	Typhoon Marge	Philippines	August 1951	895	8
10	Hurricane Allen	Caribbean	August 1980	899	9
11	Hurricane Linda	Baja Peninsula	September 1997	900	10
12	Hurricane Camille	Gulf of Mexico	August 1969	905	12
13	Hurricane Mitch	Caribbean	October 1998	905	12
14	Typhoon Babe	Philippines	September 1977	906	14
15	Cyclone Vance	Australia western	March 1999	910	15
16	Typhoon Viola	Philippines	November 1978	911	16
17	Cyclone Cossack	Australia	June 1881	914	17
18	Hurricane Janet	Mexico	September 1955	914	17
19	Cyclone Mahina	Australia	March 1899	914	17

recorded event had a different magnitude. Currently, with so many online tools being developed and made available for general consumption and convenience, probability of a particular return period event can be calculated by plugging in the required parameters. One such example is available on the website of the National Weather Service (NWS 2015). A screen capture is shown in Fig. 7.9. In this tool, the user needs to plug in a return period value (in years) for an anticipated (design) event and the number of years toward the time period over which the prediction is sought. These numbers are then converted into percent chance of occurrence of the anticipated event. As indicated in the tool, although the calculator is worded for flood events, it would work for any extreme weather event.

7.5 Exercise

Apply each of the methods discussed in this chapter to the best of your ability using any recent disaster. Suggest modification and new ideas as you see fit.

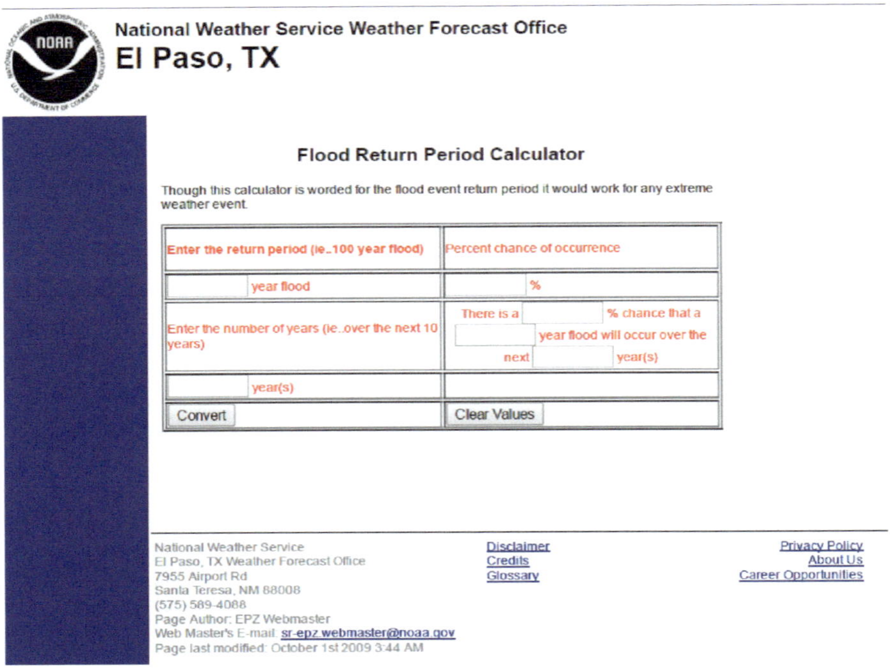

Fig. 7.9 Flood return period calculator (NWS 2015)

References

Armenakis C, Nirupama N (2014) Flood risk mapping for the City of Toronto. Proced Econ Financ, Elsevier 18:320–326

Birkmann J, Welle T (2015) Assessing the risk of loss and damage: exposure, vulnerability and risk to climate related hazards for different country classifications. Int J Glob Warm 8(2):191–212

Brayant E (2005) Natural hazards. Cambridge University Press, Cambridge, 330p

CDEM (2015) National Civil Defence Emergency Management Plan 2015. Ministry of Civil Defence & Emergency Management http://www.civildefence.govt.nz/cdem-sector/cdem-framework/national-civil-defence-emergency-management-plan/. Retrieved 20 Jan 2016

Chatham Islands Council (CIC) (2015) Chatham Islands Civil Defence and Emergency Management Plan 2013–2016. http://www.cic.govt.nz/documents/2013/10/chatham-islands-civil-defence-and-emergency-management-plan-2013-2016.pdf. Accessed 20 Jan 2016

CSRC (2015) NIST special publications. Computer Security Resource Center. http://csrc.nist.gov/publications/PubsSPs.html. Accessed 18 Jan 2016

FEMA (2015a) Risk assessment, Federal Emergency Management Agency. www.ready.gov/risk-assessment. Accessed 18 Jan 2016

FEMA (2015b) Natural disasters, Federal Emergency Management Agency. http://www.ready.gov/natural-disasters. Accessed 18 Jan 2016

FEMA (2015c) Pandemic, Federal Emergency Management Agency. http://www.ready.gov/pandemic. Accessed 18 Jan 2016

FEMA (2015d) Implementation, Federal Emergency Management Agency, http://www.ready.gov/business/implementation. Accessed 18 Jan 2016

FEMA (2015e) Business impact Analysis, Federal Emergency Management Agency. http://www.ready.gov/business-impact-analysis. Accessed 18 Jan 2016

Grance T, Nolan T, Burke K, Dudley R, Gregory W, Good T (2015) Guide to test, training, and exercise programs for IT plans and capabilities. http://csrc.nist.gov/publications/nistpubs/800-84/SP800-84.pdf. Accessed 18 Jan 2016

IIBHF (2015) Insurance Institute for Business & Home Safety. http://www.disastersafety.org/. Accessed 18 Jan 2016

Nirupama N, Armenakis C, Montpetit M (2014) Is flooding in Toronto a concern? Nat Hazards 72 (2):1259–1264

NWS (2015) Flood return period calculator. National Weather Service. http://www.srh.noaa.gov/epz/?n=wxcalc_floodperiod. Accessed 22 Jan 2016

Smith K (2004) Environmental hazards: Assessing risk and reducing disaster, Routledge 306p

Smith K, Petley DN (2009) Environmental hazards: assessing risk and reducing disaster. Routledge, New York, p 306

Swanson M, Bowen P, Phillips AW, Gallup D, Lynes D (2015) Contingency planning guide for federal information systems. NIST special publication 800-34 Rev 1. http://csrc.nist.gov/publications/nistpubs/800-34-rev1/sp800-34-rev1_errata-Nov11-2010.pdf. Accessed 18 Jan 2016

Welle T, Birkmann J (2015) The world risk index – an approach to assess risk and vulnerability on a global scale. J Extreme Events (JOEE) 2(1):34 pages

Chapter 8
Disaster Risk Evaluation – Qualitative Methods

Perceptions of risks are significant in the decision-making process. Furthermore, the level of risk acceptance varies between individuals, communities and cultures. Given recent advances in the social sciences and psychology, new theories and research in disaster management focus on understanding the root causes of individuals' changing social, physical, economic and environmental vulnerabilities. Accounting for stakeholders' perceptions is an important part of the process. This closing chapter discusses qualitative approaches to evaluating vulnerability and risk that reflect the progression of vulnerabilities; access to resources; and community perceptions.

8.1 Pressure and Release (PAR) Model

The PAR model (Fig. 8.1) provides a framework that can be used to assess the progression of vulnerability and evaluate disaster risk. The framework assists with the identification of root causes, dynamic pressures (that translate root causes into unsafe conditions), and unsafe conditions. Birkmann (2006, 2007) asserts that overall the PAR model is an important approach and one of the best known conceptual frameworks worldwide that focuses on vulnerability and its underlying driving forces. The PAR model suggests that addressing the underlying (root) causes may prove to be an effective way of reducing vulnerability and risk. Figure 8.2 is an adaptation of the original PAR for the purpose of simplifying the illustration.

N. Agrawal, *Natural Disasters and Risk Management in Canada*, Advances in Natural and Technological Hazards Research 49,
https://doi.org/10.1007/978-94-024-1283-3_8

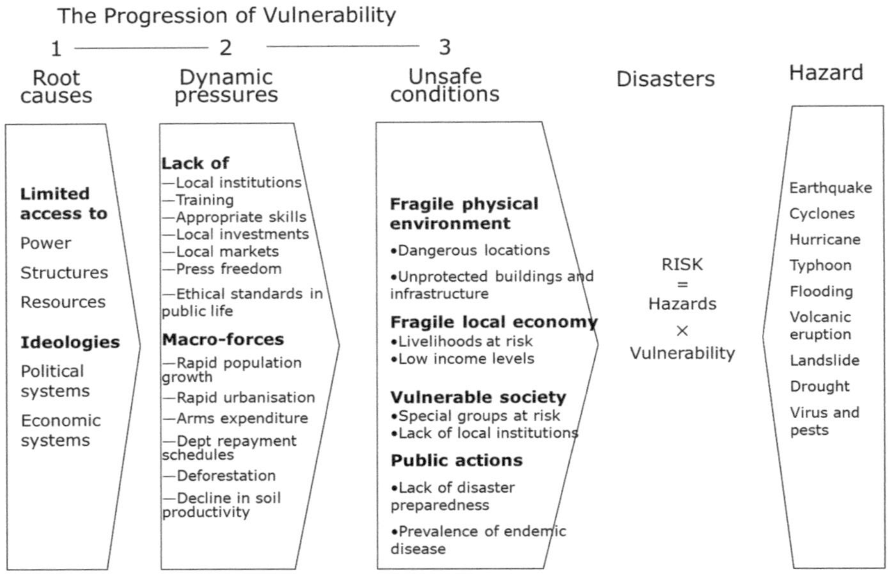

Fig. 8.1 Pressure and Release model in its original format (Wisner et al. 2004)

8.1.1 Application of Pressure and Release (PAR) Model to the Case of Hurricane Katrina

The PAR model is a useful tool to analyze a disaster for better understanding. Using the example of Hurricane Katrina that caused havoc in several cities in the United States, particularly, New Orleans in August 2005, the application of PAR model has been demonstrated. Hurricane Katrina devastated the State of Louisiana, USA, especially the City of New Orleans to an unprecedented degree. Figure 8.3 shows the powerful hurricane system in the Gulf of Mexico fast approaching the Gulf coast impact region. Figure 8.4 shows the flooding in New Orleans and Fig. 8.5 shows an image of the flooding in New Orleans approximately 1 week after.

First step in the process of applying the two models is to always gather information from reliable media resources including online news media, TV, radio, and social media, as well as scholarly articles. Documentaries produced by a variety of organizations such as CBC (Canadian Broadcasting Corporation) of Canada and PBS (Public Broadcasting Station) of the USA are also used in information gathering for this section. One particular documentary that aired on 21 November 2005 has been very helpful in this regard, produced by PBS/NOVA, titled as Storm that Drowned a City. This documentary can be accessed at (https://www.youtube.com/watch?v=zBZuPmi4i-U). The PAR model construct is shown in Fig. 8.6.

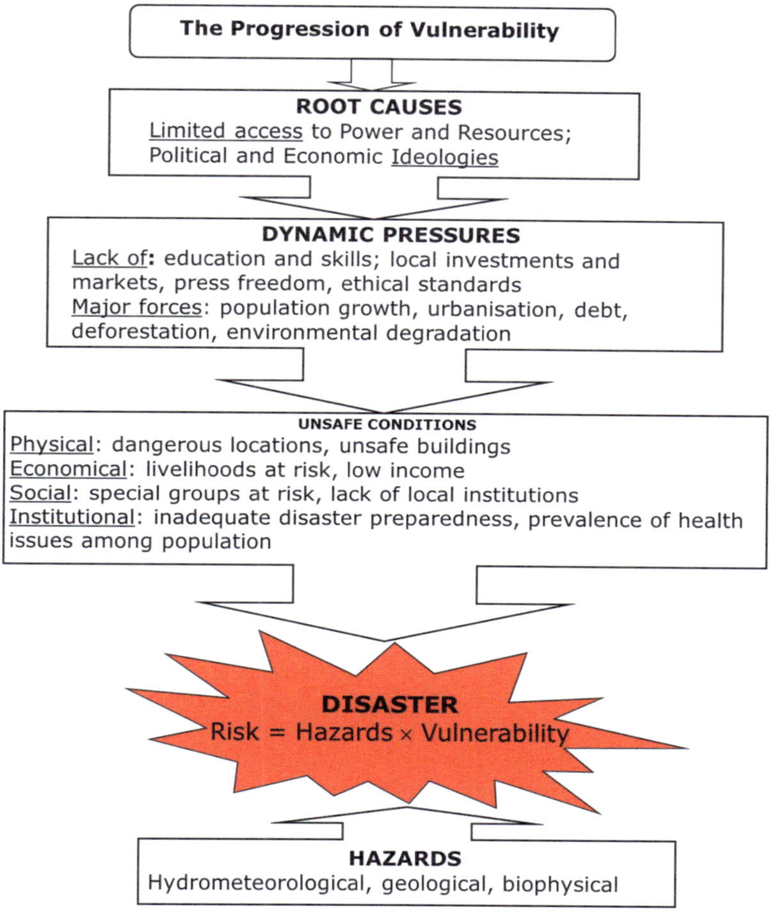

Fig. 8.2 Pressure and Release Model (Adapted from Wisner et al. 2004)

8.1.2 Application of PAR Model to the 2008 Sichuan, China Earthquake

In May 2008, the ground in China's Sichuan province shuddered and cracked open. Buildings, roads, and lives were torn apart in seconds. The massive earthquake left over 80,000 dead or missing and millions homeless. *China's Earthquake: The People in the Pictures* tells the story of four people lives that were changed forever by this disaster – a young reporter, a 9-year-old hero, a schoolgirl who lost her leg after being buried alive, and one of the many parents whose children died in a poorly constructed school. (https://curio.ca/en/video/chinas-earthquake-the-people-in-the-pictures-1154/). The Host of the documentary series is Ann-Marie MacDonald, the Director is Susan Teskey, and the Producer is Xiaoping Diana Dai.

Fig. 8.3 Hurricane Katrina August 28, 2005. The image was taken by GOES satellite, obtained from http://www.nnvl.noaa.gov/hurseas2005/Katrina1515z-050828-4kg12.jpg

Fig. 8.4 NOAA Corps (Commander Mark Moran, Lt. Phil Eastman, and Lt. Dave Demers) flew more than 100 h from August 31 to September 19, 2005 surveying Katrina's devastation

Fig. 8.5 The flooding in New Orleans nearly a week after Hurricane Katrina hit, taken by NASA's EO-1 satellite on Sept. 6, 2005. Credit: NASA

PAR Model Construct: the format of the model construct has been simplified here for the purpose of demonstrating that what is most important is that the model components are understood correctly, and that the actual shape or diagram is not necessary to draw (Fig. 8.7).

8.2 Access to Resources (ATR) Model

The Access to resources (ATR) model was developed by Wisner et al. (2004) as shown in Fig. 8.8, in its original form. Fig. 8.9 is also a part of the original model, showing Box 1 of Fig. 8.8 in its unpacked form. The concept of 'access to resources' which is mostly explained by the unpacked Box 1 is what makes this model dynamic

ROOT CAUSES

LIMITED ACCESS TO:

Power

- Federal government's focus is on fighting terrorism, not hazard mitigation
- FEMA (Federal Emergency Management Agency) is not a first responder, it doesn't have its own emergency response vehicles – forced to rely on other federal agencies, State workers, National Guard, private contractors, and US military
- Decision making problems at the local level regarding evacuation plans and emergency supply needs

Resources

- Income opportunities, education and training among the poor
- Government assistance such as subsidized home insurance for the poor available but not sufficient
- Information to build hurricane and flood resistance structures
- Limited sources of information/funds available on how to prepare for a hurricane
- Transportation - 100,000 people did not have personal transportation access

IDEOLOGIES

Political Systems

- Belief that infrastructural protection was sufficient for storm surges
- Lack of cooperation between levels of Government (Federal/State/Municipal). Bureaucracy slowed down relief effort
- Despite scenarios previously conducted, federal officials seem ignorant of possibility of level of devastation possible in New Orleans
- After 9/11, FEMA downgraded from a cabinet-level agency to part of the Homeland Security Office. Taking it out of the White House made it less effective
- Washington's reliance on media for crucial updates

Economic Systems

- US economy not oriented to safety nets for the poor and/or unemployed
- Financial power exists more with the white majority than with the black minority - unequal income distribution
- Budget shortfalls resulted in the levees not up to safety standards
- Lack of financial security creates a cycle of poverty and vulnerability whereby the poor cannot rebuild or start over in a reasonable time

DYNAMIC PRESSURES

Lack of:

- Designated large shelters, and emergency supplies
- Local investment in major industries
- Formal training of general population and police; 249 police officers deserted their post
- Emergency plans, transportation agreements
- Official media/communication EOC (Emergency Operation Centre) experts as the media reporters gave conflicting reports
- Local Institutions: safe places and shelters for more vulnerable population
- Ethical Standards: promises to deliver shelter, gasoline, diesel, and generators etc. not kept
- Training: hurricane drills, education, and awareness
- Local institution - communication and coordination between response agencies caused primarily by bureaucracy and red tape
- Business Continuity Planning: with the exception of Wal-Mart and Home Depot, many businesses were not adequately stocked with hurricane supplies

Macro forces:

- Population growth in flood zones (lower 9th ward) as a result of existing levee system that gave people a false sense of security and encouraged development
- Removal of marshes and wetlands by developers, which are natural protection against storm surges
- Local government officials not keeping up with the funding and maintenance of levee systems
- Deforestation contributed to soil instability
- Rapid urbanization of New Orleans promoted as a tourist hot spot and entertainment centre

Fig. 8.6 PAR model construct for Hurricane Katrina that drowned New Orleans, USA in August 2005

UNSAFE CONDITIONS

Physical Environment:

- Dangerous Location: large areas with streets and houses 1.2m below sea level
- Levee system not maintained
- Population centres in Louisiana and Mississippi are locatedin coastal flood plains
- City is built on the Mississippi delta and Greater New Orleans is surrounded on three sides by water (Gulf Coast, Mississippi River, and Lake Pontchartrain)
- Natural protection of marshes and wetlands are removed
- Houses not adequately retrofitted to withstand floods, built with cheap materials

Local Economy:

- Low income level earners living in French quarter and lower 9th ward
- 23% chronic unemployment rate in New Orleans so, end of the month occurrence of a disaster can hit hard
- Livelihoods are seasonal, at risk, low income jobs
- Lack of insurance

Social Relations:

- High percentage of population belong to high risk groups (minority groups, women, children, elderly)
- Mistrust of authority, politicians and media
- Expectation that authorities will 'take care' of populace
- Lack of allocation of land above sea level to poor/minorities

Public institutions/Actions:

- Planning resulting from 2004 simulation exercise not followed through
- Budgetary constraints delayed levees' repairs
- Lack of appropriate shelters
- Lack of communication means, as some of the poor do not own a television or radio
- Lack of local institutions to deal with supplies and assistance in emergency or evacuation situations
- 25,000 people take shelter at Conference Center without any relief supplies; FEMA was unaware people were taking refuge in the Conference Center
- Lack of disaster preparedness
- Poor support from local, regional, and federal government for social protection.

DISASTER

- Confirmed US fatalities of 1836, mostly in Louisiana, 705 missing
- A total of 20,000 people take shelter, thousands fell ill due to lack of food and water
- 53 levees between New Orleans and Lake Pontchartrain breached
- Extensive property damage = US$ 81.2 Billion
- Flooding of 9th ward and other areas of downtown, 80% of New Orleans flooded
- 30 oil platforms in the Gulf damaged or destroyed

HAZARD(S)

Hurricane Katrina hit U.S. States of Louisiana, Mississippi and Alabama as a Category 3 storm with winds reaching up to 200 km/hr, storm surge of 3.5-4.5m, and causing significant flooding and airborne debris due to high winds

Fig. 8.6 (continued)

Progression of Vulnerability

Root Causes:

Limited access to:

Power

- Victims became powerless once the government banned reporters from broadcasting the devastation
- Negative attitude and stigma towards people of lower economic status and thus no power to voice needs or concerns
- No political voice to stand up for parents of dead school children
- Corrupt officials did not investigate accidents

Resources

- No other aid such as social, physical, or financial, available to rescuers
- Government assistance available more to urban areas, such as Shanghai and Beijing, for schools, jobs, and health care system as opposed to rural villages in Sichuan
- Skilled labor for construction (based on speculation around the fact that a lot of buildings collapsed)
- Information regarding earthquakes resistant structures – again, speculation based on extent of damage suffered from collapse of built structures.
- Affordable and effective equipment to build safer buildings
- Investments in infrastructure as rural areas did not receive overall large investments

Ideologies:

Political Systems

- Corrupt system -government officials failed to supply grieving parents with needed answers, covered up the truth, declared accident caused by natural disaster, quashed public discussion, and resorted to harassment and threats of imprisonment
- Lack of openness of communist leaders to information
- Government trying to portray a media friendly side to China by allowing Chinese journalists to film aftermath of the disaster
- Government manipulating the media by focusing on the young heroes of the disaster rather than addressing all victims' needs

Economic Systems

- Sichuan province has a weak economy
- Little diversification amongst possible livelihoods of mostly farming and some animal rearing
- Villages of low importance to government, due to poor economic abilities
- Government does not distribute much resources to less developed rural areas
- Financial powers exist more in urban areas than in rural areas

Dynamic Pressures:

Lack of:

- Disaster response training: responders did not know how to free some of the trapped persons, and as a result one little girl lost a leg and more rapid response needed by government
- Skilled and trained builders of earthquake resistant homes (speculation due to large number of collapsed buildings)
- Ethical standards in public life by government officials: corrupt local officials unwilling to perform an investigation into the school collapse, pocketed school construction funding
- Effective land-use planning
- Employment and higher educational opportunities only available in cities such as Shanghai
- Press freedom: news stations and reporters were told to report on stories that were only positive and a benefit to the Chinese society however after one month, the government forced them not to report about the school collapse and barred foreign journalists from the school investigation sites
- Access to some essential services particularly health facilities to deal with earthquake casualties
- Public awareness of seismic risk and building codes

Fig. 8.7 PAR model applied to the 2008 China earthquake. Based on the CBC Doc Zone documentary series https://curio.ca/en/video/chinas-earthquake-the-people-in-the-pictures-1154/

Macroforces:

- Rapid population change: migration of skills and knowledge as the young people moved from rural to urban areas in search of better livelihoods and job opportunities
- Large number of people concentrated in a small area for example in the town of Beichuan, where land pressures force people to live in multi-story apartment buildings

Unsafe Conditions

Physical Environment:

- Earthquake prone region
- Building code variations led to unprotected buildings and infrastructure leading to a number of homes and schools being poorly constructed
- Unsafe critical facilities such as schools, medical buildings, etc.
- Poor communication with the rest of the Sichuan Province and China, as it is an isolated mountainous village
- Unsafe public/private buildings for all income levels due to the lack of adequate building codes prior to the disaster, the affected school buildings were described as 'iron twisted like rubber and the concrete crumbled like sand"
- Limited routes for evacuation and access for rescue equipment and response teams

Local Economy:

- Livelihoods at risk - limited to no job diversity with heavy dependence on agriculture
- Low income levels - rural community made-up of peasant farmers
- Unequal distribution of wealth and income between rural and urban areas
- Absence of social insurance

Social Relations:

- Special groups at risk – elderly, young children, and school staff, (earthquake occurred at 2:28 pm when children were still at school)
- Institutions such as hospitals were inaccessible for a small town
- Affected people being treated unequally due to political reasons

Public Actions and Institutions:

- Lack of disaster preparedness/planning
- No effective evacuation/relocation plans
- Difficult for response/recovery teams to reach impacted location
- Lack of coordinated actions to mitigate disaster
- Lack of local institutions in terms of oversight for building construction
- Inadequate land-use planning
- Lack of public education and training to manage local disasters
- Lack of emergency shelters

Disaster: *Risk = hazard x vulnerability*

- 70,000 people dead, 374,176 injured, about 5-11 million homeless
- 7,000 school buildings collapsed
- Damage to critical infrastructure (railways, bridges, roads, etc.)
- Losses greater than US $20 billion
- Destruction of millions of livelihoods
- Widespread psychological trauma and panic - some survivors committed suicide, children had difficulties sleeping in the dark and afraid to be alone

Hazards

- Earthquake (7.9 on Richter scale) - primary
- Landslides - secondary

Fig. 8.7 (continued)

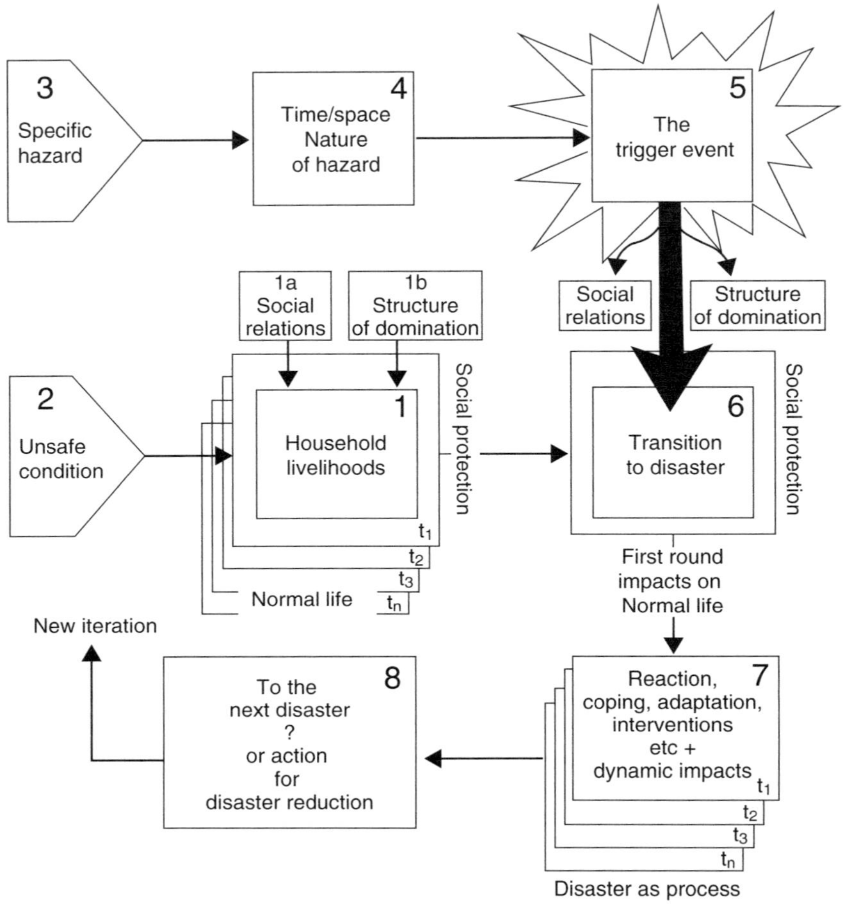

Fig. 8.8 Access to Resources model as proposed by Wisner et al. (2004)

in nature. The original model is quite complex for lay persons, therefore, a simpler version has been developed using only the core component of the model (Fig. 8.10), which is the unpacked form of Box 1. The framework facilitates an understanding of how people make decisions based on their ability to access resources – temporally and spatially – in the face of a disaster. The model looks at people's 'access profile', which is based on whether they seek opportunities, both prior to and after a disaster. The model identifies the dynamics of changing decisions, options, budgets, access profiles, and the choices made by the impacted population. The ATR model approach allows for a deeper look at disaster triggers, the extent of impact, coping mechanisms, social protection, and future regional scenarios. The model also explains whether or not recovery and

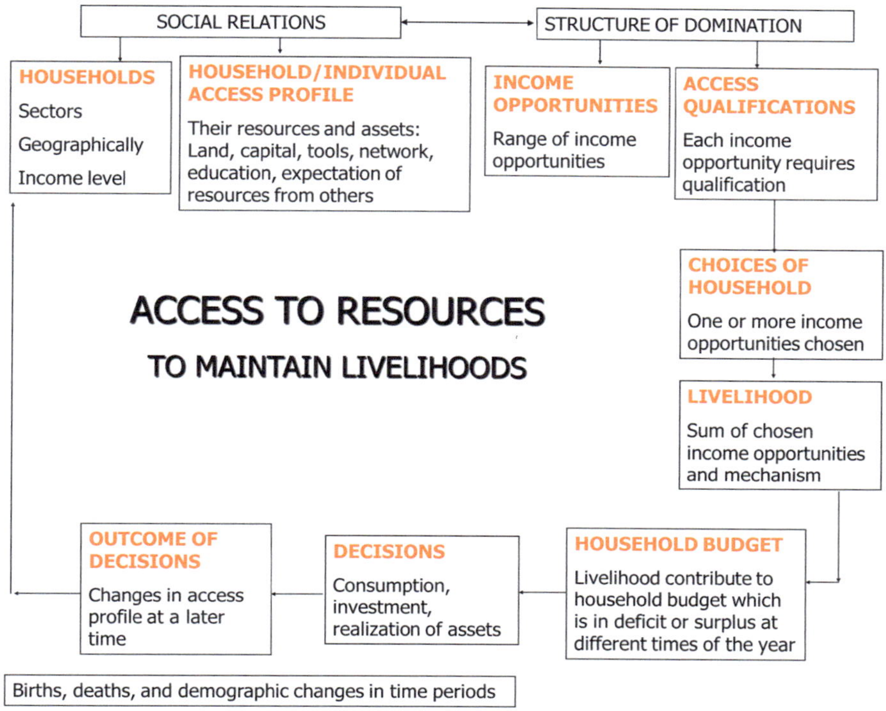

Fig. 8.9 Unpacked Box 1 of Fig. 8.10 (Wisner et al. 2004)

rebuilding efforts made a difference in terms of enhancing disaster resilience, as individual decisions are always made in economic-political environments.

8.2.1 Application of Access to Resources Model to the Case of Hurricane Katrina

Using the same documentary as the one used to develop a PAR model framework, an ATR model framework is presented here. ATR model (shown in Fig. 8.11) has been developed for the case of Hurricane Katrina using the same information as was used for the application of the PAR model shown earlier. This exercise focuses on New Orleans and demonstrates salient components of the society, social relations, structure of domination, choices made by affected households, and how they moved on with complex decision making and making livelihoods for themselves.

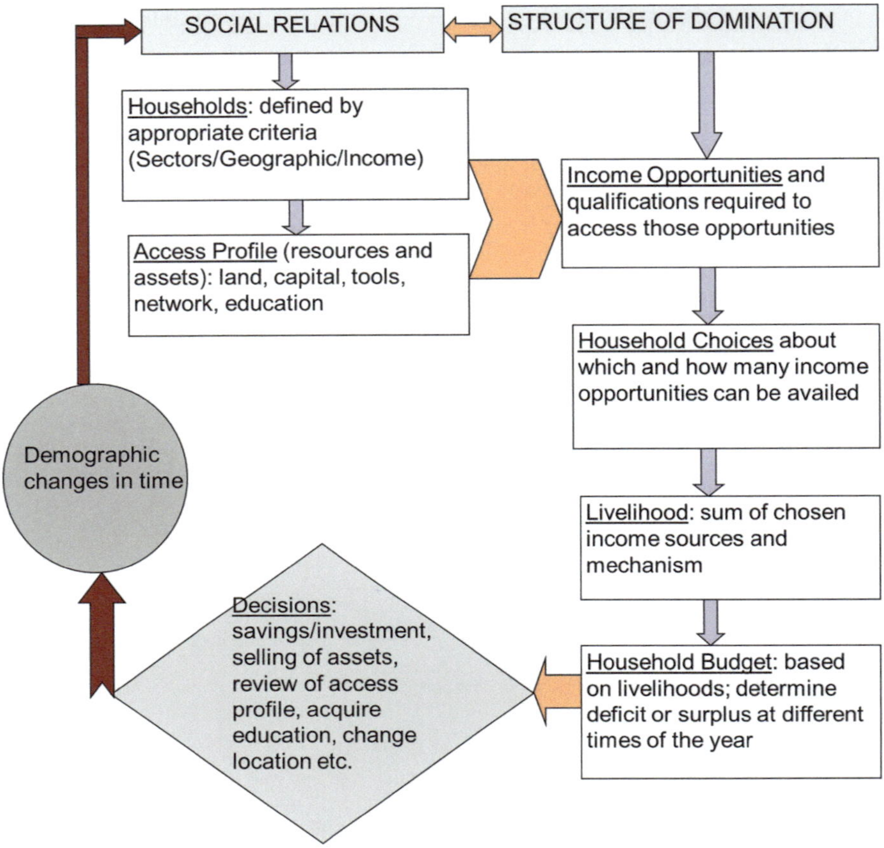

Fig. 8.10 Access to resources model in modified form based on unpacked Box 1 of Fig. 5.6.3 (Modified from Wisner et al. 2004)

8.3 Community Perception Model

Risk perception is characterised as the intuitive judgement of individuals and groups on risks in the context of limited and uncertain information (Slovic 2000). Community perception of disaster risk, disaster impact, resilience, risk reduction measures, and the role of public institutions in disaster preparedness are vital for the successful development and implementation of disaster mitigation strategies, measures, and policies. According to an ongoing study by the author, which involves a group of immigrant women in Toronto, Canada, it was observed that as a special group at risk, their ability to cope during emergencies and disasters is extremely poor. In addition, these women felt that the mitigation measures in place were not adequate in terms of 'their' needs. They expressed their concern over lack of community participation in the process of disaster risk mitigation planning and priority decision-making. In

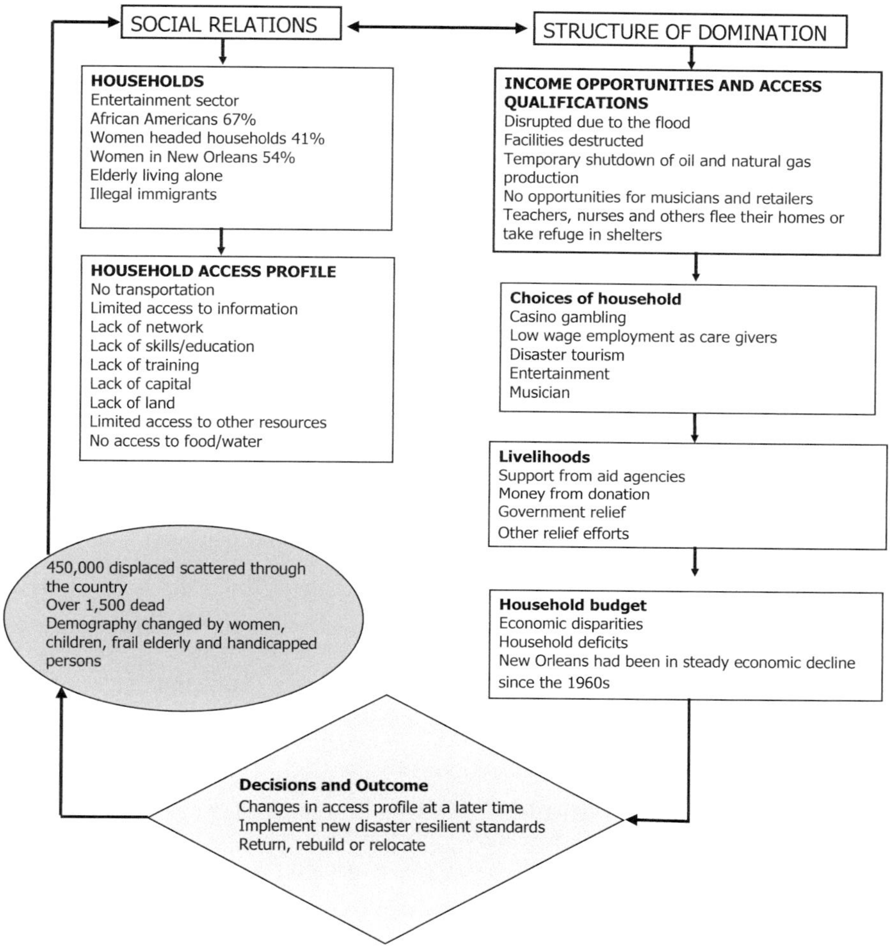

Fig. 8.11 Application of ATR model to the case study of Hurricane Katrina

other words, understanding the various perspectives of immigrants with different cultural backgrounds is necessary before beginning an evaluation of 'their' needs in the face of crises. It is noteworthy that nearly all the participants in the study had low to medium levels of education, were unemployed, faced language barriers (could not follow instructions if there was a notice to evacuate), had low annual household income, and relied on public transit for medical and other transportation needs. Providing appropriate and adequate support to vulnerable families is fundamental (Nirupama 2008; Armenakis and Nirupama 2013) for a socially strong nation. It is also important to identify specific intervention strategies (thoughtful of culturally sensitive issues) that vulnerable immigrant families are expecting to receive during emergencies (Olanubi 2009). Marginalized groups are often negatively affected in the aftermath of a hazardous event because inequity makes them largely invisible to aid agencies and governments in times of crisis. As a result, they experience the greatest levels of impoverishment, disability, and fatality after a disaster (ISDR 2004; GTZ 2004; ADRC 2005).

There is importance in the idea of building resiliency (Twigg 2007; Stewart 2007) for a vulnerable population, as it is a means that they are better able to equip themselves for emergencies. Various groups of people who are especially exposed to different hazards and risks happen to be economically challenged in most cases (HRVA 2004; Mathew and Kelly 2008). In order to build resilient communities, identifying, evaluating, and assessing people's vulnerabilities and the risks they are exposed to, is imperative. Social (minority groups, women, disabled, illiterate, etc.), physical (exposure to hazardous situation), economic (low income, unemployment), and environmental (land degradation, toxic deluge) vulnerabilities must be identified and assessed in consultation with the main stakeholders in the community. Some experts have tried including the risk perceptions of stakeholders or lay people within a given social system (Raaijmakers et al. 2008; Whyte and Burton 1982). A modified risk formula, which takes into account people's perception, is shown in Eq. (8.1), and was suggested by Whyte and Burton (1982).

$$R = p \cdot L^x \tag{8.1}$$

Where R is risk; p is probability; L denotes loss; and x (> *1*) is people's perception, which depends on number of factors. Those details have not been clearly defined by the literature. Ferrier and Haque (2003) have presented a methodology to implement the Whyte and Burton (1982) method, but they have used community perceptions of the 'likelihood' of hazard occurrence rather than the impact of disasters. Thus, there is a need for a clearly defined methodology that incorporates people's perception of the impact of disasters. Equation (8.2) suggests a method for this, and is proposed as below as:

$$R = H \times (V \times cp) \tag{8.2}$$

Where, H is hazard or likelihood (or probability); V is vulnerability/impact/ severity; and cp is community perception of the impact of disasters. A 5-point scale as shown in Table 8.1 is proposed to quantitatively describe the community perception component, cp. A community includes all concerned members of a community, neighbourhood, city, or region. It is assumed that the community under consideration will have sufficient access to relevant information regarding past occurrences of hazards in the region as well as their impact on people, property, and the environment. The PAR framework (Fig. 8.1) is constructed for a particular event/case, which feeds into the ATR model (Fig. 8.9). The detailed input from the PAR model includes unsafe conditions such as physical exposure, local economy scenario, and non-existent public institutions. The construction of the ATR model provides a clearer picture of past disasters, their consequences in terms of changing

Table 8.1 Scale to quantitatively determine Community Perception (*cp*)

Description of criteria	Rank
Very severe occurrence(s) with severe impact on a large number of people, critical infrastructure, and/or the environment hence community consent to positively invest in protection and mitigation measures as a priority	5
Severe occurrence(s) with severe impact on some people, few essential services, and/or the environment hence cautious community consent to invest in hazard protection and disaster mitigation measures	4
Somewhat severe occurrence(s) with major impact on few people and/or essential services, and/or the environment hence community consent to invest in hazard protection and disaster mitigation measures in selected areas	3
Occurrence(s) with significant impact on some people, and/or essential services, and/or the environment hence community consent to invest somewhat in hazard protection and disaster mitigation measures – not a priority	2
Occurrence(s) with minor impact on some people, and/or essential services, and/or the environment hence community consent to invest in hazard protection and disaster mitigation measures	1

dynamics among people, economy, opportunities, livelihood options, and budget decisions with regards to location and duration. Thus, the ATR framework becomes an additional and important resource to the community so they can develop their perceptions based on past disasters. Any disruption to critical infrastructure (essential services) such as hydro, water supply, natural gas, communication, and hospitals may carry an additional weight in determining the impact of past events. It is important to point out that the proposed 5-point scale approach is likely to evolve over time depending on the diversity, relevance, and applicability in communities.

To recap, the conventional risk assessment approach:

Risk Index (*RI*) is estimated using Eq. (8.1). Equations (8.1, 8.2) are sim`ilar in that 'hazard' is same as 'likelihood', and 'vulnerability' is synonymous with 'impact' or 'severity'. In the context of disasters, 'hazard' can also be considered to be synonymous with 'probability', and the vulnerability can be either personal (vulnerable communities) or impersonal (vulnerable coastline). However, the direct consequences of a disaster can be accounted for using a notational variation of the conventional formula as:

$$Risk\ Index\ (RI) = P(Likelihood) \times Impact(I) \tag{8.3}$$

Tables 8.2 and 8.3 show scales used to measuring impact and likelihood of disasters using the conventional risk formula (Eq. 8.3).

8.3.1 Application of Community Perception Model

The proposed risk assessment method is applied (Table 8.4) and compared with the conventional approach of risk assessment. Historical hazards' data of the province of

Table 8.2 Numerical ranking scheme

	Category	Details	Assessment	Description	Rank
1	Fatality		0–4	Very low	1
			4–10	Low	2
			10–50	High	3
			50 +	Very high	4
2	Injury	includes homeless, missing	0–4	Very low	1
			4–50	Low	2
			50–2000	High	3
			2000 +	Very high	4
3	Critical Infrastructure and Resources	e.g. Hospitals, schools, utilities, transportation	Temporary interruption	Very low	1
			Closure of few days	Low	2
			Loss of 50% of capability	High	3
			Permanent loss	Very high	4
4	Property Damage	Public, commercial, private	Minimal damage	Very low	1
			Localized damage	Low	2
			Localized & severe	High	3
			Widespread & severe damage	Very high	4
5	Environmental Impact	green/park, asbestos exposure, toxic releases	Minimal damage	Very low	1
			Localized damage	Low	2
			Localized & severe	High	3
			Widespread & severe damage	Very high	4
6	Economic and Social Impact	Industries, businesses and employers	Temporary impact	Very low	1
			Temporary & widespread	Low	2
			Extended & widespread	High	3
			Permanent impact	Very high	4

(Adapted from HRVA 2004) for measurement of disaster impact. This table is also used in the chapter on HRVA method in this book

Ontario, Canada has been used for the application. Impact of each hazard type has been determined using historical data (PSC 2015; Etkin and Haque 2003) on fatalities, injuries, property loss, disruption to essential services, extent of impact

Table 8.3 Likelihood ranking

Return period in years	Measure of likelihood	Rank
1–5	Very likely	5
5–10	Likely	4
10–30	Slight chance	3
30–100	Unlikely	2
>100	Very unlikely	1

Modified from HRVA (2004)

Table 8.4 Risk assessment using conventional approach (Eq. 8.3) for Ontario, Canada data (hazard information compiled from Public Safety Canada disaster database)

SN	Hazard	Likelihood (1)	Impact (2)	Risk Index (*RI*) (1) × (2) = (3)	*RI* (%) (3)÷20* × 100 = (4)
1	Winter storm	5	3	15	75
2	Wildfire	4	1	4	20
3	Land subsidence	4	2	8	40
4	Tornado	4	3	12	60
5	Epidemic/ pandemic	3	4	12	60
6	Extreme heat	3	3	9	45
7	Landslide	2	2	4	20
8	Expansive soil	2	3	6	30
9	Hurricane	2	4	8	40
10	Earthquake	1	3	3	15
11	Hail storm/wind storm	3	1	3	15
12	Flash flood from snowmelt	3	4	12	60

Note: max value of *RI* = *20* (based on max rank of Likelihood =5 and Impact =4

(localized or widespread), and the magnitude, duration, and cost of damage (Table 8.4). In addition, more realistic and holistic estimations were derived from the PAR (root causes, dynamic pressures, and unsafe conditions) and the ATR (social relations and structure of domination) frameworks. Scenarios of livelihood effects on affected populations, their decisions regarding moving, acquiring qualifications, etc., in order to access new livelihood opportunities provided a sense of people's perception during those events. Finally, based on factual information, ranks were assigned to disaster impact using a 4-point scale as given in Table 8.2 and likelihood using a 5-point scale (Table 8.3); and to community perception using a 5-point scale (Table 8.1). Tables 8.4 and 8.5 are the results of using conventional and community perception approaches respectively.

Table 8.5 Risk assessment using community perception model (Eq. 8.2) for the Province of Ontario, Canada

SN	Hazard	Likelihood (1)	Impact (2)	Community Perception (cp) (3)	Risk Index (RI_{cp}) (1) × (2 × 3) (4)	RI_{cp} (%) (4)÷100[a] × 100(5)
1	Winter storm	5	3	5	75	75
2	Wildfire	4	1	3	12	12
3	Land subsidence	4	2	1	8	8
4	Tornado	4	3	1	12	12
5	Epidemic/ pandemic	3	4	5	60	60
6	Extreme heat	3	3	5	45	45
7	Landslide	2	2	1	4	4
8	Expansive soil	2	3	1	6	6
9	Hurricane	2	4	3	24	24
10	Earthquake	1	3	5	15	15
11	Hail storm/ wind storm	3	1	1	3	3
12	Flash flood from snowmelt	3	4	5	60	60

[a]max value of $RI_{cp} = 100$ (based on max Likelihood = 5, Impact = 4, and $cp = 5$)

8.3.2 Comparison of Community Perception Model with Conventional Approach

As evident from Tables 8.4 and 8.5, the proposed community perception risk assessment method produces different results than the conventional risk assessment approach. For demonstration purposes, scores assigned to 'community perception' as given in Table 8.1 have been used. For a fair comparison, each type of Risk Index has been represented by a percent value. Figure 8.12 illustrates a comparison between Risk Index (RI) using the conventional approach and Risk Index using the community perception model (RI_{cp}), confirming that people's perception may influence risk evaluation, meaning decisions on disaster mitigation measures for a given region. For example, in Fig. 8.12, the evaluated risk for tornadoes varies from 60% using the conventional approach to 12% using the community perception model. In the real world, for a given community or region, details on a community's physical location, economic standing, and institutional support status would allow for an accurate assignment of community perception ranking. Prioritization of risks

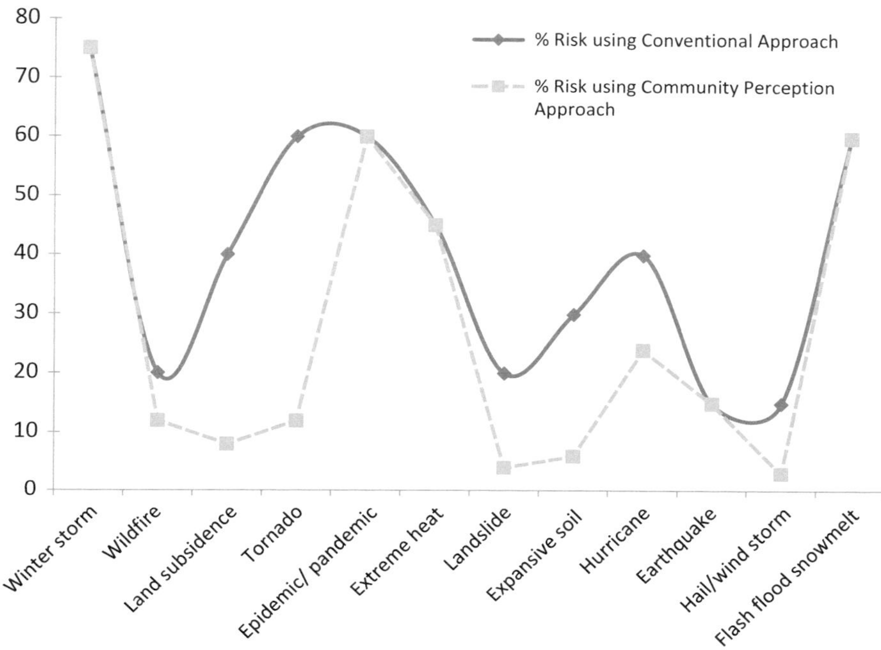

Fig. 8.12 Comparison of the two approaches, the proposed community perception and the convention one

and vulnerabilities according to the assessed values of risk using the community perception model (RI_{cp}) is essential, not only for budget and resource allocation, but also for the future implementation of disaster mitigation measures.

8.4 Risk Aversion Concept

Many people make many risk aversion decisions not only at the personal level but also on the level of regulatory agencies. Risk aversion is a phenomenon of real life and not a theoretical construct. If we don't take risk aversion into consideration when evaluating risk, we may come to inadequate or ill-informed decisions. A better approach would be to discuss risk aversion explicitly with absolute transparency than introducing it implicitly and arbitrarily. According to the principle of risk aversion as defined by John M. Keynes, "It is better to be roughly right than exactly wrong!" and we tend to propose and apply pragmatic solutions rather than ones supported by solid research. A conversation on risk aversion is necessary because it would promote a much needed dialogue between science and politics. Our knowledge and the skills of experts and professionals are growing rapidly but the quality of our societal decision-making is not growing in similar proportions. The ability for effective implementation of increasing scientific knowledge and professional skills are increasingly being bottleneck and limiting progress. One of the great challenges regarding risk aversion is fostering a dialogue and the transfer of knowledge from science to the political arena. But this also means that professionals, scientists and researchers have to be sensible with respect to the mechanisms and processes in our society for being able to produce societally relevant and adequate results. The issue of risk aversion is a good example (Schneider et al. 2006).

8.5 Exercise

Apply each of the methods discussed in this chapter to the best of your ability using any recent disaster. Suggest modification and new ideas as you see fit.

References

ADRC (2005) Major disasters in FY2005 http://www.adrc.asia/publications/stats/MajorDisasters2005.html. Accessed 10 Jan 2016

Armenakis C, Nirupama N (2013) Estimating spatial disaster risk in urban environments, geomatics, natural hazards and risk. Taylor & Francis 4(4):289–298

Birkmann J (ed) (2006) Measuring vulnerability to natural hazards: towards disaster resilient societies. United Nations University Press, New York, 524p

Birkmann J (2007) Risk and vulnerability indicators at different scales: applicability, usefulness and policy implications. Environ Hazard 7(2007):20–31

Etkin D, Haque EC (2003) An assessment of natural hazards and disasters in Canada. Kluwer, Dordrecht

Ferrier N, Haque CE (2003) Hazards risk assessment methodology for emergency managers: a standardized framework for application. Nat Hazards 28(2):271–290

GTZ (2004) Guidelines risk analysis – a basis for disaster risk management. Federal Ministry for Economic Cooperation and Development, Eschborn

HRVA (2004) Hazard, Risk, and Vulnerability Assessment toolkit, Emergency Management British Columbia, Canada http://hrva.embc.gov.bc.ca/toolkit.html

ISDR (2004) Living with risk: a global review of disaster reduction initiatives. The United Nations Office for Disaster Risk Reduction www.unisdr.org 429p

Mathew A, Kelly K (2008) disaster preparedness in urban immigrant communities: lessons learned from recent catastrophic events and their relevance to Latino and Asian Communities in Southern California, The Tomás Rivera Policy Institute University of Southern California. School of Policy, Planning and Development, Los Angeles

Nirupama N (2008) Disaster risk management: theory, practice and policy, Workshop report: public safety Canada. Government of Canada, Ottawa, 149p

Olanubi T (2009) Crisis management and psychosocial intervention: perception of immigrants, Major Research Paper, MA in Disaster & Emergency Management, York University

PSC (2015) The Canadian disaster database, public safety Canada. http://www.publicsafety.gc.ca/cnt/rsrcs/cndn-dsstr-dtbs/index-eng.aspx. Accessed 11 Mar 2016

Raaijmakers R, Jörg K, Van der Veen A (2008) Flood risk perceptions and spatial multi-criteria analysis: an exploratory research for hazard mitigation. Nat Hazards 46(3):307–322

Schneider Th, Basler E, Partner AG (2006) A delicate issue in risk assessment, In Proceedings RISK21 Workshop, Monte Verità, Ascona, Switzerland, 28 Nov–3 Dec 2004, (Eds.) W.J. Ammann, S. Dannenmann, L. Vulliet, Taylor & Francis

Slovic P (2000) The perception of risk. Earthscan Publications, London, 473 pp

Stewart RM (2007) Community perspectives of flood risk and social vulnerability reduction: the case of the Red River Basin, Doctoral Thesis, The Natural Resources Institute, The University of Manitoba, Canada

Twigg J (2007) Characteristics of a Disaster-resilient Community. DFID Disaster Risk Reduction Interagency Coordination Group, 39p

Whyte AV, Burton I (1982) Perception of risk in Canada. In: Burton I, Fowle CD, McCullough RS (eds) Living with risk. University of Toronto, Toronto, pp 39–69

Wisner B, Blaikie P, Cannon T, Davis I (2004) At Risk: Natural Hazards, People's Vulnerability and Disasters, 2nd edn. Routledge, London, 471p

Appendices

Appendix 1 (NHC 2015)

N. Agrawal, *Natural Disasters and Risk Management in Canada*, Advances in Natural and Technological Hazards Research 49,
https://doi.org/10.1007/978-94-024-1283-3

Category	Winds (1-min sustained winds in mph, kt, and km/h)	Summary	People, livestock, and pets	Mobile homes	Frame homes	Apartments, shopping centers, and industrial buildings	High-rise windows and glass	Signage, FENCES, AND CANOPIES	Trees	Power and Water	Example
1	74–95 mph 64–82 kt 119–153 km/h	*Very dangerous winds will produce some damage*	People, livestock, and pets struck by flying or falling debris could be injured or killed.	Older (mainly pre-1994 construction) mobile homes could be destroyed, especially if they are not anchored properly as they tend to shift or roll off their foundations. Newer mobile homes that are anchored properly can sustain damage involving the removal of shingle or metal roof coverings, and loss of vinyl siding, as well as damage to carports, sunrooms, or lanais.	Some poorly constructed frame homes can experience major damage, involving loss of the roof covering and damage to gable ends as well as the removal of porch coverings and awnings. Unprotected windows may break if struck by flying debris. Masonry chimneys can be toppled. Well-constructed frame homes could have damage to roof shingles, vinyl siding,	Some apartment building and shopping center roof coverings could be partially removed. Industrial buildings can lose roofing and siding especially from windward corners, rakes, and eaves. Failures to overhead doors and unprotected windows will be common.	Windows in high- rise buildings can be broken by flying debris. Falling and broken glass will pose a significant danger even after the storm.	There will be occasional damage to commercial signage, fences, and canopies.	Large branches of trees will snap and shallow rooted trees can be toppled.	Extensive damage to power lines and poles will likely result in power outages that could last a few to several days.	Hurricane Dolly (2008) is an example of a hurricane that brought category 1 winds and impacts to south Padre Island, Texas.

					soffit panels, and gutters. Failure of aluminum, screened-in, swimming pool enclosures can occur.						
2	96–110 mph 83–95 kt 154–177 km/h	*Extremely dangerous winds will cause extensive damage*	There is a substantial risk of injury or death to people, livestock, and pets due to flying and falling debris.	Older (mainly pre-1994 construction) mobile homes have a very high chance of being destroyed and the flying debris generated can shred nearby mobile homes. Newer mobile homes can also be destroyed.	Poorly constructed frame homes have a high chance of having their roof structures removed especially if they are not anchored properly. Unprotected windows will have a high probability of being broken by flying debris. Well-constructed frame homes could sustain major roof and siding damage. Failure of aluminum, screened-in, swimming	There will be a substantial percentage of roof and siding damage to apartment buildings and industrial buildings. Unreinforced masonry walls can collapse.	Windows in high-rise buildings can be broken by flying debris. Falling and broken glass will pose a significant danger even after the storm.	Commercial signage, fences, and canopies will be damaged and often destroyed.	Many shallowly rooted trees will be snapped or uprooted and block numerous roads.	Near-total power loss is expected with outages that could last from several days to weeks. Potable water could become scarce as filtration systems begin to fail.	Hurricane Frances (2004) is an example of a hurricane that brought category 2 winds and impacts to coastal portions of Port St. Lucie, Florida with category 1 conditions experienced elsewhere in the city.

(continued)

Category	Winds (1-min sustained winds in mph, kt, and km/h)	Summary	People, livestock, and pets	Mobile homes	Frame homes	Apartments, shopping centers, and industrial buildings	High-rise windows and glass	Signage, FENCES, AND CANOPIES	Trees	Power and Water	Example
					pool enclosures will be common.						
3	111–129 mph 96–112 kt 178–208 km/h	*Devastating damage will occur*	There is a high risk of injury or death to people, livestock, and pets due to flying and falling debris.	Nearly all older (pre-1994) mobile homes will be destroyed. Most newer mobile homes will sustain severe damage with potential for complete roof failure and wall collapse.	Poorly constructed frame homes can be destroyed by the removal of the roof and exterior walls. Unprotected windows will be broken by flying debris. Well-built frame homes can experience major damage involving the removal of roof decking and gable ends.	There will be a high percentage of roof covering and siding damage to apartment buildings and industrial buildings. Isolated structural damage to wood or steel framing can occur. Complete failure of older metal buildings is possible, and older unreinforced masonry buildings can collapse.	Numerous windows will be blown out of high-rise buildings resulting in falling glass, which will pose a threat for days to weeks after the storm.	Most commercial signage, fences, and canopies will be destroyed.	Many trees will be snapped or uprooted, blocking numerous roads.	Electricity and water will be unavailable for several days to a few weeks after the storm passes.	Hurricane Ivan (2004) is an example of a hurricane that brought category 3 winds and impacts to coastal portions of gulf shores, Alabama with category 2 conditions experienced elsewhere in this city.
4	130–156 mph	*Catastrophic damage will occur*	There is a very high risk of injury or death to	Nearly all older (pre-1994) mobile homes will be	Poorly constructed homes can sustain	There will be a high percentage of structural	Most windows will be blown out of high-rise	Nearly all commercial signage, fences, and	Most trees will be snapped or uprooted	Power outages will last for weeks to possibly	Hurricane Charley (2004) is an example of a

			people, live-stock, and pets due to flying and falling debris.	destroyed. A high percent-age of newer mobile homes also will be destroyed.	complete col-lapse of all walls as well as the loss of the roof struc-ture. Well-built homes also can sus-tain severe damage with loss of most of the roof structure and/or some exterior walls. Extensive damage to roof cover-ings, win-dows, and doors will occur. Large amounts of windborne debris will be lofted into the air.	damage to the top floors of apartment buildings. Steel frames in older industrial buildings can collapse. There will be a high per-centage of collapse to older unreinforced masonry buildings.	buildings resulting in falling glass, which will pose a threat for days to weeks after the storm.	canopies will be destroyed.	and power poles downed.	months. Long- term water short-ages will increase human suffer-ing. Most of the area will be uninhabitable for weeks or months.	hurricane that brought category 4 winds and impacts to coastal por-tions of Punta Gorda, Florida with category 3 conditions experienced elsewhere in the city.
	113–136 kt 209–251 km/h				Windborne debris damage will break most unpro-tected win-dows and penetrate some				Fallen trees and power poles will isolate res-idential areas.		

(continued)

Category	Winds (1-min sustained winds in mph, kt, and km/h)	Summary	People, livestock, and pets	Mobile homes	Frame homes	Apartments, shopping centers, and industrial buildings	High-rise windows and glass	Signage, FENCES, AND CANOPIES	Trees	Power and Water	Example
					protected windows.						
5	157 mph or higher	*Catastrophic damage will occur*	People, livestock, and pets are at very high risk of injury or death from flying or falling debris, even if indoors in mobile homes or framed homes.	Almost complete destruction of all mobile homes will occur, regardless of age or construction.	A high percentage of frame homes will be destroyed, with total roof failure and wall collapse. Extensive damage to roof covers, windows, and doors will occur. Large amounts of windborne debris will be lofted into the air.	Significant damage to wood roof commercial buildings will occur due to loss of roof sheathing. Complete collapse of many older metal buildings can occur.	Nearly all windows will be blown out of high-rise buildings resulting in falling glass, which will pose a threat for days to weeks after the storm.	Nearly all commercial signage, fences, and canopies will be destroyed.	Nearly all trees will be snapped or uprooted and power poles downed.	Power outages will last for weeks to possibly months. Long- term water shortages will increase human suffering. Most of the area will be uninhabitable for weeks or months.	Hurricane Andrew (1992) is an example of a hurricane that brought category 5 winds and impacts to coastal portions of cutler ridge, Florida with category 4 conditions experienced elsewhere in south Miami- Dade County.
	137 kt or higher 252 kph or higher				Windborne debris damage will occur to nearly all unprotected windows and many	Most unreinforced masonry walls will fail which can lead to the collapse of the buildings. A			Fallen trees and power poles will isolate residential areas.		

					protected windows.	high percentage of industrial buildings and low-rise apartment buildings will be destroyed.					

Appendix 2: Focus Group Discussion Themes and Questions

Demographic

The purpose of this section is to understand the structure and composition of the studied population.

- Age bracket
- Education level
- Employment status
- Job satisfaction
- Single/Married/Relationship
- Members in the household and relationship with them
- Type of dwelling
- Years in Canada
- English language proficiency

Theme 1: Risk Exposure

- Exposure to risk (river, power plant, railway, chemical plant etc.)?
- Level of safety

Theme 2: Vulnerability

- Personal experienced with hazardous event
- Importance of preparedness for unforeseen hazards
- In case of emergency, do you have the resources to cope?
- Main and most frequent source of transportation
- Challenges faced as an immigrant

Theme 3: Resiliency/Capacity

- Health status
- Disability if any
- List of items at home to cope with an emergency
- Awareness of community programs that are beneficial to my healthy living, happiness, and safety
- Participate in cultural events within the community
- Sense of belonging within the community

- Importance of socializing
- Interest in local government
- Interest in voting in municipal elections
- Satisfaction with municipal political representatives

Appendix 3[1]: Questionnaires to Assess the Impact of the CTLC Pilot Project in Pakistan

Questionnaire for CTLC Graduates

- Which one of the following would you select?
- Does the course help you in your daily life?
- What do you believe about getting support from your family and friends in moving forward?
- What was the attitude of your parents and friends when you told them that you wanted to do something for yourself and your family?
- Select any one which you think is appropriate and suits your thinking:
- Are you using computer skills in your daily life learned from this course?
- If you are offered a job in another city what will be your reaction?
- If you wanted to do anything to achieve something in your life and you think you are 100% right but your family discourages you, what would be your next step?
- What is your opinion about getting support of parents/husband/brother for women to move forward?
- You are helpful for girls like yourself; who don't have eagerness and potential to do something in life.
- Has this course changed your personality?
- Has this course changed your way of life?
- Select any one which you think is appropriate.
- *Questionnaire for Parents/Relatives (as Respondents) of the CTLC Graduates*
- Do you think that the level of confidence of your daughter/sister/relative has increased after this course?
- What is the learner's level of confidence after this course?
- Do you observe any change in her attitude and behavior?
- Do you feel any change in her knowledge and information?
- Can she now speak up and express herself? Select any one which you think is appropriate and suits your thinking:
- Is there any change in her motivation and commitment towards her studies after this course?
- Is she sensitive and demanding about her rights?

- Can she take initiatives independently?
- Do you think that without your support she can move forward and grow?
- Do you think she can achieve her goals in life?
- Is she utilizes her time productively and positively after this course?
- Does she now motivate and guide her younger brother/sister after this course?
- Is there any change in her capabilities, abilities, intelligence?
- Has she become a source of inspiration for other girls in your surrounding/family after this course?
- What do you think about C.T.L.C?

Appendix 4: Disaster Myth and Reality (Alexander 2002)

- ***Myth**: After a disaster, survivors tend to be dazed and apathetic.*
  - **Reality:** Survivors rapidly get to work on the clear-up. Activism is much more common than fatalism. In the worst possible cases only 15–30% of victims show passive and dazed reactions.
- ***Myth: Looting is a common and serious problem after disasters***
  - **Reality: The phenomenon of looting is rare and limited in scope. It mainly occurs when there are strong preconditions . . . as when a community is already deeply divided.**
- ***Myth: Disasters give rise to spontaneous displays of antisocial behavior.***
  - **Reality: Generally, they are characterized by great social solidarity, generosity and self-sacrifice, perhaps even heroism.**
- ***Myth: Any kind of aid and relief is useful after disaster, provided that it is supplied quickly enough.***
  - **Reality: Hasty and ill considered relief initiatives tend to create chaos. Only certain types of technical assistance, goods and services will be required. Not all useful resources that existed in the area before the disaster will be destroyed. Donation of unusable materials or manpower consumers' resources of organization and accommodation that could more profitably be used to reduce the toll of the disaster.**
- ***Myth: People will flee in large numbers from a disaster.***
  - **Reality: Usually there is a "convergence reaction" and the area fills up with people. Few of the survivors will leave and even obligatory evacuations will be short-lived.**

Appendix 5

Values for *Frequency* (F), *Consequences* (C), and *Changing Risk* (CR) for most hazards listed in the HIRA document (HIRA 2012)

Hazard	F	C	CR	Hazard	F	C	CR
Hazardous materials incident	6	6	4	Nuclear facility emergency	2	6	2
Flood	6	5	4	Terrorism/CBRNE	3	4	2
Forest/wildland fire	5	6	4	Windstorm	6	2	2
Freezing rain	5	6	4	Building/structural collapse	5	2	1
Snowstorm/blizzard	6	5	4	Drought/low water	5	2	3
Tornado	5	6	3	Radiological emergency	2	4	1
Drinking water emergency	6	4	4	Cyber attack	2	3	2
Human health emergency	4	6	2	Earthquake	1	6	2
Oil/natural gas emergency	6	4	2	Fog	3	2	1
Explosion/fire	6	3	4	Hail	3	2	2
Geomagnetic storm	3	5	3	Mine emergency	3	2	1
Transportation emergency	5	3	4	Natural space object crash	1	6	1
Agricultural and food emergency	4	3	3	Special event	6	1	1
Dam failure	4	3	3	Lightning	6	1	1
Civil disorder	6	2	3	Energy emergency (supply)	4	1	2
Critical infrastructure failure	6	2	2	Land subsidence	2	2	1
Extreme temperatures	4	3	2	Sabotage	3	1	1
Hurricane	4	3	4	War and international emergency	2	1	2
Landslide	4	3	2	Erosion	1	1	2

Appendix 6

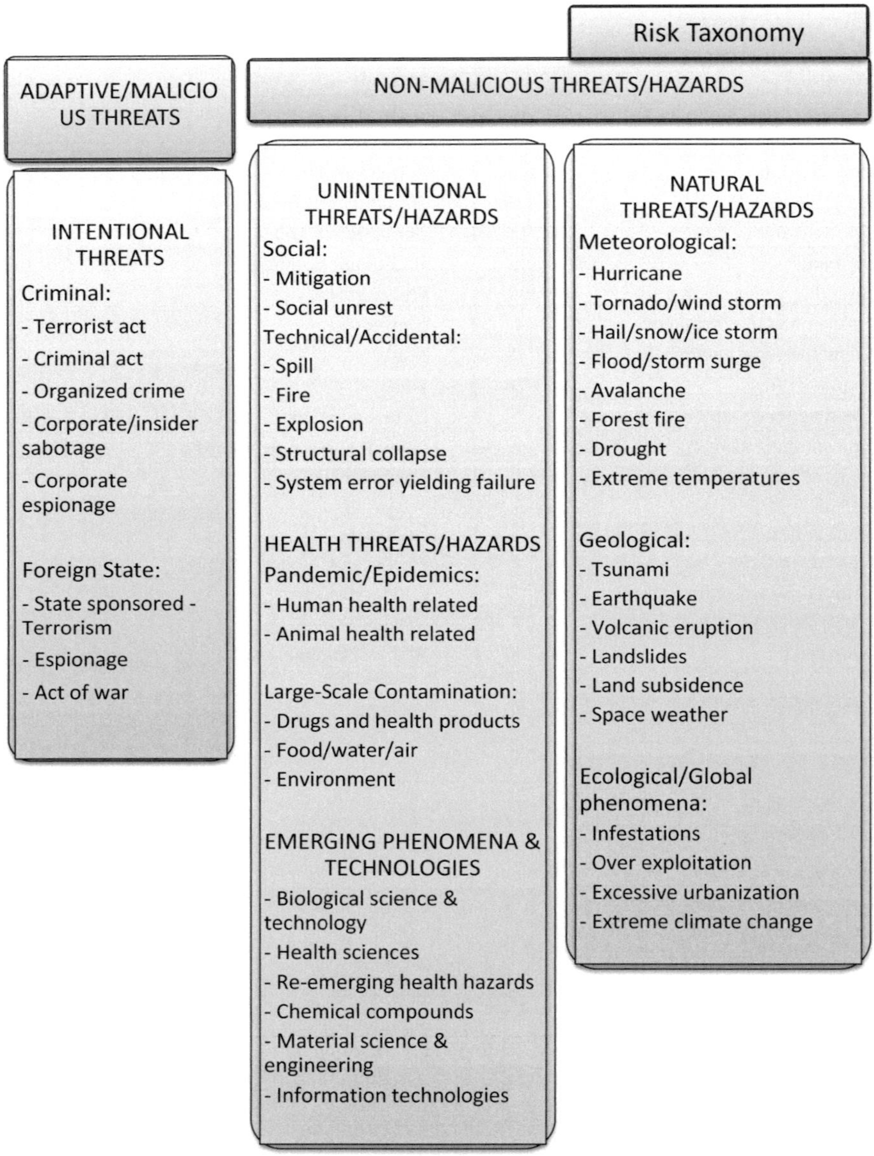
Risk Taxonomy
ADAPTIVE/MALICIOUS THREATS
NON-MALICIOUS THREATS/HAZARDS
INTENTIONAL THREATS
Criminal:
- Terrorist act
- Criminal act
- Organized crime
- Corporate/insider sabotage
- Corporate espionage
Foreign State:
- State sponsored - Terrorism
- Espionage
- Act of war
UNINTENTIONAL THREATS/HAZARDS
Social:
- Mitigation
- Social unrest
Technical/Accidental:
- Spill
- Fire
- Explosion
- Structural collapse
- System error yielding failure
HEALTH THREATS/HAZARDS
Pandemic/Epidemics:
- Human health related
- Animal health related
Large-Scale Contamination:
- Drugs and health products
- Food/water/air
- Environment
EMERGING PHENOMENA & TECHNOLOGIES
- Biological science & technology
- Health sciences
- Re-emerging health hazards
- Chemical compounds
- Material science & engineering
- Information technologies
NATURAL THREATS/HAZARDS
Meteorological:
- Hurricane
- Tornado/wind storm
- Hail/snow/ice storm
- Flood/storm surge
- Avalanche
- Forest fire
- Drought
- Extreme temperatures
Geological:
- Tsunami
- Earthquake
- Volcanic eruption
- Landslides
- Land subsidence
- Space weather
Ecological/Global phenomena:
- Infestations
- Over exploitation
- Excessive urbanization
- Extreme climate change

Appendix 7

Risk event scenario description template

Risk event scenario description	
Risk event scenario description	
Risk event name/title:	*Baseline description used to evaluate both likelihood and impact. In areas where likelihood and impact should be considered and scored, text should be marked with (L) for likelihood and (I) for impact. This is suggested for text that is embedded in descriptions and not obvious to the reader.*
Applicable risk code(s) for the principal constituent threat or hazard (including the category (ies) of the standard AHRA risk taxonomy affected):	*Please refer to the AHRA taxonomy, annex 3.*
Applicable risk code(s) for the secondary threat (s) or hazards (s) (including the category(ies) of the standard AHRA risk taxonomy affected):	*This field is optional and related to risks that have secondary effects, such as floods that occur after a hurricane.*
Primary department (for response):	*The Federal Emergency Response Plan (2011) describes the primary department as a federal government institution with a mandate related to a key element of an emergency. Several federal government institutions may be designated as primary departments, depending on the nature or severity of the emergency.*
Supporting department (for all EM components):	*According to the Federal Emergency Response Plan (2011), a supporting department is a federal government institution that provides general or specialized assistance to a primary department in response to an emergency.*
Key information sources for the risk event scenario description - please tag the information as*Unclassified* (U) or *Classified* (C; S; TS; TS SA):	*Identification of supporting documentation is important, especially in cases where qualitative and/or quantitative data supports scores decided upon during the risk scoring workshop. This ensures credibility and legitimacy of risk scores. In addition, reference can be made back to decision points at any point in time and by anyone. Clearly identify unclassified and classified information, for ease of reference when assessing likelihood components for malicious threats.*
Risk event description	
Description (context, setting, cause, source, nature, scale), of the risk event:	*The description entered here must be plausible in that factual information would support such an occurrence. The considered time-frame from which events are considered in the AHRAprocess is short-term (within the next 5 years) threats/hazards. Long-term threats/ hazards (that span 5–25 years into the future)*

(continued)

	are not currently considered in the AHRA. Background information leading up to the risk event provides context to the scenario without making broad assumptions which may skew results during the risk scoring workshop. Information inserted in this area should take into consideration the assessment of the following impact categories: People, Environment, Economic, Territorial Security, Canada's Reputation and Influence and Society and Psycho-Social.
Description of the lead-up to the incident, consisting of the (underlying) cause and any underlying insidious process:	*This section is optional.*
Geographical considerations (location, geographical extent, region):	*This section is optional. Geographical coordinate system (latitudinal and longitudinal lines), country, province, territory or region is to be included in this section.*
Natural environment:	*Relevant physical or environmental characteristics are inserted in this area facilitating the assessment of the environmental impact category.*
Meteorological conditions:	*Relevant meteorological condition(s) that influence the outcome of the scenario should be inserted in this area. If applicable variants may be inserted in this area.*
Seasonal:	*This section is optional and left to the discretion of the scenario developers. Dependent on the scenario, seasonal changes may influence the outcome of assessment of a particular risk.*
Hazard characteristics:	*Characteristics of chemical, biological, radiological and/or nuclear agent(s) involved in the scenario are inserted in this area. Elements captured should relate to: Toxicity, transmissibility, behaviour, fate and persistency to indicate a hazard severity and duration.*
Nature and vulnerability of the affected area (context, population density, degree of urbanisation, key infrastructure, economic considerations, political considerations, etc.):	*This area is important to note as it provides relevant information from which subject matter experts score risks. Population density, degree of urbanisation and key infrastructure influence the people and possibly the society and psycho-social impact category. Economic considerations affect the economic impact category. Political, geographical and territorial considerations influence Canada's reputation and influence and territorial security impact categories.*
Any other relevant assumptions made in describing the risk event scenario:	*If assumptions relating to the risk event description can be identified or isolated they should be inserted in this area. Although this*

(continued)

	field is considered optional, the information may still be required in the risk scoring tool.
Uncertainty or variability in the risk event description:	*If there are areas of uncertainty or unpredictability, it should be inserted in this area. Although this field is considered optional, information may be required in the risk scoring tool.*
Other relevant information, notes or comments:	*Any other relevant information relating to the risk event description should be identified in this area.*
Likelihood assessment	
Time period/time horizon during which the risk eventmight be realised:	*The translation of the data in likelihood of occurrence on a yearly basis will be done in the risk scoring tool.*
Uncertainty in the likelihood assessment:	*Unknown factors which would influence the likelihood assessment should be inserted in this area.*
Other relevant information, notes or comments:	*Any other relevant information relating to the primary likelihood assessment should be inserted in this area.*
Impacts/consequences assessment	
Impact categories: Nature and scale	
1. People:	*Specific indicators have been selected to evaluate the effect of hazards and threats on people. Estimated figures should be inserted in this box e.g. the number of fatalities, serious injuries, etc.*
2. Economy:	*Based on the Department of Finance Canada's criteria of risks and hazards on the economy. This impact category captures direct and indirect loses. Direct loses are immediate economic damage as a result of a risk event. Losses are measured based on repair or replacement costs. Indirect loses refer to the flow of goods and services which will not be produced as a result of damage to productive assets and infrastructure.*
3. Environment:	*Based on the indicators developed by public safety Canada, in close collaboration with Environment Canada, on the effects of hazards and threats on the environment.*
4. Territorial security:	*Based on indicators that capture conditions in which there is a loss in the ability of the Government of Canada to secure the territory or the border and to secure the safety of citizens.*
5. Canada's reputation and influence:	*Based on expert assessment of the potential international reaction to an emergency event occurring in Canada, or involving Canadians abroad.*

(continued)

6. Society and psychosocial:	*Based on indicators regarding public outrage and public anxiety, as well as social actions, such as protests, civil disturbances or vandalism, can be provoked by a risk event.*
Uncertainty in the impacts/consequences assessment:	*Uncertainty, unpredictability or areas of doubts relating to the impacts/consequences assessment should be inserted in this area.*
Other relevant information, notes or comments:	*Other relevant information relating to the impacts/consequences assessment should be inserted in this area.*
Preliminary risk treatmentplanning	
Baseline risk treatment plan (treatment actions, timeframe(s), readiness, etc.):	*This area is optional. Federal institutions may choose to fill it out after the completion of the risk scoring workshop. This area would assess the capacity of the emergency support functions (ESF).*
Risk treatment measures already in place	*As the AHRAtakes into consideration mitigation measures in place when assessing the likelihood of occurrence and the impacts of a risk, (all or some of) these measures should be clearly captured somewhere in the risk event scenario template. This will force divisions which "own" mitigation measures (usually Program divisions) to share their information with EM divisions (usually under GOC, Operations or Corporate Branches).*
Degree to which the risk (likelihood, impacts) can be reduced by risk treatment.	*This area may be completed by departments and agencies. This area would assess the capacity of the ESF.*
Additional risk treatment resources required.	*Additional information relating to risk treatment may be inserted in this area.*
Other relevant information, notes or comments:	*Other relevant information should be inserted in this area.*

Appendix 8

Economic category assessment tool – direct and indirect economic loss for repair or replacement

Direct Economic Loss (those involving damages to stock and assets occurring at the time of the disaster or soon after)
Buildings: e.g. industrial, commercial, institutional (plants, offices, recreational facilities, hospitals).
Infrastructure: e.g. roads, water systems, irrigation, docks, terminals, other transportation, electric power, oil and gas engineering.
Machinery and equipment: e.g. computers and software, agricultural and industrial machinery, furniture, trucks, etc.

(continued)

Residential housing and contents.
Raw materials: e.g. coal, crude oil, natural gas, grains, animals and animal products, wood, ferrous, non-ferrous, non-metallic.

Indirect economic loss (those involving a loss in the flow of production of goods and services which begin after the disaster and may extend through the reconstruction period)
Production or service provision losses due to the full or partial paralysis of productive activities: e.g. loss in industrial production due to damage to factories or shortages of raw materials/energy supplies, loss in agricultural production due to flooding or prolonged drought, loss of profits in the fishing and tourism industry following an oil spill, loss of production due to illness following a pandemic or listeriosis outbreak, etc.
Higher operational costs due to destruction of physical infrastructure and inventories or losses to production or income: e.g. a ban on beef and cattle exports would first translate into higher maintenance costs due to rising inventory of live animals.
Lost production due to linkage effects: e.g. destruction of a factory reduces the economic activities of suppliers who have no alternative markets.
Additional ***costs incurred due to the need to use alternative means*** of production or provision of essential services: e.g. costs arising out of need to use alternative roads or transportation means due to damage to principal routes and critical infrastructures.
Costs of required government response due to emergency and rescue operations: e.g. overtime payments to provide emergency assistance and repair critical infrastructure, additional expenses incurred to accommodate evacuees or for investigation, productivity loss induced from distortion of government resources and time allocation, etc.

Mitigating factors, if applicable (disasters sometimes involve indirect benefits or adjustments over the short-medium term which we may want to flag)
Shift in consumer demand/spending: e.g. following a BSE outbreak, demand for other types of meat may increase.
Change in the productivity of assets: e.g. following a flood, land productivity sometimes rises.
Labour reallocation: e.g. some workers could work longer or harder to make up for the shortfall in labour supply due to a pandemic outbreak.
Reconstruction activity: e.g. rebuilding activities after a hurricane.

Appendix 9

Rating of the impact on Canada's reputation and influence

Level	Actions	Political relations	Non-political relations
0–1.0 **insignificant**	Canadian missions abroad are not affected. Concentrated and short-lived condemnation of Canada/Government of Canada.	Trade regulations slow Canadian exports into some minor foreign markets but are not stopped. Canadian mission staff are not affected.	No effect on international events. International travel is discouraged to one region within Canada by foreign governments.

(continued)

Level	Actions	Political relations	Non-political relations
		Bilateral trade agreements are temporarily suspended.	
	Canadians abroad are not affected.	International working level meetings are delayed.	
1.0–2.0 **minor damage to Canadian reputation**	Canadian missions abroad receive threats but none materialize.	Temporary trade bans and/or sanctions are imposed by a few minor trading partners.	International conferences see fewer participants.
	Short-lived condemnation of Canada internationally.	Canadian mission staff exercises increased levels of vigilance.	International travel is discouraged to several regions within Canada by foreign governments.
		Canadians delayed at border crossings but visas are not imposed.	
	Threats issued to Canadians abroad but are unlikely.	Cancellation of meetings with minor international partners.	
		Minor trade agreements are temporarily suspended.	
2.0–3.0 **significant damage to Canadian reputation/ prestige**	Canadian missions abroad receive serious threats and are forced to close.	Trade bans and/or sanctions imposed by a few major and minor trading partners and trading blocs (United States, Japan, United Kingdom, China not included).	International events are forced to reschedule.
	Significant condemnation of Canada and/or the Government of Canada internationally.	Canadian mission staff leaves host country due to insecurity.	International travel to Canada is discouraged by foreign governments.
		Entry visa requirement imposed on Canadians travelling abroad.	
	The Government of Canada encourages citizens not to travel due to threats abroad.	Cancellation of bilateral meetings with major and minor international partners.	
		Minor trade agreements are cancelled.	
3.0–4.0 **major damage to Canadian reputation/ prestige**	Attempted invasion, occupation, and/or destruction of Canadian missions abroad.	Trade bans, embargoes, blockades and regulations imposed by some major and minor trading partners and trading	Significant international events are cancelled.
			International travel to Canada is discouraged

(continued)

Level	Actions	Political relations	Non-political relations
		blocs (United States not included).	by international organizations such as the World Health Organization, the United Nations, American government, etc.
		Forced deportation of Canadian mission staff.	
	Wide-spread condemnation of Canada and/or the Government of Canada nationally and internationally.	Denial of entry visas to a number of countries or the imposition of extreme fees (e.g. United Arab Emirates visa).	
		Temporary suspension of trade agreements such as the north American free trade agreement.	
	Persistent threats to Canadians abroad.	Cancellation of major and minor international delegations to Canada or the rejection of Canadian delegations to other countries.	
4.0–5.0 **severe damage to Canadian reputation/ prestige**	Invasion, occupation, and/or destruction of Canadian missions abroad.	Trade bans, embargoes, blockades and regulations imposed by major and minor trading partners and trading blocs (ex: United States, Japan, United Kingdom, China, etc.).	Refusal by major and minor Canadian partners to attend significant international events, such as the G8/20, Olympics, etc.
			Cancellation of major international events in Canada by event organizers (ex: International Olympic Committee, International Federation of Association
			Football, la Francophonie, the commonwealth, etc.).
	Wide-spread and continuous condemnation of Canada and/or the Government of Canada nationally and internationally.	Deportation, arrest and/or killing of Canadian mission staff.	Ban on international travel to Canada. Ban on Canadians travelling overseas.
		Denial of entry visas to many countries.	
	Threats to Canadians abroad materialise.	Cancellation of Canadian trade agreements such as the north American free trade agreement.	
		Canada is expelled from major security	

(continued)

Level	Actions	Political relations	Non-political relations
		organizations such as the North Atlantic Treaty Organization.	
		Relations between the Government of Canada and foreign governments cease.	
		Public cancellation of major international visits (ex: State visit by the president).	

Appendix 10

Risk Assessment Table

(1) Asset or Operation at Risk	(2) Hazard	(3) Senario (Location, Timing, Magitude)	(4) Oportunities for Prevention or Mitigation	(5) Probability (L, M, H)	Impacts with Existing Mitigation (L, M, H)					(11) Overall Hazard Rating
					(6) People	(7) Property	(8) Operations	(9) Environment	(10) Entity	

Ready Business.

Risk Assessment Table

INSTRUCTIONS

Column 1: Compile a list of assets (people, facilities, machinery, equipment, raw materials, finished goods, information technology, etc.) in the left column.

Column 2: For each asset, list hazards (review the "Risk Assessment" page from Ready Business) that could cause an impact. Since multiple hazards could impact each asset, you will probably need more than one row for each asset. You can group assets together as necessary to reduce the total number of rows, but use a separate row to assess those assets that are highly valued or critical.

Column 3: For each hazard consider both high probability/low impact scenarios and low probability/high impact scenarios.

Column 4: As you assess potential impacts, identify any vulnerabilities or weaknesses in the asset that would make it susceptible to loss. These vulnerabilities are opportunities for hazard prevention or risk mitigation. Record opportunities for prevention and mitigation in column 4.

Column 5: Estimate the probability that the scenarios will occur on a scale of "L" for low, "M" for medium and "H" for high.

Columns 6-10: Analyze the potential impact of the hazard scenario in columns 6 - 10. Rate impacts "L" for low, "M" for medium and "H" for high.

Column 8: Information from the business impact analysis should be used to rate the impact on "Operations."

Column 10: The "entity" column is used to estimate potential financial, regulatory, contractual, and brand/image/reputation impacts.

Column 11: The "Overall Hazard Rating" is a two-letter combination of the rating for "probability of occurrence" (column 5) and the highest rating in columns 6 – 10 (impacts on people, property, operations, environment, and entity).

Carefully review scenarios with potential impacts rated as "moderate" or "high." Consider whether action can be taken to prevent the scenario or to reduce the potential impacts.

Appendix 11

Ready Business.

Business Impact Analysis Worksheet

Department / Function / Process ______________________________

Operational & Financial Impacts

Timing / Duration	Operation Impacts	Financial Impact

Timing: Identify point in time when interruption would have greater impact (e.g., season, end of month/quarter, etc.)

Duration: Identify the duration of the interruption or point in time when the operational and or financial impact(s) will occur.

- < 1 hour
- >1 hr. < 8 hours
- > 8 hrs. <24 hours
- > 24 hrs. < 72 hrs.
- > 72 hrs.
- > 1 week
- > 1 month

Considerations (customize for your business)

Operational Impacts

- Lost sales and income
- Negative cash flow resulting from delayed sales or income
- Increased expenses (e.g., overtime labor, outsourcing, expediting costs, etc.)
- Regulatory fines
- Contractual penalties or loss of contractual bonuses
- Customer dissatisfaction or defection
- Delay executing business plan or strategic initiative

Financial Impact

Quantify operational impacts in financial terms.

ready.gov/business

Appendix 12

Ready Business.
ready.gov/business

Business Continuity Resource Requirements

Resource Category	Resource Details	Normal Quantity	24 hours	72 hours	1 week	Later (specify)
Managers						
Staff	Primary site, relocation site and recovery site					
Office space						
Office equipment	Furniture, phone, fax, copiers					
Office technology	Desktops and laptops (with software), printers with connectivity; wireless devices (with email access)					
Vital records, data, information	Location, backups, and media type					
Production Facilities	Owned, leased, or reciprocal agreement					
Production machinery & Equipment	Especially custom equipment with long replacement time					
Dies, patterns, molds, etc. for machinery & equipment						
Raw Materials	Single or sole source suppliers and possible alternates					
Third party services						

Instructions: Identify resources required to restore business operations following a disaster. Estimate the resources needed in the days and weeks following the disaster. Also review information technology disaster recovery plan for restoration of hardware and software.

Appendix 13

Prioritization of hazards for the Chatham Islands identified using the SMUG model (CDEM 2011)

Hazards	Risk analysis			Risk evaluation											
				Seriousness					Manageability					Growth	
DOC #1351593	Likelihood	Consequence	Rating	Social	Built	Economic	Natural	Sub-total	Reduction	Readiness	Response	Recovery	Sub-total	Sub-total	Total
Natural															
Tsunami-local	Likely	Major	VH	5	4	3	3	8.5	2	3	3	2	2.5	5	16.0
Tsunami-distance	Likely	Moderate	H	4	4	3	3	7.5	2	3	3	2	2.5	5	15.0
Storm surge	Possible	Minor	M	3	2	2	2	5	4	4	3	3	3.5	4	12.5
Wind storm	Possible	Minor	M	2	2	2	1	3.8	4	4	3	3	3.5	5	12.3
Earthquake	Unlikely	Minor	L	2	2	2	2	4	4	4	3	3	3.5	4	11.5
Volcanic	Rare	Minor	VL	2	2	1	1	3.5	4	4	3	3	3.5	4	11.0
Erosion-lagoon	Certain	Minor	M	1	2	1	3	2.9	4	4	3	3	3.5	2	8.4
Erosion-coastal	Certain	Moderate	M	1	2	1	3	2.9	4	4	3	3	3.5	2	8.4
Drifting sand dunes	Certain	Minor	M	1	1	2	3	2.7	2	4	3	3	3	3	8.7
Land slide	Rare	Very minor	M	1	1	2	3	2.7	3	3	3	3	3	3	8.7
Fire-rural	Likely	Moderate	M	4	3	2	1	6.3	3	3	3	3	3	4	13.3
Icebergs	Unlikely	Insignificant	VL	1	1	1	1	2	4	4	4	4	4	1	7.0
Sea level rise	Possible	Insignificant	L	1	1	1	1	2	4	4	3	3	3.5	1	6.5
River flood	Likely	Minor	M	1	2	1	2	2.7	2	3	3	3	2.75	2	7.5
Drought	Possible	Minor	M	1	1	2	2	2.5	4	4	4	4	4	2	8.5
Slope stability	Rare	Insignificant	VL	1	1	1	1	2	4	4	3	3	3.5	3	8.5
Power failure	Unlikely	Major	M	3	1	2	2	4.5	3	3	3	4	3.25	3	10.8
Water failure	Unlikely	Minor	H	2	1	3	1	3.6	3	3	3	3	3	3	9.6

(continued)

Hazards	Risk analysis			Risk evaluation										Growth	
				Seriousness					Manageability						
DOC #1351593	Likelihood	Consequence	Rating	Social	Built	Economic	Natural	Sub-total	Reduction	Readiness	Response	Recovery	Sub-total	Sub-total	Total
Wastewater failure	Likely	Minor	M	1	1	2	2	2.5	3	3	3	3	3	3	8.5
Communication	Likely	Minor	M	3	1	3	1	4.6	3	3	3	3	3	4	11.6
Air transport	Possible	Catastrophic	VH	4	3	4	1	6.9	3	3	3	3	3	4	13.9
Road transport	Possible	Catastrophic	VH	4	3	4	1	6.9	3	3	3	3	3	4	13.9
Civil unrest	Rare	Insignificant	VL	4	2	2	1	5.8	3	3	3	3	3	4	12.8
Fire urban	Possible	Moderate	M	3	2	2	3	5.2	4	3	3	3	3.25	4	12.5
Criminal damage	Rare	Insignificant	VL	3	2	3	1	5.1	3	3	3	3	3	2	10.1
Hazards substance	Possible	Minor	M	3	2	2	2	5	4	4	3	3	3.5	2	10.5
Biological															
Public health	Possible	Moderate	M	3	1	3	1	4.6	4	3	3	3	3.25	3	10.9
Pandemic	Possible	Moderate	M	3	1	3	1	4.6	4	3	3	3	3.25	4	11.9
Plant & animal pests	Possible	Moderate	M	1	1	3	3	3	3	3	3	3	3	3	9.0
Marine oil spill	Possible	Moderate	M	3	2	3	4	5.7	3	3	3	3	3	1	9.7

References

Alexander D (2002) Principles of emergency planning and management. Oxford University Press

AMS (2000) American Meteorological Society, Glossary of meteorology. http://glossary.ametsoc.org/wiki/Return_period

CDEM (2011) Civil Defence Emergency Management Group Plan Chatham Islands. http://www.cic.govt.nz/documents/2013/10/chathamislands-civil-defence-and-emergency-management-plan-2013-2016.pdf/

Glossary of Terms (2011) Definitions, emergency management Ontario. www.emergencymanagementontario.ca/english/emcommunity/ProvincialPrograms/hira

IPCC (2007) Contribution of Working Group II to the Fourth Assessment Report of the Intergovernmental Panel on Climate Change. In: Parry ML, Canziani OF, Palutikof JP, van der Linden PJ, Hanson CE (eds) . Cambridge University Press, Cambridge/New York

HIRA (2005, 2012) Hazard identification and risk assessment for the Province of Ontario, Emergency Management Ontario, Ministry of Community Safety and Correctional Services, Ontario, Canada

NHC (2015) National Hurricane Center. http://www.nhc.noaa.gov/pdf/sshws_table.pdf. Accessed 21 Mar 2016

Glossary and Definitions of Terms

ACE	Accumulated Cyclone Energy
AHRA	All Hazard Risk Assessment
ATR	Access to Resources
BIA	Business Impact Analysis
CBC	Canadian Broadcasting Corporation
CBS	Columbia Broadcasting Systems
CCA	Climate Change Adaptation
CDC	Center for Disease Control
CDD	Canadian disaster database
CDEM	Civil Defence Emergency Management
CEG	Coordinating Executive Group
CF	Critical facilities
CFFDRS	Canadian Forest Fire Danger Rating System
CI	Critical infrastructure
CME	Coronal Mass Ejections
CRED	Center for Research on the Epidemiology of Disasters
CSS	Centre for Security Science
CTLC	Community Technology Learning Centers
DALY	Disability-Adjusted Life Year
DRDC	Defense Research and Development Canada
DRR	Disaster Risk Reduction
EC	Environment Canada
EF	Enhanced Fujita
EM	Emergency Management
EMO	Emergency Management Ontario
EOC	Emergency Operation Centre
EVD	Ebola virus disease
FEMA	Federal Emergency Management Agency

(continued)

N. Agrawal, *Natural Disasters and Risk Management in Canada*, Advances in Natural and Technological Hazards Research 49,
https://doi.org/10.1007/978-94-024-1283-3

FERP	Federal Emergency Response Plan
FPEM	Federal Policy for Emergency Management
GC	Government of Canada
GO	GO Public Transit, a division of Metrolinx
GTA	Greater Toronto Area
HIRA	Hazard Identification and Risk Assessment
HRVA	Hazard Risk and Vulnerability Assessment
IAU	International Astronomical Union
ICT	Information and Communication Technologies
IOS	International Organization for Standardization
IPCC	Intergovernmental Panel on Climate Change
IRAWG	Interdepartmental Risk Assessment Working Group
IRAWG	Interdepartmental Risk Assessment Working Group
MDG	Millennium Development Goals
MMI	Modified Mercalli Intensity Scale
MMS	Moment Magnitude Scale
MPAC	Municipal Property Assessment Corporation (Ontario, Canada)
NASA	National Aeronautics and Space Administration
NATECH	Natural and Technological Hazards
NCHD	National Commission for Human Development (Pakistan)
NHC	National Hurricane Centre
NOAA	National Oceanic and Atmospheric Administration
NRC	National Research Council of Canada
NRCan	National Research Council of Canada
NWS	National Weather Service
OECD	Organization for Economic Co-Operation and Development
PAR	Pressure and ReleasE
PBS	Public Broadcasting Station
PEP	Provincial Emergency Program
PSC	Public Safety Canada
SARS	Severe Acute Respiratory Syndrome
SC	Statistics Canada
SME	Subject Matter Expert
SMUG	Seriousness, Manageability, Urgency and Growth
TRCA	Toronto and Region Conservation Authority
TTC	Toronto Transit Commission
USGS	United States Geological Survey
UTR	Upper Thames River
WFO	World Food Organization
WHO	World Health Organization
WMO	World Meteorological Organization

Definitions from Emergency Management Ontario http://www.emergencymanagementontario.ca/stellent/groups/public/@mcscs/@www/@emo/documents/abstract/ec159132.pdf

Acceptable Risk The level of potential losses that a society or community considers acceptable given existing social, economic, political, cultural, technical and environmental conditions (Glossary of Terms, 2011).

Assessment The evaluation and interpretation of available information to provide a basis for decision-making (Glossary of Terms, 2011).

Building Code A set of ordinances or regulations and associated standards intended to control aspects of the design, construction, materials, alteration and occupancy of structures that are necessary to ensure human safety and welfare, including resistance to collapse and damage (Glossary of Terms, 2011).

Business/Financial Impact The negative economic consequences of the occurrence of a hazard.

Changing Risk A variable in the HIRA methodology that allows for the inclusion of information on changes in the likelihood and vulnerability of the hazard.

Climate Change "A change in the state of the *climate* that can be identified (e.g., by using statistical tests) by changes in the mean and/or the variability of its properties, and that persists for an extended period, typically decades or longer. Climate change may be due to natural internal processes or *external forcing*, or to persistent *anthropogenic* changes in the composition of the *atmosphere* or in *land use*." (IPCC, 2007).

Community A generic term that includes both municipalities and First Nations (Glossary of Terms, 2011).

Comprehensive Emergency Management It is an all-encompassing risk-based approach to emergency management that includes prevention, mitigation, preparedness, response and recovery measures (Glossary of Terms, 2011).

Consequence The outcome of an event or situation expressed qualitatively or quantitatively, being a loss, injury or disadvantage (Glossary of Terms, 2011).

Critical Infrastructure (CI) Interdependent, interactive, interconnected networks of institutions, services, systems and processes that meet vital human needs, sustain the economy, protect public safety and security, and maintain continuity of and confidence in government (Glossary of Terms, 2011).

Critical Infrastructure Impact The negative consequences of the occurrence of a hazard on the interdependent, interactive, interconnected networks of institutions, services, systems and processes that meet vital human needs, sustain the economy, protect public safety and security, and maintain continuity of and confidence in government.

Current Risk The present level of risk associated with a hazard.

Damage Assessment An appraisal or determination of the effects of a disaster on people, property, the environment, the economy and/or services (Glossary of Terms, 2011).

Declared Emergency A signed declaration made in writing by the Head of Council or the Premier of Ontario in accordance with the *Emergency Management and Civil Protection Act.* This declaration is usually based on a situation or an impending situation that threatens public safety, public health, the environment, critical infrastructure, property, and/or economic stability and exceeds the scope of routine community emergency response (Glossary of Terms, 2011).

Emergency A situation or an impending situation that constitutes a danger of major proportions that could result in serious harm to persons or substantial damage to property and that is caused by the forces of nature, a disease or other health risk, an accident or an act whether intentional or otherwise (*Emergency Management and Civil Protection Act)* (Glossary of Terms, 2011).

Emergency Area A geographic area within which an emergency has occurred or is about to occur, and which has been identified, defined and designated to receive emergency response actions (Glossary of Terms, 2011).

Emergency Management Organized activities undertaken to prevent, mitigate, prepare for, respond to and recover from actual or potential emergencies (Glossary of Terms, 2011).

Emergency Management Program A risk-based program consisting of prescribed elements that may include prevention, mitigation, preparedness, response and recovery activities (Glossary of Terms, 2011).

Emergency Plan A plan developed and maintained to direct an organization's external and/or internal response to an emergency (Glossary of Terms, 2011).

Environmental Damage The negative consequences of the occurrence of a hazard on the environment, including the soil, water, air and/or plants and animals.

Frequency How often a hazard occurs at an intensity that may result in an emergency, disaster or service disruption.

Hazard A phenomenon, substance, human activity or condition that may cause loss of life, injury or other health impacts, property damage, loss of livelihoods and services, social and economic disruption, or environmental damage. These may include natural, technological or human-caused incidents or some combination of these (Glossary of Terms, 2011).

Hazard Identification A structured process for identifying those hazards which exist within a selected area and defining their causes and characteristics (Glossary of Terms, 2011).

Historical Risk The level of risk associated with a hazard in the past. The level of risk may have been altered by changes in consequence, frequency or prevention, preparedness, mitigation, response or recovery practices.

Human-Caused Hazard Human-caused hazards are hazards which result from direct human action or inaction, either intentional or unintentional. This includes hazards that arise from problems within organizational structure of a company, government etc.

Impact The negative effect of a hazardous incident on people, property, the environment, the economy and/or services (Glossary of Terms, 2011).

Incident An occurrence or event that requires an emergency response to protect life, property, or the environment (Glossary of Terms, 2011).

Land Use Planning The process undertaken by public authorities to identify, evaluate and decide on different options for the use of land to help mitigate and prevent disasters by discouraging settlements and construction of key installations in hazard-prone areas (Glossary of Terms, 2011).

Mitigation Actions taken to reduce the adverse impacts of an emergency or disaster (Glossary of Terms, 2011).

Monitor and Review The part of the HIRA process in which the HIRA is reviewed and changes in the likelihood and consequences of the hazards is updated.

Municipality "Municipality" means a geographic area whose inhabitants are incorporated (*Municipal Act*) (Glossary of Terms, 2011).

Natural Hazard Natural hazards are those which are caused by forces of nature (sometimes referred to as 'Acts of God'). Human activity may trigger or worsen the hazard; (for example deforestation may increase the risk of a landslide) but the hazard ultimately is viewed as a force of nature.

Preparedness Actions taken prior to an emergency or disaster to ensure an effective response. These actions include the formulation of emergency response plans, business continuity/continuity of operations plans, training, exercises, and public awareness and education (Glossary of Terms, 2011).

Prevention Actions taken to avoid an emergency or disaster and the associated impacts of a hazard (Glossary of Terms, 2011).

Property Damage The direct negative consequences of the occurrence of a hazard on buildings, structures and other forms of property.

Psychosocial Impact The negative response of community or a subset of the community to a hazard caused by their perception of risk. This includes human responses such as self-evacuation, mass hysteria, hoarding and other potential undesirable responses.

Recovery The process of restoring a stricken community to a pre-disaster level of functioning (Glossary of Terms, 2011).

Resources These are personnel and major items of equipment, supplies, and facilities available or potentially available for assignment to incident operations and for which status is maintained. Resources are described by kind and type and may be used in operational or support capacities (Glossary of Terms, 2011).

Response The provision of emergency services and public assistance or intervention during or immediately after an incident in order to protect people, property, the environment, the economy and/or services (Glossary of Terms, 2011).

Return Period The average time between occurrences of a defined event (AMS, 2000).

Risk The product of the probability of the occurrence of a hazard and its consequences (Glossary of Terms, 2011).

Risk Analysis The process by which hazards are prioritized for emergency management programs at that particular point in time based on their frequency and potential consequences.

Risk Assessment A methodology to determine the nature and extent of risk by analyzing potential hazards and the evaluation of vulnerabilities and consequences (Glossary of Terms, 2011).

Severity The extent of disruption and/or damages associated with a hazard (Glossary of Terms, 2011).

Site A geographical location of an incident (Glossary of Terms, 2011).

Social Impact The direct negative consequences of the occurrence of a hazard on people, such as fatalities, injuries or evacuations.

Technological Hazard Technological hazards are hazards which arise 'from the manufacture, transportation, and use of such substances as radioactive materials, chemicals, explosives, flammables, modern technology and critical infrastructure' (HIRA, 2005).

Threat A person, thing or event that has the potential to cause harm or damage (Glossary of Terms, 2011).

Vulnerability The susceptibility of a community, system or asset to the damaging effects of a hazard (Glossary of Terms, 2011)

Definitions from AHRA Canada https://www.publicsafety.gc.ca/cnt/rsrcs/pblctns/ll-hzrds-ssssmnt/index-eng.aspx

Accident An unintended, unplanned and unexpected event that interrupts an activity and sometimes causes injury or damage. Note: Examples of accidents include transportation accidents, hazardous material spills or releases, fire and accidental explosions.

All Hazards Referring to the entire spectrum of hazards, whether they are natural or human-induced. Note: For example, hazards can stem from industrial accidents, national security events or cyber events.

All Hazards Approach An emergency management approach that recognizes that the actions required to mitigate the effects of emergencies are essentially the same, irrespective of the nature of the incident, thereby permitting an optimization of planning, response and support resources. Note: The intention of an all-hazards approach is to employ generic emergency planning methodologies, modified as necessary according to the circumstances.

All Hazards Risk Assessment The process of identifying, analyzing and evaluating risks using an all-hazards approach.

Disaster An event that results when a hazard impacts a vulnerable community in a way that exceeds or overwhelms the community's ability to cope and may cause serious harm to the safety, health or welfare of people, or damage to property or the environment. Note: A disaster may be triggered by a naturally occurring phenomenon that has its origins within the geophysical or biological environment or by human action or error, whether malicious or unintentional, including technological failures and terrorist acts.

Emergency A present or imminent event that requires prompt coordination of actions concerning persons or property to protect the health, safety or welfare of people, or to limit damage to property or the environment.

Emergency Management The management of emergencies concerning all-hazards, including all activities and risk management measures related to prevention and mitigation, preparedness, response and recovery.

Frequency The number of occurrences of an event in a defined period of time.

Hazard A potentially damaging physical event, phenomenon or human activity that may cause the loss of life or injury, property damage, social and economic disruption or environmental degradation.

Hazard Identification The process of identifying, characterizing and validating hazards. Note: Hazard identification looks at the type, the properties and the potential effects of hazards and is part of hazard assessment.

Likelihood The chance of an event or an incident happening, whether defined, measured or determined objectively or subjectively.

Mitigation Actions taken to reduce the impact of disasters in order to protect lives, property and the environment, and to reduce economic disruption. Note: Mitigation includes structural mitigative measures (e.g. construction of floodways and dykes) and non-structural mitigative measures (e.g. building codes, land-use planning and insurance incentives). Prevention and mitigation may be considered independently or one may include the other.

Natural Hazards A source of potential harm originating from a meteorological, environmental, geological or biological event. Note: Examples of natural hazards include tornadoes, floods, glacial melt, extreme weather, forest and urban fires, earthquakes, insect infestations, infectious diseases.

Probability In statistics, a measure of the chance of an event or an incident happening.

Qualitative Assessment A risk assessment method that assigns non-statistical values to risks. Note: A qualitative assessment produces narrative, descriptive or comparative information about risks. It can be based on limited information, numerically incomparable data or complex non-linear relationships.

Quantitative Assessment A risk assessment method that assigns statistical values to risks.

Residual Risk Risk that remains after implementing risk mitigation measures.

Resilience The capacity of a system, community or society to adapt to disruptions resulting from hazards by persevering, recuperating or changing to reach and maintain an acceptable level of functioning. Note: Resilience is built through a process of empowering citizens, responders, organizations, communities, governments, systems and society to share the responsibility to keep hazards from becoming disasters.

Risk The combination of the likelihood and the consequence of a specified hazard being realized; refers to the vulnerability, proximity or exposure to hazards, which affects the likelihood of adverse impact.

Risk Analysis A process to comprehend the nature of a risk and to determine its level. Note: Risk Analysis provides the basis for Risk Evaluation and decisions about Risk Treatment.

Risk Assessment The overall process of Risk Identification, Risk Analysis and Risk Evaluation.

Risk Avoidance An informed decision to avert or to withdraw from, an activity in order not to be exposed to a particular risk.

Risk Communication The imparting, exchanging and/or receiving of clear, credible and timely information about the existence, nature, form, likelihood, severity, acceptability, treatment or other aspects of risk to improve decision-making in risk management. Note: Risk communication is carried out among public authorities, risk assessors, risk managers, the public and all other interested parties. It is intended to achieve a better understanding of risks and risk management.

Risk Identification The process of finding, recognizing and recording risks.

Risk Management The use of policies, practices and resources to analyze, assess and control risks to health, safety, environment and the economy.

Risk Perception A stakeholder's view on a risk. Note: Risk perception reflects the stakeholder's needs, issues, knowledge, beliefs and values.

Risk Profile A description of an entity's existing management practices, common vulnerabilities, tolerance and key interdependencies concerning its particular risks, as well as an assessment of their relative likelihood, consequences and priority.

Risk Register A register that contains a list of identified risks and related information used to facilitate the monitoring and management of risks. Note: The risk register is generally in the form of a table, spreadsheet or database and may contain the following information: statement or description of the risk, source of risk, areas of impact, cause of the risk, status or action of sector network, existing controls, risk assessment information and any other relevant information.

Risk Taxonomy A comprehensive and common set of risk categories that is used within an organization.

Risk Tolerance The willingness of an organization to accept or reject a given level of residual risk. Note: Risk tolerance may differ across an organization, but must be clearly understood by those making risk-related decisions.

Threat The presence of a hazard and an exposure pathway. Note: A threat may be natural or human-induced, accidental or intentional.

Threat Assessment A process consisting of the identification, analysis and evaluation of threats.

Vulnerability A condition or set of conditions determined by physical, social, economic and environmental factors or processes that increases the susceptibility of a community to the impact of hazards. Note: Vulnerability is a measure of how well prepared and equipped a community is to minimize the impact of or cope with hazards.

Vulnerability Assessment The process of identifying and evaluating vulnerabilities, describing all protective measures in place to reduce them and estimating the likelihood of consequences.

Definitions from EMDAT http://www.emdat.be/explanatory-notes

EM-DAT data include the main following information:

Disaster Number A unique disaster number for each event (8 digits: 4 digits for the year and 4 digits for the disaster number - i.e.: 19,950,324).

Country Country(ies) in which the disaster has occurred.

Disaster Group Two main groups of disasters are distinguished in EM-DAT: natural disasters, and technological disasters. A third category complex disasters has been added in order to include specific event (famine) which are not directly linked to a natural hazard.

Disaster Type Description of the disaster according to a pre-defined classification.

Date When the disaster occurred. The date is entered as follow: Month/Day/Year.

Death Number of people who lost their life because the event happened.

Missing The number of people whose whereabouts since the disaster is unknown, and who are presumed dead (official figure when available).

Total Deaths Sum of death and missing.

Injured People suffering from physical injuries, trauma or an illness requiring medical treatment as a direct result of a disaster.

Homeless Number of people whose house is destroyed or heavily damaged and therefore need shelter after an event.

Affected People requiring immediate assistance during a period of emergency, i.e. requiring basic survival needs such as food, water, shelter, sanitation and immediate medical assistance.

Total Affected Sum of injured, homeless, and affected.

Estimated Damage The amount of damage to property, crops, and livestock. In EM-DAT estimated damage are given in US$ ('000). For each disaster, the registered figure corresponds to the damage value at the moment of the event, i.e. the figures are shown true to the year of the event.

For a disaster to be entered into the database at least some criteria must be fulfilled.

- Ten (10) or more people reported killed.
- Hundred (100) or more people reported affected.
- Declaration of a state of emergency.
- Call for international assistance.

Learning Objectives

Post-secondary students are expected to achieve the learning objectives as outlined under the following six categories used in North American universities.

1. **Depth and breadth of knowledge**

   (a) Understanding of key concepts, theories, and methodologies in the field of natural disasters, associated risks, risk reduction, and mitigation
   (b) Understanding of types of disasters and their characteristics and data sources
   (c) Comprehending disaster management knowledge in theory and practice
   (d) Development of critical and analytical skills
   (e) Application of knowledge from other disciplines such as environmental studies, social sciences, humanities, science and engineering, and management studies to disaster management

2. **Knowledge of Methodologies**

   (a) Comprehension of various risk assessment methodologies and their application in Canadian and international disaster management context
   (b) Evaluation of disaster management frameworks for their effectiveness and efficiency in variety of scenarios
   (c) Deep understanding of disaster prevention, mitigation, preparedness, and strategic planning
   (d) Comprehension of various aspects of disasters – scientific, social, psychological, etc.
   (e) Understanding of the key research methodologies in the identification of problems and development of suitable solution approaches
   (f) Learning to design research studies to consider community engagement through participation and partnerships for policy decisions

N. Agrawal, *Natural Disasters and Risk Management in Canada*, Advances in Natural and Technological Hazards Research 49,
https://doi.org/10.1007/978-94-024-1283-3

3. **Application of Knowledge**
   (a) Comprehension of the applicability of theoretical concepts, methods, and tools
   (b) Ability to grasp step by step workable examples with clear explanations
   (c) Demonstration of the applicability of concepts such as risk perception, vulnerability resilience, and coping capacity
   (d) Learn to implement research studies in communities with ethical considerations
   (e) Critical analysis of past disasters for their physical dynamics and societal impacts

4. **Communication Skills**
   (a) Understanding of the importance of clear and timely risk communication in pre, during, and post disaster situations to stakeholders using variety of media
   (b) Comprehension and accurate analysis of past events for developing awareness campaigns, knowledge-sharing strategies, and decision-making tools
   (c) Recognizing cultural sensitivities and social complexities
   (d) Development of networking skills for effective communication

5. **Awareness of Limits of Knowledge**
   (a) Understanding of the limits of advances in science and technology in the context of disaster management
   (b) Awareness of limits of disaster simulation, modelling, and early warning systems
   (c) Recognizing the ever evolving environment and diversity in social, economic, political, and geomorphological context
   (d) Understanding of the limits in sharing information within and outside of the discipline

6. **Autonomy and Professional Capacity**
   (a) Ability to demonstrate transferable skills to foster universal understanding of the subject matter
   (b) Ability to develop leadership skills based on sound understanding of fundamentals and independent thinking
   (c) Ability to develop collaborative aptitudes and professional integrity
   (d) Ability to practice consistency and competence in decision making

Index

A
Access to resources (ATR) models, 85
 access profile, 304
 Hurricane Katrina, 305, 307
 modified form, 304, 306
 original form, 299, 304
 unpacked form, 299, 305
Accumulated Cyclone Energy (ACE), 35
Adaptation
 definitions, 152
Adaptive resilience (AR), 151
Agricultural drought, 7
Agro-Climate Consulting, 132
Alaska tsunami warning center (ATWC), 128, 129
Alberta fires, 200
All Hazard Risk Assessment (AHRA), 258–264, 266, 329, 332
 business cycle, 254–268
 critical infrastructure sectors, 254
 objectives, 253
 risk analysis
  Canada's reputation and influence, 262, 263
  duration modifiers, 260
  duration of environmental damage, 261
  economy impact category, 259
  environment impact category, 259–261
  geographical extent assessment modifier approach, 261
  impact analysis, 258
  impact ratings and adult fatalities, 259
  likelihood analysis, 258
  likelihood assessment, 266
  modifier chart, 260, 261
  public mood, 263
  society and psycho-social impacts, 262–264
  territorial security, 261–262
 risk evaluation, 266
 risk identification, 256
 risk treatment, 266–268
 setting the context, 255–256
Andhra Pradesh, India, 161
ArcGIS, 140
Atlantic hurricanes, 33
Avian influenza (AI), 3
Avian influenza A (H7N9) virus, 3

B
BC Emergency Program Act, 226
Blizzard, *see* Snow storm
Brampton, Canada, 174, 184, 205, 207
Business continuity plan, 274
Business impact analysis (BIA), 108, 271, 273

C
Canada, 154–157, 167, 174, 175, 193, 205, 210
Canadian Census program, 168
Canadian disaster database (CDD), 231, 255
Canadian disaster database interactive map, 232
Canadian earthquake, 154, 155
Canadian Forest Fire Danger Rating System (CFFDRS), 24
Canadian Standards Association (CSA), 106, 107

N. Agrawal, *Natural Disasters and Risk Management in Canada*, Advances in Natural and Technological Hazards Research 49,
https://doi.org/10.1007/978-94-024-1283-3

Canada's Reputation and Influence, 333–335
CD ROM project, 164, 165
Center for Research on the Epidemiology of Disasters (CRED), 24
Charity Shoal, 71
Chatham Islands Civil Defence Emergency Management (CDEM), 278
Citizen participation, 158, 159
Citizen Power, 159
Climate change adaptation (CCA), 151
Cloudburst, 86–88
Coastal erosion, 41
Command & Control hierarchical model, 210
Community participation programs, 158
Community perception model
- *vs.* conventional approach, 312–313
- disaster impact, 308
- disaster risk, 306
- framework, 309
- immigrants, 307
- likelihood ranking, 311
- mitigation measures, 306
- modified risk formula, 308
- numerical ranking scheme, 310
- *vs.* PAR, 308
- 5-point scale, 308, 309
- resilience, 308
- risk assessment, 309, 311, 312
- risk index, 309
- vulnerabilities, 308

Community resilience
- behavioral dimension, 154
- content analysis, 155
- earthquake events, 154
- emergency preparedness and risk, 158
- hazard effects, 154
- management, 155
- Microsoft Excel, 157
- natural hazards, 155
- NVivo qualitative software, 157
- physical and psychological resilience, 155
- productivity and socio-economic status, 164
- public transit, 157
- self-activation, 155
- social and psychological harm, 153
- social resource, 153
- statistical analysis, 163
- sufficiency and responsibility, 155
- while floods, 154

Community Technology Learning Centres (CTLC), 162
Constructionist approach, 194
Coordinating Executive Group (CEG), 278
Coronal Mass Ejections (CMEs), 73
Cost/benefit analysis, 213
Critical infrastructure, 138, 139
CTLC Pilot Project in Pakistan, 325–326
Cyclone, 112–117
- shelters, India
  - for cattle, 116
  - coastal states, 112
  - components, 117
  - design and construction, 116
  - floods and tsunamis, 114
  - local motivation, 113
  - modifications, 117
  - multi-purpose shelter, 114, 115
  - National Cyclone Risk Mitigation Project, 117
  - number and location, 116
  - Orissa State Disaster Mitigation Authority, 113
  - proper maintenance, 116
  - severe damage, 113, 114
  - sustainable uses, 114

D

Data transcription, 163
Decision support tool, resilience, 167, 168, 170, 172, 173
Deforestation, 166
Derecho, 64, 66
Determinate perception, 195
Digital archive system, 153
Digital elevation model (DEM), 100
Digital literacy, 161
Direct economic loss, 259
Disaster, definition, 232
Disaster management, 208
- apathy, 209
- cultural attitudes, 209
- institutional ambiguities, 209
- inter-governmental paradox, 209
- models of, 210
- overdependence upon technology, 209
- overestimation of capacity, 209
- and planning, 209
- social pressures, 209
- underestimation of risk, 209

Disaster Myth and Reality, 326
Disaster resilience
- management, 167
- quantitative and qualitative approaches, 174

Disaster risk management (DRR), 85–104, 106, 107, 109–133, 136, 147, 148, 151, 161
elements, 84
historical records, 82
natural, technological/human sources, 81
preparedness, response and recovery plan
in California, 117–118
Fiji, 111
Indian Ocean Tsunami 2004, 119–126
post-earthquake reconstruction, Nepal, 118
vertical evacuation shelters, India, 112–117
rehabilitation, 111
risk assessment
FEMA model, 85
HRVA method, 85
multi-tier hazard, 86–91
NOAA, 85
qualitative frameworks, 85
quantitative frameworks, 85
SMUG model, 85
risk control options
London, Ontario, 92–94
market-like tools, 92
non-structural measures, 91
Red River Basin, 95–104
structural measures, 91
risk identification, 82
risk management strategies, 82
risk perception, 82, 83
risk reduction, 82
standard risk formula, variations, 82
strategic planning
assessment, 107
CSA, 106, 107
financial assistance, 106
hazards, 107
impact on businesses, 107
2004 Indian Ocean Tsunami, 109–111
sustainable development
Arctic Canada, 127, 128
Atlantic Canada, 129–131
climate change and sunspots, 131–133
Pacific Canada, 128–129
propane explosion, Toronto, 136
resilient communities, 126
RMI, 126
threat recognition, 84
Disaster risk reduction, 160
Disaster scenario simulations, 214
Dissemination areas (DA), 134, 135, 175
Dissonant perception, 195
Don River, 18
Don Valley Parkway (DVP), 250
Drainage infrastructure, 172
Drought
acute events, 6
agricultural, 7
chronic events, 6
drainage ditch west, 7, 8
historical drought events, 9
historical events, 9
hydrological, 7
large economic losses, 7
meteorological, 7
orange and almond orchards/trees, 9
socioeconomic, 7
water shortage emergencies, 7
Dry forest fire, 200

E

Earthquake, 195, 196
adverse effects, 10
epicenter, 9
faultline, 9
foreshocks, 9
hypocenter, 9
mainshock, 10
measurement and monitoring, 10, 11
Northern Ontario, 12, 13
potential impacts, 12, 13
Southern Great Lakes, 12
Western Quebec Zone, 11
Ebola virus disease (EVD)
outbreak in 2014, 3–5
Economic category assessment tool, 332, 333
Economic vulnerability (EV), 136, 137, 235
El Niño/Southern Oscillation (ENSO), 132
Emergency crews, 197
Emergency management, 204, 209
Emergency Management and Civil Protection Act, 220, 221
Emergency Management community, 209
Emergency Management Cycle, 210
Empowerment Framework, 162
Empowerment process, 148, 161
Engineering infrastructure data, 168
Enhanced Fujita (EF) Scale, 58, 60
Environmental vulnerabilities, 235
Erosion, 41–43
Expert analysis, 195
Expert judgement of risk, 194
Extreme temperature

Extreme temperature (*cont.*)
- cold wave, 16, 17
- heat wave, 14–16
- historical cold wave events, 14

F

Federal Emergency Management Agency (FEMA), 85
- BIA report, 273, 274
- business continuity plan, 274
- business disruption scenarios, 273
- disasters, impact of, 273
- disruption, timing and duration, 273
- electronic information, 274
- exercises, 277
- hazards, impacts, 273
- information collection, 271
- IT DRP, 274
- manufacturing recovery strategies, 275, 276
- Operational & Financial Impacts worksheet, 275
- physical assets, 272
- recovery time objective, 275
- resources, 276
- risk assessment process, 272
- risk assessment table, 271
- technology recovery strategies, 274
- testing, 277

Fire raging, 201
Flash flooding, 195
Flood emergency management, 94
Flood forecasting tools, 170
Flood generation processes, 168
Flood hazard perception, 195
Flood risk mapping methodology, 202, 203
Floods, 18–22
- EM-DAT database, 87
- international database EMDAT, 17
- risk and urbanization of London, 22
- Toronto, Canada
  - Don River, 18
  - failures of infrastructure, 21
  - flood damage mitigation, 22
  - local weather, 18
  - major flooding events, 19–21
  - population, 18
  - topography, 18, 19
- in Uttarakhand, 87, 89, 90

Focus group discussion, 324–325
Fog, 73
Foreshocks, 9
Forest fires, 23, 24, 200
Fort McMurray fire, 201
Freezing rain, 26, 27
Fukushima Daiichi nuclear power plant, 199

G

Geographic information system (GIS), 85, 94, 95
Globalization, 148
Global tsunami
- amplitudes and inundation, 110, 111
- of 26 December 2004, 109
- early warning system, 109–110
- event on 27 August 1883, 109
- scientific parameters, 109
- survey results, 109

Google map, 203
Government of Canada (GoC), 14
Great East Japan Earthquake, 197
Greater Toronto Area (GTA), 18, 19, 174, 178, 179
- rainfall record, 285, 286
- rain gauge network, 285, 286
- Toronto rainfall amounts, 285, 287

Great Hanshin Earthquake, Japan, 156
Great Kanto Earthquake, Japan, 156
Great Plains tornado, 58
Group perceptions
- hazardous events, 195
- internal control, 195
- locus of control, 195
- social/cultural factors, 195
- types, 195

Gulf of Mexico, 167
Gust front, 64

H

H1N1, 5
Hailstorm, 74, 76
Hazard Identification and Risk Assessment (HIRA), 214, 222–227, 327
- change in frequency values, 224, 227
- change in vulnerability value, 224, 228
- changing risk, 224
- consequence estimation
  - business/financial Impact, 224, 227
  - change in frequency values, 227
  - critical infrastructure service disruptions/impact, 223, 226
  - environmental damage, 224, 226
  - historical winter storms, 222, 223
  - property damage, 223, 226

psychosocial impacts, 224, 227
reclassification, 227
social impacts, 223, 225
conventional and risk equation, 229
disaster databases, 221
emergency management program, 220
frequency, 222
hazard levels, 228
monitoring and review process, 221
natural, technological and human-induced hazards, 220
Provincial level, 221
risk analysis, 221
risk equation, 222
structure, 220
Hazard, Risk and Vulnerability Analysis (HRVA) method, 85, 236–239, 244–253
Canadian disaster database, 231, 232
critical facility, 239
data fields, 233
disaster-resilient communities, 230
economic and social impact, 239
economic vulnerabilities, 235
environmental impact, 239
environmental vulnerabilities, 235
Executive committee, 230
hazard categories, 232, 234
hazard types with frequency, 233, 235
ice storm in Montreal, Canada
consequences severity assessment, 246
dairy, 244
hazard likelihood, 245, 247
ice accumulation, 245
information collection, 238
manufacturing, 244
nation's economic output, 244
risk profile, 245, 247
visual information adds value, 244
vulnerabilities, 246
impact assessment and ranking
critical facility, 236, 237
economic impact, 237
economic and social impact, 236
environmental impact, 236
explosion, 238
fatality, 236, 239
injury, 236, 239
lifelines, 236
property damage, 236
lifeline, 239
likelihood ranking, 240
online HRVA Web tool, 230
online tool, 238, 241, 242
physical vulnerabilities, 234
propane explosion in Toronto, Canada
consequences severity, 251
economic vulnerability and damage, 250
hazard likelihood, 251–253
information collection, 246
map, 248
physical vulnerability and damage, 250
risk profile, 251, 252
social vulnerability and impact, 249
sources, 251
property damage, 239
purpose, 228
risk profile, 238, 240
social vulnerabilities, 234
step process, 230–232
HEC-RAS hydraulic model, 101, 102
Himalayan Mountains, 166
Human-nature interaction, 147
Human security, 148
Humidex, 14–16
Hurricane Hazel, 37
Hurricanes
classification, 32
closed low-level circulation, 32
Hurricane Hazel, 37
Hurricane Sandy, 37
intensity, 33–35
return periods, 33
tropical storm season, 32, 33
Hurricane Sandy, 37
Hydrological drought, 7

I

Ice fog, 73
Ice storm, 25–32
accumulated ice, 24
St. Lawrence River Valley, 25
Toronto, Canada, urban impacts
emergency warming centers, 31
extent of, 25, 26
freezing rain, 26, 27
freezing rain formation, 26
historical winter storms, 27, 28
ice accumulation, 27, 29
power distribution system, 27, 29, 30
precipitation, 27, 29
public transit system, 31
telecommunications, 29, 31
tree canopy management, 32
underground electrical system, 32
urban forestry, 27, 30
Immediate relief package, 125
Incident Management System, 210
Indirect economic loss, 259

Information and Communication Technologies (ICT), 160
Information sharing, 211, 212
Information technology disaster recovery plan (IT DRP), 274
Inherent resilience (IR), 151
International Astronomical Union (IAU), 69
International Nuclear Events Scale (INES), 200

K
Kedarnath, India, 165, 166
Knowledge network, 126

L
Lac-Mégantic train derailment and explosion, 204
Lahars, 43
Landslides, 165
  Brazil, 45, 49
  Canadian history, 44, 46, 47
  causes of, 43, 44
  Hindu shrine in Kedarnath, 89, 91
  impact region, 89, 90
  India, 45–48
  movements and materials, 42
  Quebec, Canada, 44
  southern California, 44, 47, 48
Landuse
  classification, 96
Large scale natural hazards, 2–22, 24–33, 37
  definition, 2
  drought
    acute events, 6
    agricultural drought, 7
    chronic events, 6
    drainage ditch west, 7, 8
    historical drought events, 9
    historical events, 9
    hydrological drought, 7
    large economic losses, 7
    meteorological drought, 7
    orange and almond orchards/trees, 9
    socioeconomic drought, 7
    water shortage emergencies, 7
  earthquake
    adverse effects, 10
    epicenter, 9
    faultline, 9
    foreshocks, 9
    hypocenter, 9
    mainshock, 10
    measurement and monitoring, 10, 11
    Northern Ontario, 12, 13
    potential impacts, 12, 13
    Southern Great Lakes, 12
    Western Quebec Zone, 11
  extreme temperature
    cold wave, 16, 17
    heat wave, 14–16
    historical cold wave events, 14
  floods
    international database EMDAT, 17
    risk and urbanization of London, 22
    Toronto, Canada, 18–22
  forest fires/wildfires, 24
  health and biophysical hazards
    epidemics, 3–5
    factors, 2
    pandemic, 3–6
  history, 2
  hurricanes
    classification, 32
    closed low-level circulation, 32
    Hurricane Hazel, 37
    Hurricane Sandy, 37
    intensity, 33
    return periods, 33
    tropical storm season, 32, 33
  ice storm
    accumulated ice, 24
    St. Lawrence River Valley, 25
    urban impacts, Toronto, 25–32
Lay judgment of risk, 194
Light Detection and Ranging (LIDAR), 100
Lightning, 76–78
Local Authority Emergency Management Regulation, 226
London, Ontario, 167, 168
Long term relief package, 125
Lot-level protection measures, 172

M
Mainshocks, 10
Media networks, 47
Medium scale hazards, 42–68
  erosion, 41–43
  landslide
    Brazil, 45, 49
    Canadian history, 44, 46, 47
    causes of, 43, 44
    India, 45–52
    movements and materials, 42
    Quebec, Canada, 44

southern California, 44, 47, 48
sinkholes, 53, 55, 57
snow storm
events, 53
Ontario, Canada, 52, 55
sub disaster type, 51
warning types, 51, 54
subsidence, 53, 56
tornado
characteristics, 58, 59
formation, 56, 58
Kawacatoose First Nation, 62
Leamington, Ontario, 60
Midland, Ontario, 62, 63
season, 56
severity and strength, 58–61
Tornado Alley, 56, 57
warnings, 58
windstorms
classification, 62
damages, 63, 64
derechos on May 8th, 2009, 64, 66, 67
downdraft, 64, 65
gust front, 64
haboob, 66, 67
microburst, 64, 65
Midwest and mid-Atlantic on June 29, 2012, 67, 68
straight-line wind, 64
Mesocyclone, 58
Meteor, 69
Meteorological drought, 7
Methods for the Improvement of Vulnerability Assessment in Europe (MOVE), 151
Metropolitan Toronto and Region Conservation Authority (MTRCA), 20
Millennium Development Goals, 161
Modified Mercalli (MM) Intensity Scale, 10
Moment Magnitude Scale (MMS), 10
Municipal Property Assessment Corporation (MPAC), 168, 175

N

National Commission for Human Development (NCHD), 161
National Disaster Mitigation Strategy, 213
National Oceanic and Atmospheric Administration (NOAA), 85
National Science & Engineering Research Council (NSERC) of Canada, 184
Natural hazards, 1, 41
forest fires/wildfires, 23
large scale (*see* Large scale natural hazards)
medium scale hazards (*see* Medium scale hazards)
small scale hazards (*see* Small scale hazards)
Natural protective barriers, 209
Natural resource base, 149
Natural Resources Canada (NRCan), 168
Nonparticipation, 158
North American snow cover (NAS), 132
Nuclear accidents, 197
Nuclear power plants, 197

O

Objective resilience, 177, 178
Oklahoma, 58
Ontario, Canada, 204, 205, 209
Ontario Propane Safety Review, 134
Open Data source, 168
Operational & Financial Impacts worksheet, 275

P

Pacific tsunami warning center (PTWC), 128
People's perception of risk and vulnerability, 205, 207
Perception of people, 202, 204, 205, 207
Perception of vulnerability, 200, 202
Perceptions of danger, 200
Physical vulnerability, 137–138, 234
Post-disaster recreation of vulnerability, 208
Power distribution system, 27, 30
Pressure and release (PAR) model, 85, 296, 298, 299
in China's Sichuan construction, 299
disaster risk evaluation, 295
flooding, in New Orleans, 299
framework assists, 295
Hurricane Katrina
application, 296
Gulf coast impact region, 296, 298
NOAA Corps flew, 298
vulnerability, progression of, 295
Probabilistic perception, 195
Propane explosion
critical infrastructure, 138, 139
data acquisition, 134, 135
economic vulnerability, 136, 137
evacuation distance, 134
explosion site, 133

Propane explosion (*cont.*)
physical vulnerability, 137–138
results, 134
risk management, 133
social vulnerability, 136
spatial disaster risk estimation, 138–140
vulnerability estimation, 135
Proper governance, 204
1-psi overpressure distance, 134
Public transit system, 31

Q
Qualitative methods, 295, 299, 304–307
ATR model
access profile, 304
Hurricane Katrina, 305, 307
modified form, 304, 306
original form, 299, 304
unpacked form, 299, 305
community perception model (*see* Community perception model)
PAR model (*see* Pressure and release (PAR) model)
risk aversion, 313, 314
Quantitative methods, 219–229, 253, 271
AHRA (*see* All Hazard Risk Assessment (AHRA))
FEMA (*see* Federal Emergency Management Agency (FEMA))
HIRA process
change in frequency values, 224, 227
change in vulnerability values, 224, 228
changing risk, 224
consequence estimation, 222–228
conventional and risk equation, 229
disaster databases, 221
emergency management program, 220
frequency, 222
hazard levels, 228
monitoring and review process, 221
natural, technological and human-induced hazards, 220
Provincial level, 221
risk analysis, 221
risk equation, 222
structure, 219, 220
HRVA (*see* Hazard, Risk and Vulnerability Analysis (HRVA))
SMUG (*see* Seriousness, Manageability, Urgency and Growth (SMUG))
WRI (*see* World Risk Index (WRI))
Quasi-biennial wind oscillation (QBO), 132
Quebec, Canada, 203

R
Rationalist approach, 194
Reactor systems test, 197
Real-time adaptation, 170
Recovery time objective, 275
Relational pathway, 163
Remote sensing technology, 92, 94
Resilience
binary values, 176
capital assets, 153
at community level, 148
cost benefit analysis, 148
data type and sources, 182, 184
definitions, 150
lack of coping capacity, 181
learning and reorganization processes, 151
maps, 184–187
qualitative estimation, 175
social, economic and environment, 148
susceptibility, 180
Resilience metric, 168
Resilience of built environment, 165, 167
ResilSIM concept, 167, 169–171, 173
Risk
definition of, 82
Risk assessment, 86–91
FEMA model, 85
HRVA method, 85
multi-tier hazard
cloud burst, 86–87
landslides, 89–91
river flooding, 87–89
NOAA, 85
qualitative frameworks, 85
quantitative frameworks, 85
risk analysis, 91
SMUG model, 85
Risk aversion, 313, 314
Risk control, 92–108
London, Ontario
flood emergency management, 94
landuse and landcover, 92, 94–96
rainfall and runoff response pattern, 92, 97
remote sensing and GIS technology, 94
surface water availability, 92
Upper Thames River Watershed, 92, 93
market-like tools, 92

non-structural measures, 91
Red River Basin, 104
buildings, roads and agriculture, 102, 103
computational procedures, 101
decision makers, 104
deterministic analysis, 104–108
during the 1997 floods, 97, 98, 100
economic development, 95
flood protection alternatives, 99–100
fuzzy analysis, 104
hydrologic and hydraulic analysis, 95
multi-objective techniques, 95
spatial data, 100–101
time and space, 96
structural measures, 91
Risk Event Scenario, 329, 331, 332
Risk management approach, 193
Risk Management Index (RMI), 126
Risk perception
communication processes, 195
cultural factors, 194
dynamic social and psychological processes, 193
expert judgement, 193
factors, 196
media influence, 200
peer influences, 195
personal experiences and prejudices, 193
personality and emotional state, 193
and risk assessment, 195, 196
role of, 205
social learning, 195
socio-economic factors, 193
subjective influences, 195
vulnerability progression, 204
Riverine flooding, 167

S

Saffir-Simpson hurricane wind scale, 33–37
SARS-associated coronavirus (SARS-CoV), 5
Seriousness, Manageability, Urgency and Growth (SMUG), 85, 341, 342
built environment, 283
CDEM, 278
definition, 278
economic environment, 284
human, economic, social, infrastructural and geographic factors, 284
information collection, 283
natural environment, 283
numeric ratings, 279–281
population, overview of, 283
results, 282
risk profile assessment, 283
Severe acute respiratory syndrome (SARS), 5, 6, 212
Sinkholes, 53, 55, 57
Small scale hazards, 69–72
extraterrestrial hazard
comets, 69
Lake Ontario, Canada, 70, 71
meteoroids, 69
risk-assessment system, 69
Russian meteor, 69, 70
Tunguska impact, 72
fog, 73
geomagnetic storm, 73, 75
hailstorm, 74, 76
lightning, 76, 78
Snow storm
events, 53
Ontario, Canada, 52, 55
sub disaster type, 51
warning types, 51, 54
Social amplification of risk, 196
Social networks, 193, 206
Social vulnerability, 136, 149, 152, 234
Socioeconomic drought, 7
Socio-economic factors, 109
Soil erosion, 42
Sound information, 202
South Asian Women's Centre (SAWC), 157
Spatial Fuzzy Compromise Programming (SFCP), 96
Spatial hazard index, 138
Spatial risk estimation model, 138–140
Spatial risk index R_{ij}, 138
Statistics Canada, 168
Subsidence, 53, 56
Survey-based studies, 193
Sustainable livelihood framework, 148, 153
Swine flu, 5

T

Technical Standards and Safety Authority (TSSA), 134
Telecommunications, 29, 31
Temporary diking measures, 171
Terrorism, 148
Three Mile Island Unit 2 (TMI-2) reactor, 197
Tohoku earthquake, 154
Tokenism, 158
Torino scale, 69, 70
Tornado
characteristics, 58, 59
formation, 56, 58
Kawacatoose First Nation, 62

Tornado (*cont.*)
- Leamington, Ontario, 60
- Midland, Ontario, 62, 63
- season, 56
- severity and strength, 58–61
- Tornado Alley, 56, 57
- warnings, 58

Toronto, Canada, 157–159, 204, 207
Toronto and Region Conservation Authority (TRCA), 20
Tree canopy management, 32
Tropical cyclones, *see* Hurricanes
Tsunami, 156, 199, *see* Global tsunami

U
United States, 197
Upper Thames River (UTR), 22
Urban forestry, 27
Urbanization, 92, 148
Urbanization of London, 22

V
Vancouver Island, 155
Vocational training for women, 160
Voluntary self-help organization, 156, 157
Vulnerability
- definitions, 149

W
Wildfire evacuation, 200
Wildfires, 23, 24
Windchill Index, 17
Windstorms
- classification, 62
- damages, 63, 64
- derechos on May 8th, 2009, 64, 66, 67
- downdraft, 64, 65
- gust front, 64
- haboob, 66, 67
- microburst, 64, 65
- Midwest and mid-Atlantic on June 29, 2012, 67, 68
- straight-line wind, 64

Women's economic independence, 163
World Bank, 117
World Meteorological Organization (WMO), 17
World Risk Index (WRI), 285–287
- annual frequency, 287
- data quality and quantity, 285
- flood magnitude with probabilities, 288, 289
- flood return period calculator, 292
- frequent events, 285
- GTA
  - rain gauge network, 285, 286
  - rainfall record, 285, 286
  - Toronto rainfall amounts, 285, 287
- hydrological and meteorological events, 287
- lowest central pressure record, 290, 291
- Mississippi River floods, 289, 290
- natural disaster types, 287
- natural hazards, 284
- naturally occurring events, 285
- return period/recurrence interval, 287
- vulnerability, 284

MIX
Papier aus verantwortungsvollen Quellen
Paper from responsible sources
FSC® C105338

If you have any concerns about our products,
you can contact us on
ProductSafety@springernature.com

In case Publisher is established outside the EU,
the EU authorized representative is:
Springer Nature Customer Service Center GmbH
Europaplatz 3, 69115 Heidelberg, Germany

Printed by Libri Plureos GmbH
in Hamburg, Germany